Computational Intelligence Techniques in Earth and Environmental Sciences

Tanvir Islam • Prashant K. Srivastava
Manika Gupta • Xuan Zhu
Saumitra Mukherjee
Editors

Computational Intelligence Techniques in Earth and Environmental Sciences

Springer

Editors
Tanvir Islam
NOAA/NESDIS Center for Satellite Applications and Research
College Park, MD, USA

Cooperative Institute for Research in the Atmosphere
Colorado State University
Fort Collins, CO, USA

Manika Gupta
Department of Civil Engineering
IIT Delhi
Delhi, India

Saumitra Mukherjee
School of Environmental Sciences Remote Sensing
Jawaharlal Nehru University
New Delhi, India

Prashant K. Srivastava
NASA Goddard Space Flight Center
Greenbelt, MD, USA

ESSIC, University of Maryland
College Park, MD, USA

Xuan Zhu
School of Geography and Environmental Science
Monash University
Melbourne, Australia

ISBN 978-94-017-8641-6 ISBN 978-94-017-8642-3 (eBook)
DOI 10.1007/978-94-017-8642-3
Springer Dordrecht Heidelberg New York London

Library of Congress Control Number: 2014931489

Printed on acid-free paper

Springer is part of Springer Science+Business Media (www.springer.com)

Preface

Computational intelligence (CI) techniques have garnered considerable interest in recent decades within the earth and environmental science research communities because of its capacity to solve and understand various complex problems achieving a sustainable planet. This book is a collection of recent developments and applications of the computational intelligence techniques in earth and environmental sciences. Topics addressed in this volume include meteorology, atmospheric modeling, climate change, environmental engineering, water resources, and hydrological modeling. The editors hope that this volume will promote increased collaboration of scientists with backgrounds in computer sciences with earth and environmental scientists.

The main focus of the contents are divided into three broad categories—classical intelligence techniques, probabilistic and transforms intelligence techniques, and hybrid intelligence techniques. Part I of the volume serves as an overview of computational intelligence techniques and their application in earth and environmental sciences. In Part II, which covers classical intelligence techniques, the contributions related to the classical methods applied to earth and environmental sciences are gathered. Part III concerns probabilistic and transforms intelligence techniques, more specifically, the application of Markov analysis and wavelet transforms to the earth and environmental sciences. Lastly, the analyses of problems that cannot be adequately solved with traditional techniques are addressed in Part IV. Hybrid approaches are highlighted.

The editors would like to thank all the contributors to this volume for their valuable contributions and their patience with the editing of the manuscript.

College Park, MD, USA — Tanvir Islam
Greenbelt, MD, USA — Prashant K. Srivastava
Delhi, India — Manika Gupta
Melbourne, Australia — Xuan Zhu
Delhi, India — Saumitra Mukherjee

Contents

Contributors

Munmun Akter Department of Computer Science and Engineering, The University of Rajshahi, Rajshahi, Bangladesh

A.T.M. Jahangir Alam Department of Environmental Sciences, Jahangirnagar University, Savar, Dhaka, Bangladesh

Md. Ayub Ali Department of Statistics, University of Rajshahi, Rajshahi, Bangladesh

A.B.M. Shawkat Ali School of Engineering and Technology, Central Queensland University, North Rockhampton, QLD, Australia

Md. Asaduzzaman Institute of Statistical Research and Training (ISRT), University of Dhaka, Dhaka, Bangladesh

Abdus Salam Azad Department of Computer Science and Engineering, Bangladesh University of Engineering and Technology, Palashi, Dhaka, Bangladesh

H. Md. Azamathulla REDAC, Universiti Sains Malaysia, Georgetown, Pulau Penang, Malaysia

Aida Hosseini Baghanam Faculty of Civil Engineering, Department of Water Resources Engineering, University of Tabriz, Tabriz, Iran

Varsha Chorsiya National Institute of Occupational Health, Ahmedabad, India

Priya Dutta Indian Institute of Public Health Gandhinagar, Public Health Foundation of India, Ahmedabad, India

Halil İbrahim Fedakar Faculty of Engineering and Natural Sciences, Civil Engineering Department, Abdullah Gül University, Kayseri, Turkey

Amrina Ferdous Department of Mathematics, Washington State University, Pullman, WA, USA

Department of Statistics, University of Rajshahi, Rajshahi, Bangladesh

Kh. Ghorbani Water Engineering Department, Gorgan University of Agricultural Sciences and Natural Resources, Gorgan, Golestan Province, Iran

Manika Gupta Department of Civil Engineering, Indian Institute of Technology Delhi, New Delhi, India

Dawei Han Department of Civil Engineering, University of Bristol, Bristol, UK

Md. Al Mehedi Hasan Department of Computer Science and Engineering, Rajshahi University of Engineering and Technology, Rajshahi, Bangladesh

Md. Kamrul Hasan Department of Computer Science and Engineering, Bangladesh University of Engineering and Technology, Palashi, Dhaka, Bangladesh

Tanvir Islam NOAA/NESDIS Center for Satellite Applications and Research, College Park, MD, USA

Cooperative Institute for Research in the Atmosphere, Colorado State University, Fort Collins, CO, USA

Özgür Kişi Faculty of Engineering and Architecture, Civil Engineering Department, Canik Başarı University, Samsun, Turkey

A.H.M. Mahbub Latif Institute of Statistical Research and Training (ISRT), University of Dhaka, Dhaka, Bangladesh

Md. Khademul Islam Molla Department of Information and Communication Engineering, The University of Tokyo, Bunkyo, Tokyo, Japan

Department of Computer Science and Engineering, The University of Rajshahi, Rajshahi, Bangladesh

Saumitra Mukherjee Remote Sensing, School of Environmental Sciences, Jawaharlal Nehru University, New Delhi, India

Anjali Nag National Institute of Occupational Health, Ahmedabad, India

P.K. Nag National Institute of Occupational Health, Ahmedabad, India

Farzad Hassan Nejad Faculty of Civil Engineering, Department of Water Resources Engineering, University of Tabriz, Tabriz, Iran

Vahid Nourani Faculty of Civil Engineering, Department of Water Resources Engineering, University of Tabriz, Tabriz, Iran

Aida Yahyavi Rahimi Faculty of Civil Engineering, Department of Water Resources Engineering, University of Tabriz, Tabriz, Iran

M. Arif Imtiazur Rahman Department of Computer Science and Engineering, Bangladesh University of Engineering and Technology, Palashi, Dhaka, Bangladesh

Md. Jahanur Rahman Department of Statistics, University of Rajshahi, Rajshahi, Bangladesh

Md. Mustafizur Rahman Department of Computer Science and Engineering, Bangladesh University of Engineering and Technology, Palashi, Dhaka, Bangladesh

M. Sayedur Rahman Department of Statistics, The University of Rajshahi, Rajshahi, Bangladesh

Renji Remesan Cranfield Water Science Institute, Cranfield University, Cranfield, Bedfordshire, UK

A.H.M. Sadaat Department of Environmental Sciences, Jahangirnagar University, Savar, Dhaka, Bangladesh

Md. Abu Shahin Department of Statistics, University of Rajshahi, Rajshahi, Bangladesh

Nashid Shahriar Department of Computer Science and Engineering, Bangladesh University of Engineering and Technology, Palashi, Dhaka, Bangladesh

A.R. Shoyeb Ahmed Siddique Department of Computer Science and Engineering, The University of Rajshahi, Rajshahi, Bangladesh

Prashant K. Srivastava NASA Goddard Space Flight Center, Greenbelt, MD, USA

ESSIC, University of Maryland, College Park, MD, USA

A. Zahiri Water Engineering Department, Gorgan University of Agricultural Sciences and Natural Resources, Gorgan, Golestan Province, Iran

Xuan Zhu School of Geography and Environmental Science, Monash University, Clayton, VIC, Australia

About the Editors

Dr. Tanvir Islam received the Ph.D. degree in remote sensing from the University of Bristol in Bristol, UK, in 2012. His Ph.D. research was focused on the remote sensing of precipitation through the use of radar polarimetry, especially towards algorithm developments and data quality improvements.

Following his Ph.D., he joined the University of Tokyo as a visiting scientist, more specifically, as part of the NASA/JAXA precipitation measurement missions (PMM) algorithm development team, which was supported by JAXA. Since 2013, he has been with the NOAA/NESDIS Center for Satellite Applications and Research, as a CIRA/CSU fellow, working on the development of satellite remote sensing algorithms with an emphasis on microwave variational inversion techniques. Currently, he is the scientific algorithm developer for the NOAA's Microwave Integrated Retrieval System (MiRS), where he is leading a number of efforts to enhance the performance of geophysical retrievals and extend the capability of the system to new passive microwave sensors.

Dr. Tanvir was the recipient of the Faculty of Engineering Commendation from the University of Bristol for his outstanding Ph.D. thesis (nominated for a University Prize) in 2012, the JAXA visiting fellowship award in 2012, and the CIRA postdoctoral fellowship award in 2013. He has been a member of the American Geophysical Union (AGU), International Association of Hydrological Sciences (IAHS), and American Society of Civil Engineers (ASCE). His primary research interests include satellite retrieval algorithms, cloud and precipitation system, radiative transfer modeling, data assimilation, radar polarimetry, mesoscale modeling, and artificial intelligence techniques applied to remote sensing and hydrometeorology.

Dr. Prashant K. Srivastava received his bachelors degree in Agriculture Sciences from the Institute of Agricultural Sciences, Banaras Hindu University (BHU), India, in 2004 and masters degree in Environmental Sciences from the School of Environmental Sciences (SES), Jawaharlal Nehru University (JNU), India in 2006. He worked on remote sensing applications as a Council of Scientific and Industrial Research-Junior Research Fellow in SES, JNU, for a year and then joined as an

Assistant Professor in the Department of Environmental Sciences, NVPAS, Sardar Patel University, Gujarat, in 2007. In 2010, he began work on his Ph.D. from the Department of Civil Engineering, University of Bristol. His research was sponsored and funded by India's Ministry of Human Research Development and under the United Kingdom's Commonwealth Scholarship and Fellowship Plan (CSFP). His Ph.D. research was focused on the soil moisture retrieval algorithm development from SMOS satellite and mesoscale model for hydrological applications. Following receipt of his Ph.D. in 2013, he moved to a position as a research scientist in the ESSIC/NASA Goddard Space Flight Center, Hydrological Sciences Branch, conducting work on SMAP satellite soil moisture retrieval algorithm development. Dr. Prashant K. Srivastava has been a recipient of many awards including Commonwealth fellowship, UK, CSIR fellowship (twice), MHRD, and UGC fellowships from India. He has published more than 45 peer-reviewed journal papers and more than 20 conference papers. Currently he is serving as an editorial board member of five journals related to remote sensing and environmental sciences. He is currently a member of the Indian Society of Geomatics, Indian Society of Remote Sensing, Indian Association of Hydrologists (*IAH*), International Society for Agrometeorology (INSAM), and International Association of Hydrological Sciences (IAHS).

Dr. Manika Gupta received her Master's degree in Environmental Sciences from the School of Environmental Sciences (SES), Jawaharlal Nehru University (JNU), India, in 2006. She worked as a researcher on remote sensing applications from SES, JNU, for a year and then joined the Department of Civil Engineering, Indian Institute of Technology, New Delhi, India, for her Ph.D. in 2007 as a UGC fellow. Her Ph.D. focused on numerical modeling for water resource management. Following receipt of her Ph.D. in 2012, she worked as a research consultant in Jaipur, India, on GIS and remote sensing for slum area delineation. She is currently working in collaboration with a number of pioneering global organizations such as University of Bristol, NOAA, and NASA on remote sensing advancements. Her current research involves assimilation of hyperspectral and microwave (precipitation) data in numerical models for pesticide and soil moisture prediction. Dr. Gupta has been a recipient of many awards including EIPRS (Australia), COSPAR (China), CSIR, ICMR, and UGC fellowships from India. She has published more than 25 peer-reviewed journal papers and more than 15 conference papers. She is a member of Indian Society of Geomatics, Indian Society of Remote Sensing, Indian Association of Hydrologists (*IAH*), and International Society for Agrometeorology (INSAM).

Dr. Xuan Zhu is Senior Lecturer in the School of Geography and Environmental Science at Monash University, where he is Director of Centre for GIS. He is also Guest Professor at Sichuan University. He has nearly 30 years of international experience with GIS development, application, and education in China, the UK, Ireland, Singapore, and Australia. He holds a Master's degree from Nanjing University in Cartography and Remote Sensing and a Ph.D. from the University of Edinburgh in GIS. His primary research areas are GIS for environmental management, spatial decision support systems for regional sustainable development, land use analysis and modeling, and remote sensing of the environment. He is the author of more than 50 scholarly works including a co-authored textbook on Principles and Techniques of Geographical Information Systems (East China Normal University Press, 2006). He teaches various GIS and remote sensing courses at Monash University.

Dr. Saumitra Mukherjee is Professor of Geology, Remote Sensing and Space Sciences at the School of Environmental Sciences, Jawaharlal Nehru University, Delhi. He obtained his M.Sc. and Ph.D. in Geology from the Banaras Hindu University, Varanasi, India. Prior to joining JNU in 1992, Professor Mukherjee served as Hydrogeologist and Remote Sensing Scientist for the Government of India. He has vast experience on satellites like IRS, LANDSAT, SPOT, Indian Remote Sensing Satellites, Hyperion, ResourceSat, LIDAR, and SOHO data for Sun–Earth Environment. He has authored seven books in the field of remote sensing and environmental sciences. He has published more than 80 papers in the peer-reviewed journals with nearly 60 conference contributions. He has supervised 18 doctoral theses and completed 14 collaborative projects funded by the national and international agencies, and is also involved in the projects funded by NASA on sun–earth connections. Professor Mukherjee has been designated as UGC Professor of Geology and Remote Sensing in view of his contribution in the field of Geology and Remote Sensing. He has also served as an executive commonwealth Fellow at University of Liverpool, UK. Prof. Mukherjee received many awards including the award given by the Government of India's Ministry of Environment and Forest for his academic contributions. He serves as an editorial board member of four journals. Currently, he is a member of number of pioneer organizations in the world including American Geophysical Union, European Space Agency, European Geosciences Union, SOC, JPL/CALTECH/NASA, European Fleet for Airborne Research, Canadian Remote Sensing Society, Indian Society of Remote Sensing, and Executive Fellow (Earth Sciences—India).

Part I
General

Chapter 1
Computational Intelligence Techniques and Applications

Xuan Zhu

Abstract Computational intelligence is a group of computational models and tools that encompass elements of learning, adaptation, and/or heuristic optimization. It is used to help study problems that are difficult to solve using conventional computational algorithms. Neural networks, evolutionary computation, and fuzzy systems are the three main pillars of computational intelligence. More recently, emerging areas such as swarm intelligence, artificial immune systems (AIS), support vector machines, rough sets, chaotic systems, and others have been added to the range of computational intelligence techniques. This chapter aims to present an overview of computational intelligence techniques and their applications, focusing on five representative techniques, including neural networks, evolutionary computation, fuzzy systems, swarm intelligence, and AIS.

Keywords Computational intelligence • Neural networks • Evolutionary computation • Fuzzy systems • Swarm intelligence • Artificial immune systems

1.1 Introduction

The earth and environmental systems are dynamic. They are shaped and changed continuously by complex and interrelated physical, biological, and chemical processes. For example, the physical process of weathering breaks down the rocks and soils into their constituent substances. When it rains, the water droplets absorb and dissolve carbon dioxide from the air, which causes the rainwater to be slightly acidic. The released sediment and chemicals then take part in chemical

X. Zhu (✉)
School of Geography and Environmental Science, Monash University, Building 11, Clayton Campus, Clayton, VIC 3800, Australia
e-mail: xuan.zhu@monash.edu

T. Islam et al. (eds.), *Computational Intelligence Techniques in Earth and Environmental Sciences*, DOI 10.1007/978-94-017-8642-3_1,

reactions that erode the earth's surface. The sediment may be transformed by geological forces into other rocks and soils. Living organisms also play a dynamic role through respiration, excretion, death, and decay. They recycle their constituent elements through the environment. Therefore, earth and environmental systems encompass numerous biological, physical, and chemical processes, which interact with each other and which are difficult to model and analyze. In addition, many earth and environmental systems present complex spatial and temporal patterns and behaviors. The interactions between these systems are often ill-defined and their relationships are generally nonlinear. Many earth and environmental problems have no strong theoretical understanding and, therefore, there are no full numerical models. The complexity of the earth and environmental systems has led to the need for effective and efficient computational tools to analyze and model highly nonlinear functions and can be trained to accurately generalize when presented with new, unseen data. The emerging computational intelligence techniques have some or all of these features. They provide an attractive alternative to developing numerical models to conventional statistical approaches, from which the new insights and underlying principles of earth and environmental sciences can be derived.

Computational intelligence is a group of computational models and tools devoted to solution of problems for which there are no effective computational algorithms (Konar 2005; Madani 2011). It is an offshoot of artificial intelligence, which focuses on heuristic algorithms such as neural networks, evolutionary computation, and fuzzy systems. Computational intelligence also involves adaptive mechanisms and learning ability that facilitate intelligent behavior in complex and changing environment. In addition to the three main pillars (neural networks, evolutionary computation, and fuzzy systems), computational intelligence includes swam intelligence, artificial immune systems (AIS), support vector machines, rough sets, chaotic systems, probabilistic methods, and other techniques or a combination of these techniques (Engelbrecht 2007). They have been successfully applied in a wide range of applications in the earth and environmental sciences, for example, species distribution modeling (Watts et al. 2011), site quality assessment of forests (Aertsen et al. 2010), prediction of water quality (Areerachakul et al. 2013), and forecasting of air pollution (Antanasijevia et al. 2013). This chapter aims to present an overview of computational intelligence techniques and their applications, focusing on the following five representative techniques, including neural networks, evolutionary computation, swam intelligence, AIS, and fuzzy systems.

1.2 Neural Networks

1.2.1 Basic Principles

Neural networks here refer to artificial neural networks (ANNs). They are developed to emulate biological neural systems. The basic building blocks of biological neural systems are neurons. A neuron consists of a cell body, an axon, and dendrites.

Fig. 1.1 An artificial neural network (ANN) with one hidden layer

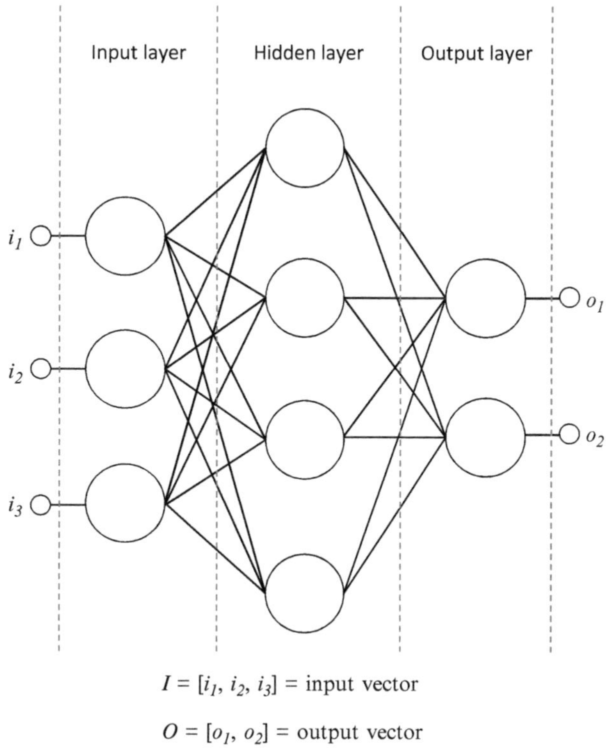

A neuron is connected through its axon with a dendrite of another neuron. The connection point between neurons is called a synapse. Basically, a neuron receives signals from the environment. When the signals are transmitted to the axon of the neuron, i.e., the neuron is fired, the cell sums up all the inputs, which may vary by the strength of the connection or the frequency of the input signals, processes the input sum, and then produces an output signal, which is propagated to all connected neurons. An artificial neuron is a simplified computational model of a biological neuron, performing the above basic functions.

In an artificial neuron, the cell body is modeled by an activation or transfer function. The artificial neuron receives one or more inputs (representing signals received through dendrites that are excited or inhibited by positive or negative numerical weights associated with each dendrite), calculates the weighted sum of the inputs, and passes the sum to the activation function. The activation function is a nonlinear function, and its output represents an axon, which propagates as an input to another neuron through a synapse. An activation function can be sigmoid, hyperbolic tangent, or linear (Konar 2005).

An ANN is a layered network of artificial neurons. It typically consists of an input layer, one or more hidden layers, and an output layer (Fig. 1.1). Each layer is composed of a number of artificial neurons, also called nodes. The artificial neurons in one layer are connected by weights to the artificial neurons in the next layer. It is essentially a model representing a nonlinear mapping between an input vector and output vector. As the output from an artificial neuron or node is a function of the sum of the inputs to the node modified by a nonlinear activation function (e.g., a logistic function), an ANN superposes many simple nonlinear activation functions used by the nodes constituting the network, which enables it to approximate highly nonlinear functions, thus introducing complex nonlinear behavior to the network (Principe et al. 2000). These functions can be trained to accurately generalize with new data. The adaptive property is embedded within the network by adjusting the weights that interconnect the nodes during the training phase. After the training phase the ANN parameters are fixed and the system is deployed to solve the problem at hand. Therefore, an ANN is an adaptive, nonlinear system that learns to perform a function from data.

There are several types of ANN, including multilayer feed-forward, recurrent, temporal, probabilistic, fuzzy, and radial basis function ANNs (Haykin 1999). The most popular one is the multilayer feed-forward neural network as shown in Fig. 1.1. It is also referred to as the multilayer perceptron network. In this type of ANN, the output of a node is scaled by the connection weights and fed forward as an input to the nodes in the next layer. That is, information flow starts from the nodes in the input layer, and then moves along weighted links to the nodes in the hidden layers for processing. The input layer plays no computational role, but provides the inputs to the network. The connection weights are normally determined through training. Each node contains an activation function that combines information from all the nodes in the preceding layer. The output layer is a complex function of the outcomes resulted from internal network transformations.

Multilayer perceptron networks are able to learn through training. Training involves the use of a set of training data with a systematic step-by-step procedure to optimize a performance criterion or to follow some implicit internal constraint, which is commonly referred to as the learning rule. Training data must be representative of the entire data set. A neural network starts with a set of initial connection weights. During training the network is repeatedly fed with the training data (a set of input–output pattern pairs obtained through sampling) and the connection weights in the network are modified until the learning rule is satisfied. One performance criterion could be a threshold value of an error signal, which is defined as the difference between the desired and actual output for a given input vector. Training uses the magnitude of the error signal to determine how much the connection weights need to be adjusted so that the overall error is reduced. The training process is driven by a learning algorithm, such as back-propagation (Rumelhart et al. 1986) and scaled conjugate gradient algorithms (Hagan et al. 1996). Once trained with representative training data, the multilayer perceptron network gains sufficient generalization ability and can be applied to new data.

1.2.2 Applications

ANNs are applicable when a relationship between the independent variables (outputs) and dependent variables (inputs) exists. They are able to learn the relationship from a given data set without any assumptions about the statistical distribution of the data. In addition, ANNs perform a nonlinear transformation of input data to approximate output data, learning from training data and exhibiting the ability for generalization beyond training data. This makes them more practical and accurate in modeling complex data patterns than many traditional methods that are linear. They are also able to deal with outlying, missing, and noisy data due to their ability to generalize well on unseen data. The capability of learning from data, modeling nonlinear relationships, and handling noises in data makes ANN particularly suitable for pattern classification, function approximation, and prediction in most earth and environmental applications. Pattern classification is to classify data into the predetermined discrete classes. Functional approximation is to formulate a function from a given set of training data to model the relationship between variables. Prediction involves the estimation of output from previous samples or forecasting of future trends in a time series of data given previous and current conditions.

ANNs have been employed to predict and assess air quality and forecast severe weather. For example, Gardner and Dorling (1999) applied multilayer perceptron networks to predict hourly NO_x and NO_2 concentrations in urban air in London. The neural networks used in this study have two hidden layers, each containing 20 nodes. They used five variables (low cloud amount, visibility, dry bulb temperature, vapor pressure, and wind speed) to predict the pollutant concentration. The activation function used in the models was the hyperbolic tangent function (Engelbrecht 2007). The networks were trained using the scaled conjugate gradient algorithm based on hourly meteorological data on the five variables. Their results suggest that ANNs outperform regression models, and can be used to resolve complex patterns of source emissions without any explicit external guidance. Sousa et al. (2007) used feed-forward ANNs to predict hourly ozone concentrations with principal components as inputs, which effectively combined principal component analysis and ANN techniques. This combination reduces the collinearity of the data sets, determines the relevant independent variables for the prediction of ozone concentrations, and thus leads to a less complex architecture of the ANN due to the decrease of input variables. It also eliminates the overfitting problem, so that mean square errors of both validation and training continuously decline. Data on precursor concentrations and meteorological variables on ozone formation were used. The principal component-based ANN was compared with multiple linear regression, principal component regression, and feed-forward ANNs based on the original data. The results suggested that the use of principal components as inputs improved the ANN prediction. Krasnopolsky (2013) provided a representative set of ANN applications in meteorology, oceanography, numerical weather prediction, and climate studies.

In geology ANNs have been used for sediment prediction, permeability prediction, simulation of chemical concentrations in minerals, and geological hazard assessment. For example, Yang and Rosenbaum (2003) built a multilayer perceptron network, integrated with a GIS, to predict distributions of sediments (sand, silt, and clay) in Gothenburg harbor, Sweden. The developed network has four input nodes (water depth, distance from the river mouth, bank, and shipping lanes) and three output nodes (one for each sediment grade) as well as two hidden layers. The network was trained using the data from 139 sample stations. Fegh et al. (2013) applied ANN to determine rock permeability of a gas reservoir from well logs at well locations, and used the results from the ANN modeling as an input to several geostatistical models through the structural model to construct 3D geological reservoir permeability models. Permeability prediction using ANN used the data sets derived from four wells of the studied gas field. Two of the wells have core permeability data, one used for constructing the ANN model and the other for evaluating the reliability of the ANN model. The model was then applied to predict permeability at the other un-cored wells. Torkar et al. (2010) used ANN to model nonlinear dependency of radon concentrations in soil gas on five environmental variables (air and soil temperature, barometric pressure of air and soil, and rainfall). A four-layer perceptron network was developed in this study with five input nodes (the environmental variables) and one output node (radon concentration) and it was trained using the back-propagation algorithm. The ANN model correctly predicted 10 seismic events out of 13 within the 2-year period. Bui et al. (2012) investigated the potentials of ANN in landslide susceptibility mapping at the Hoa Binh province of Vietnam. They built two multilayer feed-forward ANNs with back-propagation training algorithms, Levenberg–Marquardt and Bayesian regularization. Ten landslide conditioning factors were used as input nodes: slope, aspect, relief amplitude, lithology, land use, soil type, rainfall, and distance to roads, rivers, and faults. A landslide inventory over 10 years derived from satellite images, field surveys, and existing literature was utilized as training data. The connection weights obtained in the training phase were applied to the entire study area to produce landslide susceptibility indexes. The prediction accuracy of landslide susceptibility mapping by the Bayesian regularization neural network and the Levenberg–Marquardt neural network was 90.3 % and 86.1 %, respectively. The study suggested that the ANNs have good predictive capability, but the Bayesian regularization network model appears more robust and efficient.

ANNs have also found to outperform traditional or classic modeling methods in ecological modeling (Lek and Guegan 1999). There have been numerous applications in ecological modeling in various fields of ecology since the early 1990s. For example, Ranković et al. (2010) built a feed-forward ANN model to predict the dissolved oxygen in the Gruža Reservoir, Serbia. The input variables of the neural network include water pH, water temperature, chloride, total phosphate, nitrites, nitrates, ammonia, iron, manganese, and electrical conductivity. The Levenberg–Marquardt algorithm was used to train the network. The results were compared with the measured data on the basis of correlation coefficient, mean absolute error, and mean square error, which indicated that the neural network model provided accurate results. Other examples include using ANN to relate flow conditions to fish community diversity (Chang et al. 2013), determine factors that influence the dispersal of invasive species (Pontin et al. 2011), predict water quality indicators (Kuo et al. 2007; Huo et al. 2013), predict biodiversity (Yoo et al. 2012), estimate tree height (Özçelik et al. 2013), and simulate denitrification rates in wetlands (Song et al. 2013). Zhang (2010) offered an overall and in-depth knowledge on algorithms, programs, and case studies of ANNs in ecology.

Examples of ANN applications in hydrology include hydrological time series modeling (Lohani et al. 2012), groundwater salinity forecasting (Banerjee et al. 2011), rainfall–runoff modeling (Wu and Chau 2011), reservoir inflow prediction (Okkan 2012), and suspended sediment load prediction (Kakaei et al. 2013).

Pattern recognition and image classification are among the most common applications of ANN in remote sensing. The major advantage of ANN to image classification over conventional statistical approaches, such as maximum likelihood and Mahalanobis distance classifiers, is that ANN is essentially nonparametric and nonlinear, and has no assumptions regarding the underlying distribution of values of the explanatory variables and the training data. ANNs are found to be accurate in the classification of remotely sensed data (Mas 2004; Bao and Ren 2011; Dobreva and Klein 2011; Cruz-Ramírez et al. 2012).

1.3 Evolutionary Computation

1.3.1 Basic Principles

Evolutionary computation simulates natural evolution. It was born out of the idea of evolutionary programming introduced in the 1960s (Eiben and Smith 2003). Evolutionary computation includes evolutionary algorithms for solving search and optimization problems. It was thought of as a model for machine learning in which a population of randomly created individuals goes through a process of evolution mimicking the process of natural selection and natural genetics (Yu and Gen 2010). In every generation, a new set of artificial creatures is created using bits and pieces of the fittest of the old. An artificial creature is an individual representing a point in the problem's solution search space.

Evolutionary algorithms use random choice as a tool to guide a highly exploitive search toward regions of the search space with likely improvement. A single point in the solution search space is an individual, represented by a chromosome, which consists of genes. For example, a land use pattern is a single point in the search space of a land use optimization problem. A land use pattern can be seen as a chromosome. Genes are essentially the parameters of the problem to be solved. A gene can take many forms depending on the problem definition. For the land use allocation problem, a gene can be a land parcel with a given land use and land

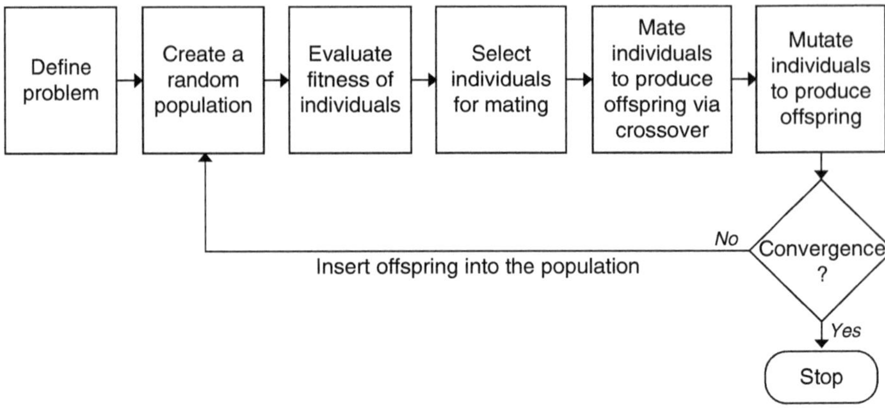

Fig. 1.2 General procedure of an evolutionary algorithm

attributes. The fitness (survival strength) of a chromosome is measured by a fitness or an objective function. A fitness function is some measure of profit, utility, or goodness to be maximized. A population is a collection of all the chromosomes being evolved. As new chromosomes are created and reinserted into the population, less fit chromosomes are replaced and only the fittest survive into the next generation. It is here that the process of evolution occurs, as the fitness of the competing chromosomes is compared in order to select parent chromosomes to reproduce.

Evolutionary algorithms generally follow the basic procedure as depicted in Fig. 1.2. The first step is to define the problem, which includes defining genes to encode the information needed for problem solving, specifying chromosomes to represent single solutions, and defining a fitness function. After the problem is defined, an initial population is randomly created as the first generation. All the chromosomes in the generation are then evaluated using the fitness function. After that, chromosomes are selected from the population according to their fitness function values to ensure that only the fittest chromosomes can survive into the next generation. The selected chromosomes are then combined in a process called crossover to create a set of children. The children are randomly mutated to create a new set of chromosomes to be reinserted into the population. Once enough children chromosomes have been created to replace a population, a generation is said to have passed. For the new generation, the evaluation, selection, crossover, mutation, and insertion process starts again. After a number of generations have elapsed, an optimal solution is converged and the process stops. The best chromosome is selected from the final population and represents the optimal solution to the problem being solved. Essentially what's happening is that a random set of solutions to a problem within a given search space is created and evolves over an amount of time to find an optimal solution.

There are different categories of evolutionary algorithms, including genetic algorithms, genetic programming, evolutionary programming, and evolution strategies. Genetic algorithms model genetic evolution (Goldberg 1989); genetic programming optimizes a population of computer programs based on genetic algorithms, where each individual is a computer program (Langdon and Poli 2002); evolutionary programming is similar to a genetic algorithm, but simulates only adaptive behavior in evolution (Yao et al. 1999); and evolution strategies are used to minimize functions of real variables, which are individuals, each having a "strategy variable" that determines the degree of mutation to be applied to the corresponding variable (Auger and Hansen 2011). They all share a basic principle of simulating the evolution of individuals through the process of evaluation, selection, crossover, mutation, and reproduction.

1.3.2 Applications

Evolutionary computation has been used successfully for optimization, scheduling, and time series approximation in earth and environmental applications.

Downing (1998) illustrated three applications of evolutionary computation in environmental modeling: optimal foraging strategies, temporal ideal-free distributions of larval emergence dates, and evolution of microscopic aquatic ecosystems. In nature, foraging animals often face difficult decisions on where to search for food. They need to make tradeoffs based on their knowledge of the relative abundance of food resources and predatory dangers as well as their current energetic condition (e.g., starvation, strong, weak). A foraging strategy consists of a list of the best foraging areas for each of the discrete energetic conditions. In optimization of animal foraging strategies, a genetic algorithm was used. It starts with an initial population of foragers that have randomly generated foraging strategies. Each forager with the initial middle energy level implements its strategy over a season. Every day over the season, the forager selects the area determined by its energy level and strategy, and it is then charged the daily metabolic cost and possibly fed and/or killed in accordance with the death and feeding-success probabilities associated with that area. The fitness of the forager and its associated strategy is measured as the average number of days it survives over the season. After all foragers and their associated strategies in a generation have obtained a fitness value, the fittest foragers are selected to produce the new generation of foragers by crossing over and mutating the strategy chromosomes of the parents. The new generation of strategies is then evaluated and the generational cycle continues until the convergence criterion is met and near-optimal strategies are found. Other examples of optimization with evolutionary computation include eco-design (Lim et al. 2013), mine planning (Riff et al. 2008), optimization of power factor and power output of wind turbines (Kusiak and Zheng 2010), and determination of the optimum pumping rates of coastal aquifers (Mantoglou et al. 2004).

The case of temporal ideal-free distributions of larval emergence dates described in Downing (1998) is an example of time series approximation. An ideal-free distribution refers to the situation where organisms diversify over space and/or time so that each gets about the same amount of resources. For species for which males often mate with several females, when male and female insects emerge from dormant stages before mating, male emergence distributions often match those of the females due to the competition for females (i.e., resources). This phenomenon is known as protandry. In the representation of a genetic algorithm, each individual is a chromosome encoding emergence times and sex. Males mate several times over several days and have either fixed active periods or age-independent mortality. Females mate only once on the emergence day. The life cycle of an insect commences with a dormant phase, which lasts until its emergence date. The insect becomes active for a fixed number of time steps, during which it shares the available food resources with other active insects. An insect's fitness is directly proportional to the amount of resources it acquires during the active period. The genetic algorithm was run on ten different cases. Each case has a different initial distribution of emergence times (the number of insects emerging on each day of the season), different frequency of the resource curve, and different postemergence life-span of the simulated insects. In all cases, the emergence distributions evolved to closely match the computed ideal-free distributions for the situation. The protandry simulations support one of the tacit assumptions of research on ideal-free distributions; that is, emergence times are genetically controlled. Other examples of time series approximation using genetic algorithms include the simulation of copepod population dynamics (Record et al. 2010) and modeling of long-term hydrological time series (Wang et al. 2009).

Timber harvest scheduling is a typical scheduling problem that can be solved using evolutionary computation. For instance, Ducheyne et al. (2004) applied genetic algorithms to develop forest harvesting plans. In the genetic algorithms, each chromosome represents a harvesting plan, encoding the felling period of each stand (gene). The fitness function maximizes the present value and minimizes deviations between successive cutting periods. It suggested that using multiple objective genetic algorithms to solve the harvest scheduling problem speeds up the optimization process and distributes the solutions evenly along the Pareto front.

1.4 Swarm Intelligence

1.4.1 Basic Principles

Swarm intelligence models the social behavior of organisms living in swarms or colonies. It is a form of agent-based modeling aiming at collective behavior of intelligent agents in decentralized, self-organized systems. Basic principles of swarm intelligence are derived from real swarms in nature, including ant colonies, bees, bird flocking, animal herding, fish schooling, and bacterial growth. A swarm intelligence system typically consists of a population of simple agents (individuals) interacting locally with one another and with their environment. These agents follow simple rules without centralized control dictating how individuals should behave. Local interactions of agent-to-agent or agent-to-environment lead to the emergence of intelligent global behavior or pattern which is unknown to the individual agents. In addition, the behavior of agents may change when the local environment is modified, which is referred to as stigmergy (i.e., the trace left in the environment by an action incites the performance of a next action by the same or different agents).

Swarm theory has led to the development of a number of algorithms for routing and optimization (Blum and Merkle 2008). Ant colony optimization (ACO) and particle swarm optimization (PSO) are two popular swarm intelligence techniques. ACO is a probabilistic technique for solving computational problems which involve finding optimum paths through graphs. It is inspired by the behavior of ants in finding paths toward food sources or their colonies. In the real world, ants initially roam randomly. Once they find foods, they return to their colony while leaving pheromone trails. When other ants find such a trail, they are likely to follow it and reinforce it with their own pheromone if they find foods. However, the pheromone along the trail evaporates over time, reducing its attractive strength. The longer it takes for an ant to travel along the trail, the weaker and less attractive the pheromone it laid down becomes. Over a short path, the strength of the pheromone remains high as it is reinforced as fast as or quicker than it can evaporate or decay. Ants tend to choose their trails with stronger pheromone concentrations. Therefore, when one ant finds a shorter path, other ants are more likely to follow that trail, and positive feedback eventually leads all the ants choose the shortest trail.

ACO mimics the behavior of ants with artificial ants walking along a graph representing the problem to solve. The graph consists of edges linking nodes. Each edge is a path from one node to another, representing a potential solution. Each edge is also assigned a pheromone value. Figure 1.3 shows a general procedure of ACO. First, the graph is initialized by assigning the same initial pheromone value to each edge and randomly selecting a node to place an artificial ant. Then the ant selects an edge to move at the current node with a probabilistic transition rule. This rule is commonly expressed as the probability (Dorigo et al. 1996)

$$p\left(e_{i,j}\right) = \frac{\left[\tau_{i,j}\right]^{\alpha}\left[f_{i,j}\right]^{\beta}}{\sum_{k\in\theta_i}\left[\tau_{i,k}\right]^{\alpha}\left[f_{i,k}\right]^{\beta}} \tag{1.1}$$

where $p(e_{i,j})$ is the probability of $e_{i,j}$ (the edge from node i to node j) being selected at node i, $\tau_{i,j}$ is the pheromone value associated with $e_{i,j}$,$f_{i,j}$ is a value of a weighting

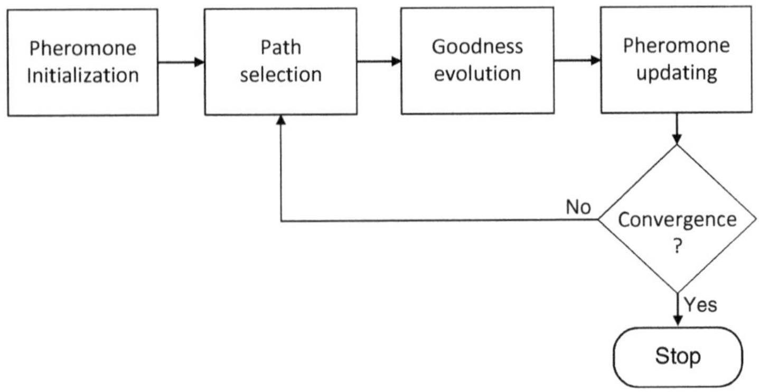

Fig. 1.3 A general procedure of ant colony optimization (ACO)

function (called a heuristic information) of $e_{i,j}$ measuring the desirability of the edge, θ_i is the set of edges available at node i, and α and β are parameters defining the relative importance of the pheromone strength and desirability. After a solution is obtained by the ant, its goodness is evaluated using an objective function. The pheromone value of each edge is then updated by uniformly decreasing all the pheromone values (pheromone evaporation) and increasing the pheromone values of one or more better solutions. The above process is repeated by applying a number of ants per iteration until a given convergence criterion (e.g., a time limit) is satisfied or all artificial ants follow the same path.

PSO simulates the social behavior of bird flocking and fish schooling. It is a search and optimization technique, similar to evolutionary computation techniques. It is initialized with a population of random solutions and searches for the optimal solutions through updating generations. However, unlike evolutionary computation techniques, it does not use evolution operators such as crossover and mutation. In PSO, individuals, referred to as particles, are grouped into a swarm. Each particle in the swarm represents a potential solution to the optimization problem. The particles are "flown" through the multidimensional search space by following the current "best" particle called guide. In each iteration, each particle is evaluated using the fitness function, and its velocity and position in the search space are updated using the following two equations (Kennedy and Eberhart 1995):

$$v(t+1) = v(t) + c_1 \times R_1 \times [\text{pbest} - p(t)] + c_2 \times R_2 \times [\text{gbest} - p(t)] \tag{1.2}$$

$$p(t+1) = p(t) + v(t+1) \tag{1.3}$$

where $v(t)$ is the current velocity of the particle, $p(t)$ is its current position, pbest and gbest are the best solutions achieved so far by the particle and population, respectively, R_1 and R_2 are the random numbers in the range [0, 1], and c_1 and c_2 are acceleration coefficients. pbest is also called the personal best, and gbest the global best. The process terminates when a time limit is reached or gbest cannot be improved further. While ACO is mainly used for combinatorial optimization, PSO is primarily employed for continuous optimization.

1.4.2 Applications

The study of swarm intelligence is providing insights that can help humans manage complex systems, from truck routing to military robots. ACO has many successful applications in discrete optimization problems such as travelling salesman problem. For example, Afshar (2010) used ACO for optimal design of sewer networks. In this study, the nodal elevations of the network were used as the decision variables of the optimization problem. The pheromone concentration over the allowable range of each decision variable was modeled with a Gaussian probability density function. It tried two alternative approaches to the implementation of ACO: constrained and unconstrained. The unconstrained approach did not take into account the minimum slope and other constraints. In the constrained approach, the elevation at downstream node of a pipe is used to define new bounds on the elevation of the upstream node, which represents the constraints on the pipe slopes. Other constraints include hydraulic radius, pipe diameter, average excavation depth, velocity, and flow depth of each link. The results from the constrained approach were compared with those of the unconstrained one. The constrained ACO was shown to be very effective in locating the optimal solution and efficient in terms of the convergence. It was also found to be relatively insensitive to the initial colony and its size when compared to the unconstrained algorithm.

PSO has few or no assumptions about the problem being optimized. It can search large solution spaces. Ma et al. (2011) applied PSO for land use optimization. In their study, each land parcel is abstracted to a particle by its centroid. Particles constantly fly to adjust their positions and velocities according to their personal best and the global best assessed in terms of the cost of land use transformation, biophysical suitability, and compactness of landscape. Chou (2012) used PSO in modeling rainfall–runoff relationships. The study compared PSO and a simple linear model in simultaneous identification of system structure and parameters of the rainfall–runoff relationship. The simple linear model combines classic models typically used in hydrology to simulate the subsystems, and transforms the system structure identification problem into a combinational optimization problem. The PSO was employed to select the optimized subsystem model with the best data fit. It found that the PSO simulates the time of peak arrival more accurately compared to the simple linear model, and it also accurately identifies the system structure and parameters of the rainfall–runoff relationship. PSO has also been applied for forecasting river stage (Chau 2007), modeling turbidity intrusion processes in flooding season (Wang et al. 2012), simulating soil moisture (Alizadeh and Mousavi 2013), optimizing greenhouse climate model parameters (Hasni et al. 2011), scheduling electric power production at a wind farm, predicting penetration rate of hard rock tunnel boring machines (Yagiz and Karahan 2011), etc.

1.5 Artificial Immune Systems

1.5.1 Basic Principles

AIS model the human immune system. The human immune system is a robust, decentralized, error-tolerant, and adaptive system. It is composed of a great variety of molecules, cells, and organs spreading all over the body. The main function of the human immune system is to search for malfunctioning cells from its own body (such as cancer cells) and foreign disease causing elements (such as bacteria and viruses). The elements that can be recognized by the immune system are referred to as antigens. The cells belonging to the body are called self or self-antigens, while the foreign cells entering the body are termed nonself or nonself-antigens. The immune system can distinguish between self and nonself. The field of AIS encompasses a spectrum of algorithms. Different algorithms mimic the behavior and properties of different immunological cells (specifically B-cells, T-cells, and dendritic cells). There are three main categories of AIS algorithms derived from the simplified immune systems: negative selection, clonal selection, and immune networks.

Negative selection algorithms simulate the negative selection process that occurs during the maturation of T-cells in the thymus. T-cells originate in the bone marrow, but pass on to the thymus to mature before they circulate the body in the blood and lymphatic vessels. Negative selection refers to the identification and elimination of those T-cells that may recognize and attack the self-antigens presented in the thymus. All T-cells that leave the thymus to circulate throughout the body are said to be tolerant to self. A typical negative selection algorithm involves a self-set S which defines the self-elements in a problem space (i.e., representative samples of self-antigen), and a detector set F which contains all elements that have been identified as nonself-antigens and do not belong to S (Dasgupta and Forrest 1999; de Castro and Timmis 2002). Basically, the algorithm first generates a random set of candidate elements C, and then compares the elements in C with the elements in S. If an element in C matches an element in S, it will be discarded; otherwise, it will be added to F. D'haeseleer et al. (1996) proposed a more efficient approach that tries to minimize the number of generated detector elements while maximizing the coverage of the nonself-space. After F is generated, it is used to detect nonself-elements in a data set (feature set) S^*, which may be composed of the set S plus other new features or a completely new set. The elements in S^* are checked against those in F. A match indicates a nonself-element is identified and an action will be followed. The follow-up action of detecting nonself varies according to the problem under investigation. The efficiency and complexity of a negative selection algorithm depend on the type of problem space (continuous, discrete, mixed, etc.), the problem representation scheme, and the matching rules. Most of the research works on the negative selection algorithm have used the binary matching rules like r-contiguous (Forrest et al. 1994). Negative selection algorithms have been applied to pattern recognition and classification.

Clonal selection algorithms are inspired by the clonal selection theory that explains how the immune system reacts when a nonself-antigen is detected by a B-cell. B-cells produce antibodies. When an antibody strongly matches an antigen, the corresponding B-cell is stimulated to generate clones of itself, which then produce more antibodies. The binding of an antibody to a nonself-antigen is a signal to destroy the invading organism on which the antigen is found. This process is called clonal selection. Clonal selection algorithms are most commonly applied to optimization and pattern recognition. They generally evolve candidate solutions by means of selection, cloning, and mutation. de Castro and von Zuben (2002) proposed a clonal selection algorithm for pattern recognition, which includes the following steps:

1. Create a random population of individuals (B-cells or antibodies).
2. Given a set of patterns (antigens) to be recognized, for each pattern, determine its affinity with each element of the population.
3. Select a number of the highest affinity individuals from the population and clone these individuals to a certain number of copies proportional to their affinity with the antigen. The greater the affinity, the larger the number of copies, and vice versa.
4. Mutate all the clones with a rate inversely proportional to their affinity with the input pattern. The higher the affinity, the smaller the mutation rate, and vice versa.
5. Add the mutated individuals to the population, and reselect the maturated (optimized) individuals to be kept as memories of the system. Delete other superfluous clones and replace them with new randomly generated individuals.
6. Repeat steps 2–5 until a stop condition is met, e.g., a minimum pattern recognition or classification error.

There are several variants of the clonal selection algorithm and clonal selection-based hybrid algorithms, which are reviewed in Berna and Sadan (2011).

Immune network algorithms assume that B-cells form a network. When a B-cell is activated as a response to an antigen, it stimulates all other B-cells to which it connects in the network. These algorithms are similar to clonal selection algorithms in that they both measure the goodness of B-cells by affinities and both involve a process of selection, cloning, and mutation. The main difference is that the immune network algorithms consider that B-cells are not isolated, but communicate with each other via collective dynamic network interactions, while clonal selection algorithms only care about the interactions between B-cells and antigens. An immune network algorithm develops a population of individuals (B-cells) that interact with data (antigens) and with each other. The interactions with antigens and between B-cells fire up the B-cells. Highly stimulated B-cells undertake cloning and mutation as they do in a clonal selection algorithm. The number of clones and mutation rate also depend on the affinity of the cell with the current stimulating antigen. This process is regulated by the interaction between B-cells, which can stimulate them in order to create a memory of observed antigens, or suppress them, in order to control the immune response. It also includes natural death of unstimulated B-cells and addition of new random B-cells to the population. A working procedure of an artificial immune network can be found in de Castro and von Zuben (2001). Immune network algorithms perform unsupervised learning. They have been typically used for clustering, but have also been adapted to optimization and classification.

1.5.2 Applications

AIS have been successfully used for optimization, classification/clustering, and pattern recognition. For example, Liu et al. (2012) applied AIS for optimizing multi-objective allocation of water resources in river basins. They integrated the macroevolution algorithm (Marin and Sole 1999), clonal selection, and an entropy-based density assessment scheme (EDAS) to perform a global optimal search. The clonal selection was based on the diversity in the evolving population and applied for solution exploitation. EDAS was used to distribute non-dominated individuals uniformly along the discovered Pareto-frontier, and the macroevolution algorithm is employed to preserve the diversity of individuals and form part of the pool solution.

AIS have been widely used for remote sensing image classification and pattern recognition (Xu and Wu 2008; Zhang et al. 2004; Zheng and Li 2007; Zhong et al. 2007). Gong et al. (2011) developed an improved artificial immune network algorithm for land cover classification from remote sensing images. It involves creation of land cover class representatives as antibodies or B-cells. Basically, an initial population of antibodies is randomly generated so that the possibility of successful recognition of a land cover class is maximized without prior knowledge of antigens. The initialized population of antibodies is then optimized by cloning and mutating the antibodies that can best recognize the antigens. The system then evolves antibodies over a number of generations until stopping criteria are met. In this study, the best antibodies for each land cover class were preserved in each generation. An adaptive mutation rate was used to adjust the learning speed in response to the difference between the classification accuracies of the current and previous generation. In addition, the Euclidean distance and spectral angle mapping distance are used as affinity measures. A genetic algorithm was also used to identify optimal weights representing contributions from different affinity measures. The artificial immune network algorithm was applied to classify land covers in a residential area in Denver, CO, with high-spatial resolution QuickBird image and LiDAR data and in a suburban area in Monticello, UT, with HyMap hyperspectral imagery. The method was compared with a decision tree, a multilayer feed-forward back-propagation neural network, and another artificial immune networks algorithm from de Castro and von Zuben (2001). The results showed that it outperformed the other methods with higher accuracy and more spatially cohesive land cover classes with limited salt-and-pepper effect.

1.6 Fuzzy Systems

1.6.1 Basic Principles

Fuzzy systems make use of fuzzy sets or fuzzy logic. Traditional set theory requires objects to be either part of a set or not. For example, the set of rice paddy fields is distinct from the set of forest stands. If a piece of land belongs to one of the two sets, it cannot belong to the other. Such sets are called crisp. Crisp sets have well-defined boundaries with no ambiguity about an object's membership. However, in earth and environmental studies, our observations and reasoning are often not this exact and usually include a measure of membership. For instance, many data collected in the field survey are often described in ambiguous words: soils can be recorded as being poorly drained, slightly susceptible to soil erosion, and marginally suitable for maize. Such sets are fuzzy. With fuzzy sets, an object belongs to a set to a certain degree of membership. Mathematically, a crisp set is described by a characteristic function whose value is always either 1 for elements of the set or 0 for those outside the set (Fig. 1.4a). A fuzzy set is defined by a membership function that takes values in the range of 0 and 1 (Fig. 1.4b). An element belongs to a fuzzy set if the value of the set's membership function at that element is nonzero. A nonzero value of the membership function indicates the degree to which an element belongs to the set.

Fuzzy logic allows approximate reasoning based on fuzzy sets. It usually uses IF-THEN rules. The following is an example of a fuzzy rule:

IF the soil depth is shallow and accumulated temperature is moderate
THEN the suitability for maize is marginal

The AND, OR, and NOT logic operators of traditional Boolean logic are also used in fuzzy logic. The AND operation is the intersection of fuzzy sets, given by the minimum of the membership functions. OR is the union of fuzzy sets, defined as the

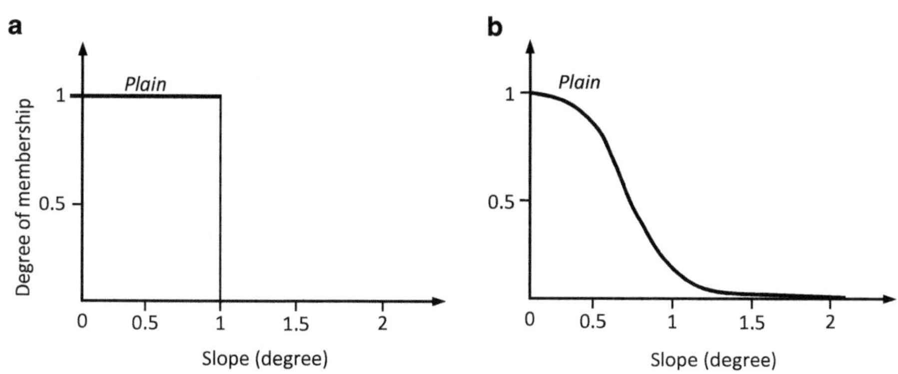

Fig. 1.4 Crisp set vs. fuzzy set defining plain. (**a**) Characteristic function, (**b**) fuzzy membership function

maximum of the membership functions. NOT gives the complement of a fuzzy set. Other operators exist in fuzzy logic (Zimmermann 2001).

A fuzzy system can be defined as a collection of IF-THEN rules with fuzzy predicates, or an algebraic or differential equation in which parameters are fuzzy numbers expressing the vagueness in the parameter values, or a system with fuzzy inputs (such as readings from unreliable sensors or quantities relating to human perception) and fuzzy outputs. In the earth and environmental systems, the majority of the phenomena are understood only partially and therefore cannot be modeled using mathematical models. A significant portion of knowledge about the earth and environmental systems is available as the heuristic knowledge or rule of thumb from experts or practitioners. Fuzzy rule-based systems can be used to represent such knowledge and make approximate reasoning using the heuristic knowledge. From this point of view, fuzzy systems can be considered as a type of expert systems or knowledge-based systems (Patterson 1990). In addition, fuzzy systems provide an alternative approach to dealing with uncertainty as not all types of uncertainty can be handled by traditional stochastic modeling framework. The uncertainty in fuzzy systems is non-statistical, based on vagueness, imprecision, and/or ambiguity. Non-statistical uncertainty is an inherent property of a system and cannot be changed or resolved by observations. Stochastic methods deal with statistical uncertainty, which is based on the laws of probability and can be resolved through observations. Moreover, like other computational intelligence techniques such as ANN, fuzzy systems can be used to model nonlinear systems and approximate other functions or measurement data with a desired accuracy. They provide a more transparent representation of the system under investigation, mainly due to the linguistic model interpretation in the way close to the one humans use for reasoning about the real world.

1.6.2 Applications

Fuzzy systems are often a choice when dealing with non-statistic uncertainties in applications including classification/clustering, function approximation, and prediction. Rezaei et al. (2013) provides an example of classification using fuzzy logic in evaluation of groundwater pollution. In this study, groundwater vulnerability was assessed by using linguistic variables to describe hydrogeological characteristics and linguistic terms to define vulnerability ratings. The Boolean logic cannot reflect the actual differences between the points in a hydrogeological setting; thus, regions that should have different vulnerability indices may be characterized by the same index. In contrast, a fuzzy system is able to adjust itself with the range of variation of input indices. The study demonstrated that fuzzy logic allows better and more logical ratings to be obtained for the values located near the classification boundaries. Other examples include using fuzzy clustering methods to characterize the physicochemical properties of groundwater and assess the impact of anthropogenic activities on the groundwater hydrology and chemistry (Güler et al. 2012), and to cluster forests according to their forest fire risk (Iliadis et al. 2010).

An example of function approximation is the use of fuzzy systems to downscale the regional wind climate to local wind conditions, considering the surrounding topography, in order to assess the wind potential available in an area for wind farms (de la Rosa et al. 2011). Wind exhibits high local-scale variability caused by the local topography, roughness, obstacles, etc. Although the regional wind climate can be extrapolated from the wind measurements at the meteorological stations, it does not reflect the local variability. The real wind is the result of local conditions. Due to the vagueness in the terrain description (e.g., down slope, up slope, and plain) in the local scale, the low quality of meteorological data (without measuring local topographic effects) recorded at the stations, and the chaotic dynamics inherent to atmospheric events, de la Rosa et al. (2011) proposed to use a fuzzy system to transform the regional wind distribution into the real one, which takes into account topographic parameters. The fuzzy system was designed to establish a link between the local wind conditions and the terrain features. It calculates the probability of possible changes in the wind direction based on the analysis of the terrain in those directions. The membership distributions of the fuzzy system were also optimized using a genetic algorithm. The fuzzy system effectively approximated the local wind conditions by transforming a regional wind climate model.

In the study by Kayastha (2012), a fuzzy system approach was used to predict landslide susceptibility. Eight causative factors were used in the predictive modeling, including slope, aspect, slope shape, relative relief, distance from drainage, land use, geology, and distance from active faults. Likelihood ratios were calculated for each class of the causative factors by comparing with past landslide occurrences. The likelihood ratios were then normalized to fuzzy membership values between 0 and 1. The study compared several different fuzzy operators and found that the fuzzy gamma (λ) operator with a λ-value of 0.70 produced the best prediction accuracy. Other instances of prediction with fuzzy systems include prediction of air quality (Fisher 2003; Carbajal-Hernández et al. 2012), ocean wave energy (Özger 2011), and habitat quality (Mocq et al. 2013).

1.7 Conclusions

The complexity and nonlinearity of most earth and environmental problems have led to the increased use of computational intelligence techniques. This chapter reviewed five representative methods of computational intelligence and their applications. The characteristics of these techniques reveal that each technique has its own merits and limitations. Fuzzy systems, for instance, are good at approximate reasoning, but do not have learning and optimization ability, while ANN is capable of adaptive learning and evolutionary computation is efficient in intelligent search and optimization. Fusion of different computational intelligence techniques thus may provide better computational models. For example, coupling ANN and fuzzy systems provides capabilities of learning from approximate data or knowledge and approximate reasoning using the knowledge derived through adaptive learning (Azar 2010). Attempts to combine evolutionary computation and ANN have a long history. For example, evolutionary computation techniques can be used to determine the weights of an ANN with the mean square error sum of the neurons at the output layer as the fitness function (Piotrowski and Napiorkowski 2011). Evolutionary computation can also be used to optimize parameters of a fuzzy system and adapt the membership distribution to optimize the performance of a fuzzy system (Sanchez et al. 1997; de la Rosa et al. 2011). An ANN can be trained with fuzzy membership distributions that have been optimized by an evolutionary computation technique. Moreover, evolutionary computation may be used to determine the best set of training data in an ANN-fuzzy system (Azar 2010). Chapter 2 of this book provides a more extensive overview of computational intelligence applications in earth and environmental sciences. Other chapters in this book offer more technical discussions on some computational intelligence techniques including those that were not reviewed in this chapter.

Computational intelligence is a collection of computational models and tools, whose classification, clusterization, optimization, prediction, reasoning, and approximation capabilities have been improved incrementally and continuously. There are already many computational intelligence techniques or combinations of the techniques. It is always possible to find alternative techniques to address a specific earth and environmental problem. However, domain-specific knowledge underlying the physical, chemical, or biological processes in an earth or environmental system under investigation should be incorporated into the problem formulation, selection of appropriate computational intelligence techniques, and evaluation of modeling results. There is also a need to incorporate computational intelligence techniques into existing modeling framework so that they can be widely accepted in the communities who are traditionally accustomed to process models. Computational intelligence provides practical tools for earth and environmental scientists to handle heterogeneous, incomplete, and noise data and to build models of uncertainty trained on the historical, current, and predicted data. We will see more advances in computational intelligence techniques and more innovative use of these techniques in earth and environmental applications in the future.

References

Aertsen W, Kint V, van Orshoven J, Özkan K, Muys B (2010) Comparison and ranking of different modelling techniques for prediction of site index in Mediterranean mountain forests. Ecol Model 221(8):1119–1130

Afshar MH (2010) A parameter free Continuous Ant Colony Optimization Algorithm for the optimal design of storm sewer networks: constrained and unconstrained approach. Adv Eng Softw 41(2):188–195

Alizadeh H, Mousavi SJ (2013) Coupled stochastic soil moisture simulation-optimization model of deficit irrigation. Water Resour Res 49(7):4100–4113

Antanasijevia DZ, Pocajt VV, Povrenovia DS, Ristia MA, Peria-Grujia AA (2013) PM.sub.10 emission forecasting using artificial neural networks and genetic algorithm input variable optimization. Sci Total Environ 443:511–519

Areerachakul S, Sophatsathit P, Lursinsap C (2013) Integration of unsupervised and supervised neural networks to predict dissolved oxygen concentration in canals. Ecol Model 261–262:1–7

Auger A, Hansen N (2011) Theory of evolution strategies: a new perspective. In: Auger A, Doerr B (eds) Theory of randomized search heuristics, foundations and recent developments. World Scientific, Singapore, pp 289–325

Azar AT (2010) Adaptive neuro-fuzzy systems. In: Azar AT (ed) Fuzzy systems. InTech, Croatia, pp 85–110

Banerjee P, Singh VS, Chatttopadhyay K, Chandra PC, Singh B (2011) Artificial neural network model as a potential alternative for groundwater salinity forecasting. J Hydrol 398(3):212–220

Bao YH, Ren JB (2011) Wetland landscape classification based on the BP neural network in DaLinor Lake Area. Procedia Environ Sci 10:2360–2366

Berna HU, Sadan KK (2011) A review of clonal selection algorithm and its applications. Artif Intell Rev 36(2):117–138

Blum C, Merkle D (2008) Swarm intelligence: introduction and applications. Springer, Berlin

Bui DT, Pradhan B, Lofman O, Revhaug I, Dick OB (2012) Landslide susceptibility mapping at Hoa Binh Province (Vietnam) using an adaptive neuro-fuzzy inference system and GIS. Comput Geosci 45:199–211

Carbajal-Hernández JJ, Sánchez-Fernández LP, Carrasco-Ochoa JA, Martínez-Trinidad JF (2012) Assessment and prediction of air quality using fuzzy logic and autoregressive models. Atmos Environ 60:37–50

Chang F, Tsai W, Chen H, Yam RS, Herricks EE (2013) A self-organizing radial basis network for estimating riverine fish diversity. J Hydrol 476:280–289

Chau KW (2007) A split-step particle swarm optimization algorithm in river stage forecasting. J Hydrol 346(3):131–135

Chou CM (2012) Particle swarm optimization for identifying rainfall-runoff relationships. J Water Resour Protect 4:115–126

Cruz-Ramírez M, Hervás-Martínez C, Jurado-Expósito M, López-Granados F (2012) A multi-objective neural network based method for cover crop identification from remote sensed data. Expert Syst Appl 39(11):10038–10048

D'haeseleer P, Forrest S, Helman P (1996) An immunological approach to change detection: algorithms, analysis, and implications. In: Proceedings of the IEEE symposium on computer security and privacy. IEEE Computer Society Press, Los Alamitos, CA, pp 110–119

Dasgupta D, Forrest S (1999) An anomaly detection algorithm inspired by the immune system. In: Dasgupta D (ed) Artificial immune systems and their applications. Springer, Berlin, pp 262–277

de Castro LN, Timmis J (2002) Artificial immune systems: a novel paradigm to pattern recognition. In: Corchado JM, Alonso L, Fyfe C (eds) Artificial neural networks in pattern recognition. SOCO-2002. University of Paisley, Paisley, England, pp 67–84

de Castro LN, von Zuben FJ (2001) aiNet: an artificial immune network for data analysis. In: Abbass HA, Sarker RA, Newton CS (eds) Data mining: a heuristic approach (Chapter XII). Idea Group Publishing, Hershey, PA, pp 231–259

de Castro LN, von Zuben FJ (2002) Learning and optimization using the clonal selection principle. IEEE Trans Evol Comput 6(3):239–251

de la Rosa JJG, Pérez AA, Salas JCP, Leo JGR, Muñoz AM (2011) A novel inference method for local wind conditions using genetic fuzzy systems. Renew Energy 36:1747–1753

Dobreva ID, Klein AG (2011) Fractional snow cover mapping through artificial neural network analysis of MODIS surface reflectance. Remote Sens Environ 115(12):3355–3366

Dorigo M, Maniezzo V, Colorni A (1996) Ant system: optimization by a colony of cooperating agents. IEEE Trans Syst Man Cybern B Cybern 26(1):29–41

Downing K (1998) Using evolutionary computational techniques in environmental modelling. Environ Model Softw 13:519–528
Ducheyne EI, de Wulf RR, de Baets B (2004) Single versus multiple objective genetic algorithms for solving the even-flow forest management problem. For Ecol Manage 201(2):259–273
Eiben AE, Smith JE (2003) Introduction to evolutionary computing. Springer, New York
Engelbrecht AP (2007) Computational intelligence: an introduction. Wiley, Chichester, England
Fegh A, Riahi MA, Norouzi GH (2013) Permeability prediction and construction of 3D geological model: application of neural networks and stochastic approaches in an Iranian gas reservoir. Neural Comput Appl 23(6):1763–1770. doi:10.1007/s00521-012-1142-8
Fisher B (2003) Fuzzy environmental decision-making: applications to air pollution. Atmos Environ 37(14):1865–1877
Forrest S, Perelson AS, Allen L, Cherukuri R (1994) Self-nonself discrimination in a computer. In: Proceedings of the IEEE symposium on research in security and privacy. IEEE Computer Society Press, Los Alamitos, CA, pp 202–212
Gardner MW, Dorling SR (1999) Neural network modelling and prediction of hourly NO_x and NO_2 concentrations in urban air in London. Atmos Environ 33:709–719
Goldberg DE (1989) Genetic algorithm in search, optimization, and machine learning. Addison-Wesley, Reading, MA
Gong B, Im J, Mountrakis G (2011) An artificial immune network approach to multi-sensor land use/land cover classification. Remote Sens Environ 115:600–614
Güler C, Kurt MA, Alpaslan M, Akbulut C (2012) Assessment of the impact of anthropogenic activities on the groundwater hydrology and chemistry in Tarsus coastal plain (Mersin, SE Turkey) using fuzzy clustering, multivariate statistics and GIS techniques. J Hydrol 414–415:435–451
Hagan MT, Demuth HB, Beale MH (1996) Neural network design. PWS Publishing, Boston
Hasni A, Taibi R, Draoui B, Boulard T (2011) Optimization of greenhouse climate model parameters using particle swarm optimization and genetic algorithms. Energy Procedia 6:371–380
Haykin S (1999) Neural networks: a comprehensive foundation, 2nd edn. Prentice-Hall, Upper Saddle River, NJ
Huo S, He Z, Su J, Xi B, Zhu C (2013) Using artificial neural network models for eutrophication prediction. Procedia Environ Sci 18:310–316
Iliadis LS, Vangeloudh M, Spartalis S (2010) An intelligent system employing an enhanced fuzzy c-means clustering model: application in the case of forest fires. Comput Electron Agric 70(2):276–284
Kakaei LE, Moghaddam NA, Ahmadi A (2013) Daily suspended sediment load prediction using artificial neural networks and support vector machines. J Hydrol 478:50–62
Kayastha P (2012) Application of fuzzy logic approach for landslide susceptibility mapping in Garuwa sub-basin, East Nepal. Front Earth Sci 6(4):420–432
Kennedy J, Eberhart RC (1995) Particle swarm optimization. In: Proceedings of IEEE international conference on neural networks, vol 4. IEEE Press, Piscataway, NJ, pp 1942–1948
Konar A (2005) Computational intelligence: principles, techniques and applications. Springer, Berlin
Krasnopolsky VM (2013) The application of neural networks in the earth system sciences. Springer, The Netherlands
Kuo J, Hsieh M, Lung W, She N (2007) Using artificial neural network for reservoir eutrophication prediction. Ecol Model 200(1):171–177
Kusiak A, Zheng H (2010) Optimization of wind turbine energy and power factor with an evolutionary computation algorithm. Energy 35(3):1324–1332
Langdon WB, Poli R (2002) Foundations of genetic programming. Springer, Berlin
Lek S, Guegan JF (1999) Artificial neural networks as a tool in ecological modelling, an introduction. Ecol Model 120:65–73

Lim S, Kim YR, Woo SH, Park D, Park JM (2013) System optimization for eco-design by using monetization of environmental impacts: a strategy to convert bi-objective to single-objective problems. J Clean Prod 39:303–311
Liu D, Guo S, Chen X, Shao Q, Ran Q, Song X, Wang Z (2012) A macro-evolutionary multi-objective immune algorithm with application to optimal allocation of water resources in Dongjiang River basins, South China. Stoch Environ Res Risk Assess 26(4):491–507
Lohani AK, Kumar R, Singh RD (2012) Hydrological time series modeling: a comparison between adaptive neuro-fuzzy, neural network and autoregressive techniques. J Hydrol 442–443:23–35
Ma S, He J, Liu F, Yu Y (2011) Land-use spatial optimization based on PSO algorithm. Geo-spat Inf Sci 14:54–61
Madani K (2011) Computational intelligence. Springer, Berlin
Mantoglou A, Papantoniou M, Giannoulopoulos P (2004) Management of coastal aquifers based on nonlinear optimization and evolutionary algorithms. J Hydrol 297(1):209–228
Marin J, Sole RV (1999) Macroevolutionary algorithms: a new optimization method on fitness landscapes. IEEE Trans Evol Comput 3(4):272–286
Mas JF (2004) Mapping land use/cover in a tropical coastal area using satellite sensor data, GIS and artificial neural networks. Estuar Coast Shelf Sci 59:219–230
Mocq J, St-Hilaire A, Cunjak RA (2013) Assessment of Atlantic salmon (Salmo salar) habitat quality and its uncertainty using a multiple-expert fuzzy model applied to the Romaine River (Canada). Ecol Model 265:14–25
Okkan U (2012) Wavelet neural network model for reservoir inflow prediction. Sci Iran 19(6):1445–1455
Özçelik R, Diamantopoulou MJ, Crecente-Campo F, Eler U (2013) Estimating Crimean juniper tree height using nonlinear regression and artificial neural network models. For Ecol Manage 306:52–60
Özger M (2011) Prediction of ocean wave energy from meteorological variables by fuzzy logic modeling. Expert Syst Appl 38(5):6269–6274
Patterson DW (1990) Introduction to artificial intelligence and expert systems. Prentice-Hall, Englewood Cliffs, NJ
Piotrowski AP, Napiorkowski JJ (2011) Optimizing neural networks for river flow forecasting—evolutionary computation methods versus the Levenberg–Marquardt approach. J Hydrol 407:12–27
Pontin DR, Schliebs S, Worner SP, Watts MJ (2011) Determining factors that influence the dispersal of a pelagic species: a comparison between artificial neural networks and evolutionary algorithms. Ecol Model 222(10):1657–1665
Principe JC, Euliano NR, Lefebvre WC (2000) Neural and adaptive systems: fundamentals through simulations. Wiley, New York
Ranković V, Radulović J, Radojević I, Ostojić A, Čomić L (2010) Neural network modeling of dissolved oxygen in the Gruža reservoir, Serbia. Ecol Model 221(8):1239–1244
Record NR, Pershing AJ, Runge JA, Mayo CA, Monger BC, Chen C (2010) Improving ecological forecasts of copepod community dynamics using genetic algorithms. J Mar Syst 82(3):96–110
Rezaei F, Safavi H, Ahmadi A (2013) Groundwater vulnerability assessment using fuzzy logic: a case study in the Zayandehrood aquifers, Iran. Environ Manage 51:267–277
Riff MC, Alfaro T, Bonnaire X, Grandon C (2008) EA-MP: an evolutionary algorithm for a mine planning problem. In: Proceedings of IEEE congress on evolutionary computation, June 2008, pp 4011–4014
Rumelhart DE, Hinton GE, Williams RJ (1986) Learning internal representations by error propagation. In: Rumelhart DE, McClelland JL (eds) Parallel distributed processing: explorations in the microstructure of cognition, vol 1. MIT Press, Cambridge, MA, pp 318–362
Sanchez E, Shibata T and Zadeh LA (1997) Genetic algorithms and fuzzy logic systems: soft computing perspectives. World Scientific Pub., Singapore; River Edge, NJ

Song K, Park Y, Zheng F, Kang H (2013) The application of Artificial Neural Network (ANN) model to the simulation of denitrification rates in mesocosm-scale wetlands. Ecol Inform 16:10–16

Sousa SIV, Martins FG, Alvim-Ferraz MCM, Pereira MC (2007) Multiple linear regression and artificial neural networks based on principal components to predict ozone concentrations. Environ Model Softw 22:97–103

Torkar D, Zmazek B, Vaupotič J, Kobal I (2010) Application of artificial neural networks in simulating radon levels in soil gas. Chem Geol 270:1–8

Wang W, Xu D, Qiu L, Ma J (2009) Genetic programming for modelling long-term hydrological time series. In: Proceedings of the fifth international conference on natural computation, Aug 2009, vol 4, pp 265–269

Wang S, Qian X, Wang QH, Xiong W (2012) Modeling turbidity intrusion processes in flooding season of a canyon-shaped reservoir, South China. Procedia Environ Sci 13:1327–1337

Watts MJ, Li Y, Russell BD, Mellin C, Connell SD, Fordham DA (2011) A novel method for mapping reefs and subtidal rocky habitats using artificial neural networks. Ecol Model 222(15):2606–2614

Wu CL, Chau KW (2011) Rainfall–runoff modeling using artificial neural network coupled with singular spectrum analysis. J Hydrol 399(3–4):394–409

Xu S, Wu Y (2008) An algorithm for remote sensing image classification based on artificial immune B-cell network. In: The international archives of the photogrammetry, remote sensing and spatial information sciences, vol XXXVII, part B6b, pp 107–112

Yagiz S, Karahan H (2011) Prediction of hard rock TBM penetration rate using particle swarm optimization. Int J Rock Mech Mining Sci 48(3):427–433

Yang Y, Rosenbaum MS (2003) Artificial neural networks linked to GIS. In: Nikravesh M, Aminzadeh F, Zadeh LA (eds) Developments in petroleum science, vol 51, Soft computing and intelligent data analysis in oil exploration. Elsevier, The Netherlands, pp 633–650

Yao X, Liu Y, Lin G (1999) Evolutionary programming made faster. IEEE Trans Evol Comput 3(2):82–102

Yoo J, Lee Y, Lee C, Kim C (2012) Effective prediction of biodiversity in tidal flat habitats using an artificial neural network. Mar Environ Res 83:1–9. doi:10.1016/j. marenvres.2012.10.001

Yu X, Gen M (2010) Introduction to evolutionary algorithms. Springer, New York

Zhang W (2010) Computational ecology: artificial neural networks and their applications. World Scientific, Singapore

Zhang X, Shan T, Jiao L (2004) SAR image classification based on immune clonal feature selection. In: Mohamed SK, Aurélio CC (eds) Proceedings of image analysis and recognition, vol 3212, Lecture notes in computer science. Springer, Berlin, pp 504–511

Zheng H, Li L (2007) An artificial immune approach for vehicle detection from high resolution space imagery. Int J Comput Sci Network Security 7:67–72

Zhong Y, Zhang L, Huang B, Li P (2007) A resource limited artificial immune system algorithm for supervised classification of multi/hyper-spectral remote sensing imagery. Int J Remote Sens 28:1665–1686

Zimmermann H (2001) Fuzzy set theory and its applications. Kluwer Academic, Boston

Part II
Classical Intelligence Techniques in Earth and Environmental Sciences

Chapter 2
Vector Autoregression (VAR) Modeling and Forecasting of Temperature, Humidity, and Cloud Coverage

Md. Abu Shahin, Md. Ayub Ali, and A.B.M. Shawkat Ali

Abstract Climate change is a global phenomenon but its implications are distinctively local. The climatic variables include temperature, rainfall, humidity, wind speed, cloud coverage, and bright sunshine. The study of behavior of the climatic variables is very important for understanding the future changes among the climatic variables and implementing important policies. The problem is how to study the past, present, and future behaviors of the climatic variables. The purpose of the present study was to develop an appropriate vector autoregression (VAR) model for forecasting monthly temperature, humidity, and cloud coverage of Rajshahi district in Bangladesh. The test for stationarity of the time series variables has been confirmed with augmented Dickey–Fuller, Phillips–Perron, and Kwiatkowski–Phillips–Schmidt–Shin tests. The endogenity among the variables was examined by F-statistic proposed by C.W.J. Granger. The order of the VAR model was selected using Akaike information criterion, Schwarz information criteria, Hannan–Quinn information criteria, final prediction error, and likelihood ratio test. The ordinary least square method was used to estimate the parameters of the model. The VAR(8) model was found to be the best. Structural analyses were performed using forecast error variance decomposition and impulse response function. These structural analyses divulged that the temperature, humidity, and cloud coverage would be interrelated and endogenous in future. Finally, temperature, humidity, and cloud coverage were forecasted from January 2011 to December 2016 using the best selected model VAR(8). The forecasted values showed an upward trend in temperature and humidity and downward trend in cloud coverage. Therefore, we must show our friendly behavior to the environment to control such trends.

M.A. Shahin (✉) • M.A. Ali
Department of Statistics, University of Rajshahi, Rajshahi 6205, Bangladesh
e-mail: shahin.hera10@gmail.com; ayubali67@yahoo.com

A.B.M. S. Ali
School of Engineering and Technology, Central Queensland University,
North Rockhampton, QLD 4702, Australia
e-mail: s.ali@cqu.edu.au

T. Islam et al. (eds.), *Computational Intelligence Techniques in Earth and Environmental Sciences*, DOI 10.1007/978-94-017-8642-3_2,

Keywords Forecast error variance (FEV) decomposition • Granger causality test • Impulse response function (IRF) • Unit root • Vector autoregression (VAR)

2.1 Introduction

Bangladesh is a developing country and likely to be one of the worst victims of the climate change. Two-third of the country stands less than 5 m above sea level, and a quarter of the country is flooded every year on average. Cyclones, floods, and droughts often pay a visit to this country, badly affecting a large number of people, especially those who live in rural areas. So, the learning of climatic variables is necessary. The time series analysis of climatic variables is a way to learn whether the climate change takes place or not. There are many climatic variables, but in this study, three variables, such as monthly temperature, humidity, and cloud coverage, have been considered to investigate their relationship and climate conditions. Time series modeling and forecasting help us to understand the long-term weather pattern. This study is very important because climatic change is now a problem of whole world. To make a solution of this problem, we have to find out its pattern first. Vector autoregression (VAR) model can play a vital role in sketching out this pattern. This method is capable of producing forecasts of interrelated variables, examining the effects of interrelated time series variable's shocks, and analyzing the dynamic impact of random disturbances. Many researchers have studied the climatic variables. Ferdous and Baten (2011) used least square method for analyzing trend of climatic data (temperature, rainfall, relative humidity, and sunshine) of Rajshahi and Rangpur Division to observe the climate variability. Shamsnia et al. (2011) used stochastic methods (autoregressive integrated moving average [ARIMA] model) for modeling of weather parameters, such as precipitation, temperature, and relative humidity. Kleiber et al. (2013) developed a bivariate stochastic model was applied to a daily temperature (minimum and maximum) data set covering the complex terrain of Colorado, USA, for studying climate impact and successfully quarters considerable temporally varying non-stationarity in both the direct-covariance and cross-covariance functions. But they didn't study the climate variability among temperature, humidity and cloud coverage for the response of one climatic variable on the other variable. The variability study among the climatic variables is essential. VAR offers such analysis. A VAR analysis is widely used in several disciplines. Khan and Hossain (2010) used VAR model for democracy and trade balance. Altaf et al. (2012) used the VAR model for macroeconomic variables of Pakistan's economic growth. Moneta et al. (2011) used structural VAR models for causal search. Awokuse and Bessler (2003) used VAR model to the US economy. Stergiou et al. (1997) analyzed monthly fisheries catches using seven forecasting techniques including regression and univariate and multivariate time series methods and finally mentioned that the univariate ARIMA and multivariate dynamic regression (MDREG) and VAR time series models all predicted persistence of catches. Durban and Glasbey (2001) used the vector autoregressive moving average (VARMA) process as a model for daily weather data and selected a vector second-order autoregressive first-order moving average process, which fits the data better and produces more realistic simulated series than existing models. Mosedale et al. (2006) used bivariate VAR time series models to fit daily winter time sea surface temperatures and North Atlantic Oscillation time series produced by a 50-year simulation of the Third Hadley Centre Coupled Ocean–Atmosphere GCM (HadCM3). Wang and Niu (2009) used the VAR model for wind speeds, temperatures, soil temperatures, and dew points for Los Angeles Long Beach area. Janjua et al. (2010) used the VAR model to investigate the impact of climate change on wheat production in Pakistan and concluded that there is no significant negative impact of climate change on wheat production in Pakistan. Liu et al. (2011) used the VAR model for examining how climate problem indicates high-profile international event, and climate science feedback influence media and congressional attention to global warming and climate change. Adenomon et al. (2013) used VAR model for analyzing dynamic relationship between time series rainfall and temperature data in Niger state, Nigeria, and found that there exists bidirectional causation between them. So, it is needed to analyze all the interrelated variables for a better forecast on climatic data. Thus, the purpose of the present study is to develop an appropriate VAR model for a better forecasting on climatic data such as monthly temperature, humidity, and cloud coverage of Rajshahi district in Bangladesh.

2.2 Materials and Methods

2.2.1 Data

Temperature is a physical property of matter that quantitatively expresses the common notions of hot and cold. Objects of low temperature are cold, while various degrees of higher temperatures are referred to as warm or hot. Humidity is a term for the amount of water vapor in the air. The relative humidity is a measure of the amount of water in air relative to the total amount of water the air can absorb, expressed as a percentage. A small part of the sky obscured by clouds, when observed from a particular location, is referred to as cloud coverage. In this study, we have used relative humidity and defined it as humidity. The measurement units of these variables are Celsius, percent, and octas. The monthly data of the climatic variables—temperature, humidity, and cloud coverage—of Rajshahi district for the period of January 1979 to December 2010 (i.e., 32 years) are used in this study and data are collected from Bangladesh Agricultural Research Council (BARC). Here the sample size (N) is 384. The data was compiled and no value for either of the variables was found to be missing.

After collecting the data we have compiled, tabulated, and analyzed them according to the objectives of the study. The data were put into MS excel in order to arrange them as time series data, and the popular software R and EViews were used for whole statistical analysis related to the objectives.

2.2.2 *Test of Stationarity*

Before applying VAR model among the temperature, humidity, and cloud coverage, a pretest for stationarity for each of the variables is needed to know the nature of the variables. If a time series is stationary, then its mean, variance, and autocovariance (at various lags) remain the same over time; that is, they are time invariant. The present study used the following three methods for testing stationarity of the variables.

2.2.2.1 Augmented Dickey–Fuller Test

The augmented Dickey–Fuller (ADF) test (Dickey and Fuller 1979) examines the presence of unit root (non-stationarity) in the autoregressive model. The ADF test here consists of estimating the following regression (Eq. 2.1):

$$\Delta y_t = \beta_1 + \beta_2 t + \delta y_{t-1} + \sum_{j=1}^{p} \gamma_j \Delta y_{t-j} + \varepsilon_t, \tag{2.1}$$

where y_t is any time series variable, y_{t-1} is the one period lag value of y_t, $\Delta y_t = y_t - y_{t-1}$, and t is the trend variable. The symbols β_1, β_2, δ, and γ are the parameters, and ε_t is a pure white noise error term. The fourth term on the right hand side of Eq. 2.1 is the augmentation term. The ADF test is based on the following hypothesis: $H_0 : \delta = 0$ (non-stationary) against the alternative hypothesis that $H_1 : \delta < 0$ (stationary). The null hypothesis is rejected if the ADF test statistic (tau statistic) is less than its critical value.

2.2.2.2 Phillips–Perron Test

Phillips and Perron (1988) propose a nonparametric statistical method to take care of the serial correlation without augmented term of the ADF equation (Eq. 2.1). In this test, the series is assumed to be non-stationary under null hypothesis. For the Phillips–Perron (PP) test, firstly δ is estimated from the non-augmented Dickey–Fuller (1979) equation (Eq. 2.2) as

$$\Delta y_t = \alpha + \beta t + \delta y_{t-1} + \varepsilon_t \tag{2.2}$$

and modifies the $t_\delta = \hat{\delta} / \mathrm{se}(\hat{\delta})$ of the δ coefficient so that serial correlation does not affect the asymptotic distribution of the test statistics. The PP test is based on the statistic:

$$t_{\delta,\mathrm{PP}} = t_{\delta}\left(\frac{\gamma_0}{f_0}\right)^{1/2} - \frac{N(f_0 - \gamma_0)\left(\mathrm{se}(\hat{\delta})\right)}{2f_0^{1/2}s}, \tag{2.3}$$

where $\hat{\delta}$ is the estimated value of δ, t_{δ} is the t-ratio of δ, $\mathrm{se}(\hat{\delta})$ is coefficient standard error, and s is the standard error of the test regression. In addition, γ_0 is a consistent estimate of the error variance in non-augmented Dickey–Fuller (DF) equation (Eq. 2.2) and calculated as $(N - K)s^2/N$, where K is the number of regressors. The remaining term, f_0, is an estimator of the residual spectrum at frequency zero.

2.2.2.3 Kwiatkowski–Phillips–Schmidt–Shin Test

Kwiatkowski–Phillips–Schmidt–Shin (KPSS) (1992) test differs from the other tests. In this test, the series is assumed to be stationary under null hypothesis. The KPSS statistic is based on the residuals from the ordinary least square (OLS) regression of y_t on the exogenous variables of lagged y_t. For one exogenous variable using one lagged value of y_t, the regression model can be written as follows:

$$y_t = \delta y_{t-1} + u_t. \tag{2.4}$$

The LM statistic is being defined as:

$$\mathrm{LM} = \sum S(t)^2/\left(N^2 f_0\right), \tag{2.5}$$

where f_0 is an estimator of the residual spectrum at frequency zero and $S(t) = \sum_{r=1}^{t} \hat{u}_r$ is a cumulative residual function based on the residuals $\hat{u}_r$ (Eq. 2.4). It is pointed out that the estimator of δ used in this calculation differs from the estimator for δ used by GLS detrending since it is based on regression involving the original data, and not on the quasi-differenced data. To specify the KPSS test, we must specify the set of exogenous regressors of lagged y_t and a method for estimating f_0. The reported critical values for LM test statistic (Eq. 2.5) are based upon the asymptotic results presented by KPSS and the whole procedure is known as KPSS test.

2.2.3 Vector Autoregression Model

VAR models are well documented through many textbooks and scientific articles (Hamilton 1994; Johansen 1995; Hatanaka 1996; Lütkepohl and Krätzig 2004; Lütkepohl 2005; Litterman 1986; Canova 1995; Sun and Ni 2004; Ni and Sun 2005; Liu and Theodoridis 2012; and so on). A VAR model can be built through the following steps (Fig. 2.1).

Fig. 2.1 Flow chart of vector autoregression (VAR) modeling strategy

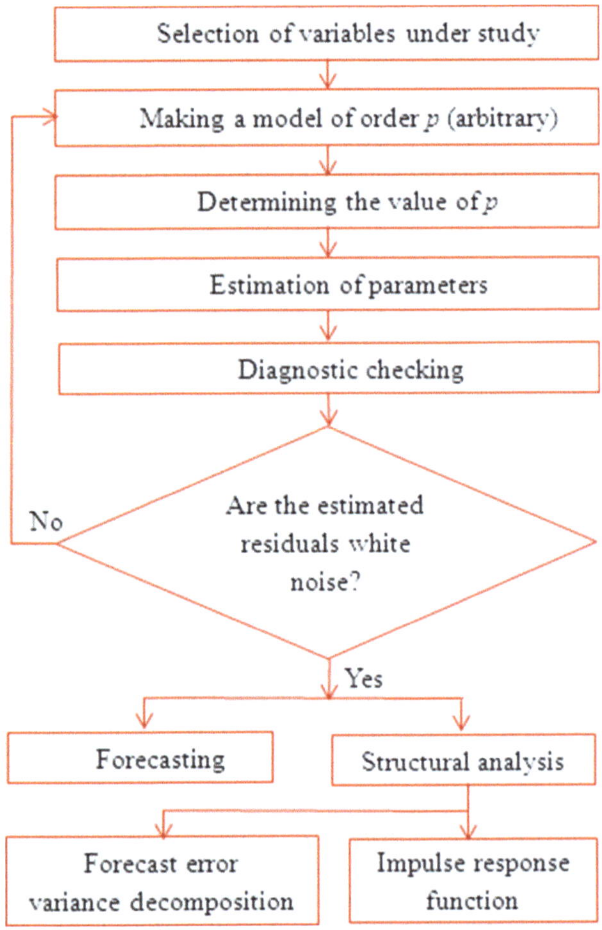

2.2.3.1 Selection of Variables Under Study

In simultaneous equation model, some variables are treated as endogenous and some as exogenous. According to Sims (1980) all the variables in a VAR system are endogenous variables. The concept of building a VAR model is that all the variables under study are endogenous and usually no one is exogenous (Gujarati 1993). Logically, our study variables: temperature, humidity, and cloud coverage, are endogenous in nature. The endogenity of the variables can be tested using Granger causality detection process that was proposed by Granger (1969), and later on, it was popularized by Sims (1972). The variables having endogenous property are finally selected for the VAR analysis.

2.2.3.2 Making a Model of Order *p* (Arbitrary)

Let us denote the temperature, humidity, and cloud coverage by T_t, H_t, and C_t; $t = 1, 2, \ldots, N$ (sample size $= N$), respectively. The three-variable VAR model of arbitrary order p can be denoted by VAR(p) and written as:

$$T_t = c_1 + a_{11}^1 T_{t-1} + \cdots + a_{1p}^1 T_{t-p} + a_{11}^2 H_{t-1} + \cdots + a_{1p}^2 H_{t-p} + a_{11}^3 C_{t-1} + \cdots + a_{1p}^3 C_{t-p} + \varepsilon_{1t}, \tag{2.6}$$

$$H_t = c_2 + a_{21}^1 T_{t-1} + \cdots + a_{2p}^1 T_{t-p} + a_{21}^2 H_{t-1} + \cdots + a_{2p}^2 H_{t-p} + a_{21}^3 C_{t-1} + \cdots + a_{2p}^3 C_{t-p} + \varepsilon_{2t}, \tag{2.7}$$

$$C_t = c_3 + a_{31}^1 T_{t-1} + \cdots + a_{3p}^1 T_{t-p} + a_{31}^2 H_{t-1} + \cdots + a_{3p}^2 H_{t-p} + a_{31}^3 C_{t-1} + \cdots + a_{3p}^3 C_{t-p} + \varepsilon_{3t}. \tag{2.8}$$

In matrix notation, we can write

$$\begin{pmatrix} T_t \\ H_t \\ C_t \end{pmatrix} = \begin{pmatrix} c_1 \\ c_2 \\ c_3 \end{pmatrix} + \begin{pmatrix} a_{11}^1 & a_{11}^2 & a_{11}^3 \\ a_{21}^1 & a_{21}^2 & a_{21}^3 \\ a_{31}^1 & a_{31}^2 & a_{31}^3 \end{pmatrix} \begin{pmatrix} T_{t-1} \\ H_{t-1} \\ C_{t-1} \end{pmatrix} + \cdots + \begin{pmatrix} a_{1p}^1 & a_{1p}^2 & a_{1p}^3 \\ a_{2p}^1 & a_{2p}^2 & a_{2p}^3 \\ a_{3p}^1 & a_{3p}^2 & a_{3p}^3 \end{pmatrix} \begin{pmatrix} T_{t-p} \\ H_{t-p} \\ C_{t-p} \end{pmatrix} + \begin{pmatrix} \varepsilon_{1t} \\ \varepsilon_{2t} \\ \varepsilon_{3t} \end{pmatrix}. \tag{2.9}$$

Therefore, the reduced form of VAR process of order p can be written as follows

$$y_t = c + A_1 y_{t-1} + A_2 y_{t-2} + \cdots + A_p y_{t-p} + \varepsilon_t, \quad t = 1, 2, \ldots, N, \tag{2.10}$$

where y_t is 3×1 vector, c is a 3×1 vector of constants (intercept), A_i is a 3×3 matrix (for every $i = 1, 2, \ldots, p$), and ε_t is a 3×1 vector of error terms and assumed $\varepsilon_t \sim \mathrm{NID}(0, \Omega)$.

2.2.3.3 Determining the Value of Order *p*

In VAR model, we need to have the number of appropriate lags to fit the model but there is no formal unique guidance for choosing this. To determine the lag length in VAR model, the Akaike information criterion (AIC) (Akaike 1974) (Eq. 2.11), Schwarz information criteria (SC) (1978) (Eq. 2.12), Hannan–Quinn information criteria (HQ) (1979) (Eq. 2.13), final prediction error (FPE) (Akaike 1969) (Eq. 2.14), and likelihood ratio statistic (LRT) (Lütkepohl 1991) (Eq. 2.15) are used. Based on the same sample size, N, the above information criteria are

$$\mathrm{AIC}(p) = \log\left|\hat{\Omega}(p)\right| + \frac{2m(p^2+1)}{N}, \tag{2.11}$$

$$\mathrm{SC}(p) = \log\left|\hat{\Omega}(p)\right| + \log(N)\frac{m(p^2+1)}{N}, \tag{2.12}$$

$$\mathrm{HQ}(p) = \log\left|\hat{\Omega}(p)\right| + 2\log(\log(N))\frac{m(p^2+1)}{N}, \tag{2.13}$$

$$\mathrm{FPE}(p) = \left(\frac{N+mp+1}{N-mp-1}\right)^m \left|\hat{\Omega}(p)\right| \tag{2.14}$$

with $\hat{\Omega}(p) = N^{-1}\sum_{t=1}^{N}\hat{\varepsilon}_t\hat{\varepsilon}_t^{'}$ and $m(p^2+1)$ is the total number of the parameters in each equation, m is the number of equation or variables in VAR model, and p assigns the lag order. The likelihood ratio test statistic is defined as:

$$\lambda_{\mathrm{N}} = -2\left[L(\hat{\theta}) - L(\hat{\theta}_0)\right]. \tag{2.15}$$

Under the null hypothesis, λ_{N} asymptotically has a χ^2 distribution with degrees of freedom equal to the number of restrictions imposed under null hypothesis (H_0).

2.2.3.4 Estimating Parameters

The likelihood function is defined as the joint density function of random variables but considered as a function of parameters. For Eq. 2.10, the sample log likelihood function (Eq. 2.16) can be written as:

$$L(\Omega, \Pi) = -\frac{Nn}{2}\log(2\pi) + \frac{N}{2}\log\left|\Omega^{-1}\right| - \frac{1}{2}\sum_{t=1}^{N}\left(y_t - \Pi^{'}x_t\right)^{'}\Omega^{-1}\left(y_t - \Pi^{'}x_t\right), \tag{2.16}$$

where $\Pi' = [c\ A_1\ A_2\ \cdots\ A_p]$ and $x_t = [1\ y_{t-1}\ y_{t-2}\ \cdots\ y_p]'$.

Using the maximum likelihood method, we get the estimated value of Π and Ω as $\hat{\Pi}^{'} = \left[\sum_{t=1}^{N} y_t x_t^{'}\right]\left[\sum_{t=1}^{N} x_t x_t^{'}\right]^{-1}$ and $\hat{\Omega} = N^{-1}\sum_{t=1}^{N}\hat{\varepsilon}_t\hat{\varepsilon}_t^{'}$, respectively, which are the same as the least squared estimator of the parameters of VAR(p) model.

2.2.3.5 Diagnostic Checking

For diagnostic checking of selected VAR(p) model, it is used unit root test for checking stationarity, the normal Q–Q plot for normality of the residual, and Durbin–Watson d test (1951) for autocorrelation. The outliers of the fitted models were checked using standardized residuals plot.

2.2.3.6 Cross Validity of the Fitted VAR Models

Cross validity predictive power (Stevens 1996, p. 100) is used for testing the validity and stability of the fitted VAR models. The fitted equations with high value of cross validity predicted power are assumed a better representation of the population. The cross validity predictive power (Eq. 2.17), denoted by ρ_{cv}^2, is defined as:

$$\rho_{\text{cv}}^2 = 1 - \frac{(N-1)(N-2)(N+1)}{N(N-k-1)(N-k-2)}\left(1 - R^2\right), \tag{2.17}$$

where N is the sample size, k is the number of predictors in the regression equation, and the cross-validated R is the correlation coefficient between observed and predicted values of the dependent variable. Using the above statistic, it can be concluded that if the prediction equation is applied to many other samples from the same population, then ($\rho_{\text{cv}}^2 \times 100$) % of the variance on the predicted variable would be explained by the regression equation (Stevens 1996, p. 100).

2.2.3.7 Forecasting

Forecasting is the prediction of values of a variable based on its known past values. In VAR model, this forecasting also depends on the lag values of other endogenous variables. Since reduced forms of the VAR models represent the conditional mean of a stochastic process, they lend themselves in forecasting. The 1-period-ahead forecasts are constructed as follows:

$$\hat{y}_{t+1} = \hat{A}_1 y_t + \hat{A}_2 y_{t-1} + \cdots + \hat{A}_p y_{t-p+1} \tag{2.18}$$

and recursively we can find the forecast values of $\hat{y}_{t+2}, \hat{y}_{t+3}, \ldots, \hat{y}_{t+h}, \ldots$.

2.2.3.8 Forecast Error Variance Decomposition

The forecast error variance (FEV) decomposition is the part of structural analysis which is a tool for investigating the impact of shocks in VAR models. Forecast error can be obtained from the variance decomposition of each VAR model. The hth step forecast error and its variance can be computed, respectively, as:

$$y_{t+h} - E(y_{t+h}) = \Psi_0\varepsilon_{t+h} + \Psi_1\varepsilon_{t+h-1} + \cdots + \Psi_{h-1}\varepsilon_{t+1} \tag{2.19}$$

and

$$\mathrm{var}_t(y_{t+h}) = \Psi_0\Psi_0' + \Psi_1\Psi_1' + \cdots + \Psi_{h-1}\Psi_{h-1}'. \tag{2.20}$$

Then, $w_{h,r} = \sum_{j=1}^{h-1}\Psi_j I_N \Psi_j'$ is the variance of h step ahead forecast errors due to the Nth shock and the variance is the sum of these components, e.g., $\mathrm{var}_t(y_{t+h}) = \sum_N w_{h,r}$.

2.2.3.9 Impulse Response Function

Impulse response function (IRF) is another part of structural analysis. The stochastic error terms are called impulses or innovations in the language of VAR. Impulse responses trace out the response of current and future values of each of the variables to a one-unit increase in the current value of one of the VAR errors, assuming that this error returns to zero in subsequent periods and that all other errors are equal to zero.

Any covariance stationary VAR(p) process has a Wold representation (vector MA(∞) process) of the form as

$$y_t = \mu + \varepsilon_t + \Psi_1\varepsilon_{t-1} + \Psi_2\varepsilon_{t-2} + \Psi_3\varepsilon_{t-3} + \cdots.$$

Thus the matrix Ψ_s has the interpretation

$$\frac{\partial y_{t+s}}{\partial \varepsilon_t'} = \Psi_s$$

That is, the row i, column j element of Ψ_s identifies the consequences of a one-unit increase in the jth variable's innovation at date t (ε_{jt}) for the value of the ith variable at time $t + s(y_{i,t+s})$ holding all other innovations at all dates constant. A plot of the row i, column j element of Ψ_s as a function of s is called the impulse response function.

2.3 Results and Discussion

2.3.1 *Descriptive Statistics*

Descriptive statistics, e.g., mean, minimum, maximum, range, standard deviation, skewness, and kurtosis, for the climatic variables temperature, humidity, and cloud coverage are calculated and shown in Table 2.1. The software EViews

Table 2.1 Descriptive statistics of temperature, humidity, and cloud coverage

Variable	Mean	Minimum	Maximum	Range	Standard deviation	Skewness	Kurtosis
Temperature	31.116	21.300	39.200	17.900	3.603	−0.563	2.623
Humidity	77.909	44.900	90.300	45.400	8.952	−0.983	3.610
Cloud overage	3.265	0.200	6.900	6.700	2.030	0.210	1.554

Table 2.2 Unit root test for the time series data of temperature, humidity, and cloud coverage

Variable	ADF	PP	KPSS
Temperature	−3.747***	−6.768***	0.202*
Humidity	−3.052**	−6.393***	1.338
Cloud coverage	−3.265**	−4.576***	0.485***

Note: The asterisks ***, **, and * represent the statistical significance at 1 %, 5 %, and 10 % levels, respectively

and R were used. The cloud coverage is more concentrated followed by the temperature and humidity. The distribution of temperature and humidity is negatively skewed but cloud coverage is positively skewed. Again the curve of the distribution of humidity is leptokurtic but temperature and cloud coverage are platykurtic.

2.3.2 Tests for Stationarity

The test for stationary has been done by the test statistics: ADF, PP, KPSS are listed in Table 2.2. From this table, it is clear that the ADF, PP, and KPSS tests show that temperature, humidity, and cloud coverage are stationary at level except humidity in KPSS test only.

2.3.3 Selection of Variables Under Study

The Granger causality test was applied for testing endogenity (dependency) of the variables temperature (T), humidity (H), and cloud coverage (C), and the results are shown in Table 2.3. From this table, it is found that the causality with lag length 2, 3, 4, and 5 is the same type but lag length 1 is different. Table 2.3 also shows the unidirectional causality in humidity to temperature, temperature to humidity, temperature to cloud coverage, cloud coverage to humidity, and humidity to cloud coverage but there is no unidirectional causality in temperature to cloud coverage at lag 1. Since the direction of causality varied with lag length we will choose the lag length for which SC is minimum. Regressing temperature on lagged itself and cloud

Table 2.3 The F-statistic for the pairwise Granger causality test among temperature, humidity, and cloud coverage

	Lag				
Null hypothesis	1	2	3	4	5
$H \nrightarrow T$	40.950***	46.144***	80.806***	68.207***	57.828***
$T \nrightarrow H$	233.820***	138.184***	93.734***	69.284***	59.395***
$C \nrightarrow T$	0.017	104.326***	141.845***	120.729***	94.346***
$T \nrightarrow C$	93.735***	53.109***	43.668***	28.603***	23.691***
$C \nrightarrow H$	113.873***	46.923***	42.874***	27.388***	26.458***
$H \nrightarrow C$	113.476***	37.452***	22.842***	11.628***	11.452***

Note: The symbol $\nrightarrow$ indicates "does not granger cause". The asterisk *** represent the statistical significance at 1 % level

Table 2.4 Schwarz information criteria (SC) for different lag lengths in regression

	SC for the lags				
Regression	1	2	3	4	5
Temperature on lagged itself and cloud coverage	4.769	3.852	3.553	3.509*	3.517

Note: * indicates lag order selected by the criterion

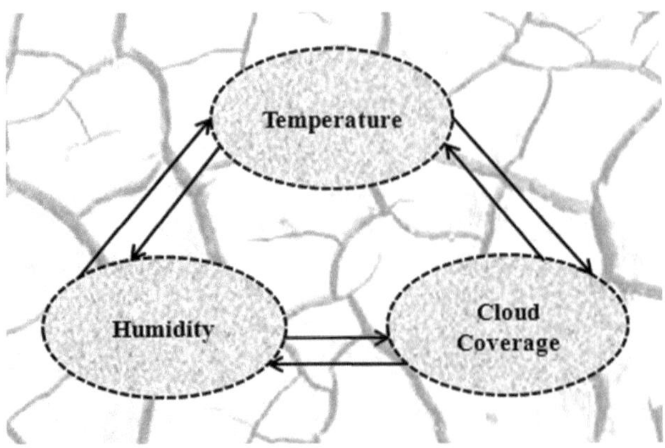

Fig. 2.2 Diagram for the direction of causality

coverage with lag 1–5, the values of SC are found and shown in Table 2.4. The value of SC is minimum at lag 4, implying unidirectional causality in cloud coverage to temperature. For example, if X is granger causes to Y and Y is granger causes to X then it is called bidirectional causality between X and Y. Therefore, the Granger causality test suggests that there exists bidirectional causality between the variables temperature, humidity, and cloud coverage. This direction of causality is diagrammatically shown in Fig. 2.2. Figure 2.2 shows that the variables are interrelated and endogenous in nature that is in favor of the VAR application.

Table 2.5 Various selection criteria for detecting lag order of VAR model

Lag	Log *L*	LRT	FPE	AIC	SC	HQ
0	−2866.897	NA	855.579	15.265	15.297	15.278
1	−2288.221	1145.040	41.330	12.235	12.361	12.285
2	−2124.355	321.631	18.135	11.411	11.631	11.499
3	−2009.098	224.382	10.305	10.846	11.160	10.971
4	−1961.723	91.475	8.403	10.642	11.050	10.804
5	−1938.443	44.578	7.789	10.566	11.068	10.765
6	−1910.218	53.599	7.033	10.464	11.060	10.700
7	−1864.722	85.666	5.793	10.270	10.960*	10.544*
8	−1848.726	29.865*	5.582*	10.233*	11.016	10.544

Note: * indicates lag order selected by the criterion

2.3.4 Selection of Order (p)

To build a VAR model, at first we need to select the order (lag length) of VAR model. This order can be determined through the procedures: (1) Likelihood ratio test statistics (LRT), (2) Final prediction error (FPE), (3) Akaike information criteria (AIC), (4) Schwarz information criteria (SC) and (5) Hannan-Quinn information criteria (HQ) or (1) LRT, (2) FPE, (3) AIC, (4) SC and (5) HQ. The results are shown in Table 2.5. From this table, the LRT statistics, FPE, and AIC indicate that the appropriate lag order of VAR is 8 but that of SC and HQ is 7. Ivanov and Kilian (2001) suggest that in the context of VAR models AIC tends to be more accurate with monthly data. Also, almost all of the selection criteria suggest that the appropriate lag order is 8. Therefore, the order of VAR model in this study is chosen as 8.

2.3.5 Estimation of Parameters

Applying OLS method, the parameters of the best selected VAR(8) model are estimated and the results are shown in Table 2.6. From this table, the first column, second column, third column, and forth column (except first and last rows) represent the predictor variables and estimated parameters of Eqs. 2.6, 2.7, and 2.8, respectively. The first and last rows represent the response variables and the coefficient of determination (R^2) for Eqs. 2.6, 2.7, and 2.8, respectively. The significance coefficient of estimated model is defined by asterisks. The R^2 of VAR(8) models are 0.884, 0.826, and 0.891 with respect to temperature, humidity, and cloud coverage, respectively; means fit the data as well. The actual and fitted values of VAR(8) model in place of temperature, humidity, and cloud coverage are shown in Figs. 2.3, 2.4, and 2.5, respectively. The actual and fitted values are presented by solid and dot line, respectively.

Table 2.6 The estimated coefficients of VAR(8) model for Eq. 2.10

Variable	T_t	H_t	C_t
Constant	$\hat{c}_1 = 41.697^{***}$	$\hat{c}_2 = -25.2660$	$\hat{c}_3 = 2.5155$
T_{t-1}	$\hat{a}^1_{11} = 0.5330^{***}$	$\hat{a}^1_{21} = 0.2428$	$\hat{a}^1_{31} = 0.0424$
T_{t-2}	$\hat{a}^1_{12} = -0.1471^{**}$	$\hat{a}^1_{22} = 0.6570^{***}$	$\hat{a}^1_{32} = 0.0030$
T_{t-3}	$\hat{a}^1_{13} = -0.348^{***}$	$\hat{a}^1_{23} = 0.6737^{***}$	$\hat{a}^1_{33} = 0.0559$
T_{t-4}	$\hat{a}^1_{14} = -0.1624^{**}$	$\hat{a}^1_{24} = 0.0425$	$\hat{a}^1_{34} = -0.0058$
T_{t-5}	$\hat{a}^1_{15} = 0.1217^{*}$	$\hat{a}^1_{25} = -0.5067^{**}$	$\hat{a}^1_{35} = -0.0327$
T_{t-6}	$\hat{a}^1_{16} = 0.0252$	$\hat{a}^1_{26} = -0.0271$	$\hat{a}^1_{36} = -0.0841^{**}$
T_{t-7}	$\hat{a}^1_{17} = -0.1245^{*}$	$\hat{a}^1_{27} = 0.0073$	$\hat{a}^1_{37} = -0.06946^{*}$
T_{t-8}	$\hat{a}^1_{18} = -0.0766$	$\hat{a}^1_{28} = 0.1336$	$\hat{a}^1_{38} = -0.0730^{**}$
H_{t-1}	$\hat{a}^2_{11} = -0.0261$	$\hat{a}^2_{21} = 0.5653^{***}$	$\hat{a}^2_{31} = 0.0166$
H_{t-2}	$\hat{a}^2_{12} = 0.05420^{*}$	$\hat{a}^2_{22} = -0.0572$	$\hat{a}^2_{32} = 0.0047$
H_{t-3}	$\hat{a}^2_{11} = -0.0943^{***}$	$\hat{a}^2_{23} = 0.2426^{***}$	$\hat{a}^2_{33} = 0.0132$
H_{t-4}	$\hat{a}^2_{14} = 0.0205$	$\hat{a}^2_{24} = -0.0218$	$\hat{a}^2_{34} = 0.0056$
H_{t-5}	$\hat{a}^2_{15} = 0.0001$	$\hat{a}^2_{25} = -0.0278$	$\hat{a}^2_{35} = 0.03216^{**}$
H_{t-6}	$\hat{a}^2_{16} = 0.0322$	$\hat{a}^2_{26} = -0.1764^{**}$	$\hat{a}^2_{36} = -0.0314^{**}$
H_{t-7}	$\hat{a}^2_{17} = -0.0936^{***}$	$\hat{a}^2_{27} = 0.2638^{***}$	$\hat{a}^2_{37} = 0.0507^{***}$
H_{t-8}	$\hat{a}^2_{18} = 0.0349$	$\hat{a}^2_{28} = 0.0746$	$\hat{a}^2_{38} = 0.0040$
C_{t-1}	$\hat{a}^3_{11} = 0.4308^{***}$	$\hat{a}^3_{21} = -0.0710$	$\hat{a}^3_{31} = 0.2107^{***}$
C_{t-2}	$\hat{a}^3_{12} = 0.3494^{***}$	$\hat{a}^3_{22} = -0.4873$	$\hat{a}^3_{32} = -0.0610$
C_{t-3}	$\hat{a}^3_{13} = -0.2395^{*}$	$\hat{a}^3_{23} = 0.4095$	$\hat{a}^3_{33} = 0.0249$
C_{t-4}	$\hat{a}^3_{14} = -0.469^{***}$	$\hat{a}^3_{24} = 0.2142$	$\hat{a}^3_{34} = -0.12357^{*}$
C_{t-5}	$\hat{a}^3_{15} = -0.0310$	$\hat{a}^3_{25} = -0.4658$	$\hat{a}^3_{35} = -0.2199^{***}$
C_{t-6}	$\hat{a}^3_{16} = -0.0072$	$\hat{a}^3_{26} = 0.4688$	$\hat{a}^3_{36} = -0.0481$
C_{t-7}	$\hat{a}^3_{17} = 0.2565^{**}$	$\hat{a}^3_{27} = -0.7400^{**}$	$\hat{a}^3_{37} = -0.1928^{***}$
C_{t-8}	$\hat{a}^3_{18} = -0.1043$	$\hat{a}^3_{28} = 0.0380$	$\hat{a}^3_{38} = -0.0788$
R^2	$R^2_1 = 0.8841$	$R^2_2 = 0.8259$	$R^2_3 = 0.8908$

Note: The asterisks ***, **, and * represent the statistical significance at 1 %, 5 %, and 10 % levels, respectively

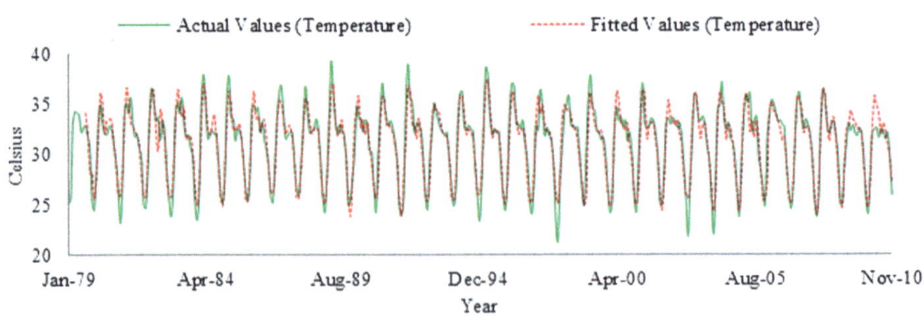

Fig. 2.3 Actual and fitted plot for temperature of VAR(8) model

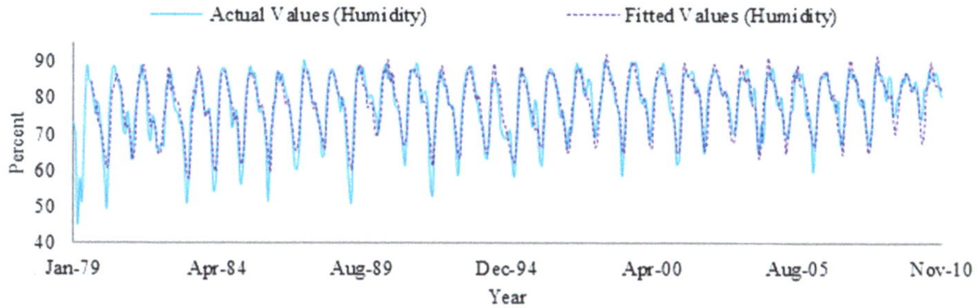

Fig. 2.4 Actual and fitted plot for humidity of VAR(8) model

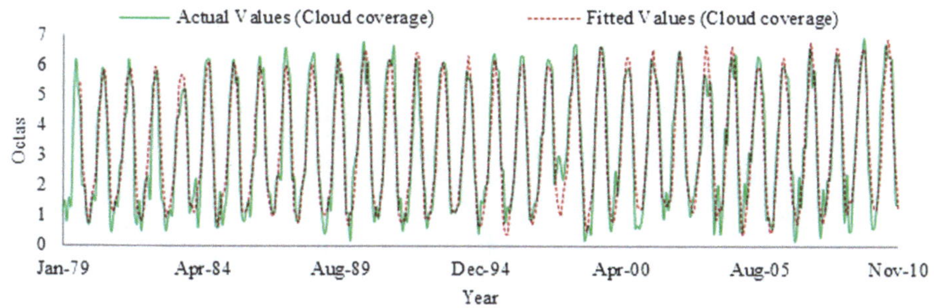

Fig. 2.5 Actual and fitted plot for cloud coverage of VAR(8) model

2.3.6 Diagnostic Checking

The diagnostic checking on residual of the selected VAR(8) model is confirmed using the *Q–Q* plot for normality, ADF test for unit root, and DW test for autocorrelation. From the results shown in Table 2.7, it is clear that all residuals series are stationary. The problem of serial correlation really fades away. The *Q–Q* plot (Fig. 2.6) of the residuals series for temperature, humidity, and cloud coverage is approximately normally distributed. Also, the histograms of the residuals series for all the variables (Figs. 2.7, 2.8, and 2.9) are roughly normally distributed.

To detect the outliers, the standardized residuals plot against the time is considered. We know that 68-95-99 rule for perfect normal frequency distribution is that 68.26 % of the samples fall between [mean ± 1 standard deviation], 95.44 % of the samples fall between [mean ± 2 standard deviation], and 99.73 % of the samples fall between [mean ± 3 standard deviation]. In case of a standard normal distribution the mean is zero and variance is one. Then we investigated whether they are between 3.0 or 3.5 and 4.0 (Pankratz 1991, p. 208) with their absolute values. We take the value 3.0 and the residuals are outlying 3.0 with their absolute values as an indication of outliers.

From the graphical representation (Fig. 2.10) we observe that there are five residuals for the series temperature falls outside 3.0 with its absolute value as

Table 2.7 Augmented Dickey–Fuller (ADF) and Durbin–Watson (DF) tests for residuals series of VAR(8) model

Series of residuals	ADF-value (probability)	5 % critical value	Comment	DW-value	Comment
Temperature	−20.197 (0.000)	−2.869	Stationary	1.999	No autocorrelation
Humidity	−19.335 (0.000)	−2.869	Stationary	2.000	No autocorrelation
Cloud coverage	−19.037 (0.000)	−2.869	Stationary	2.000	No autocorrelation

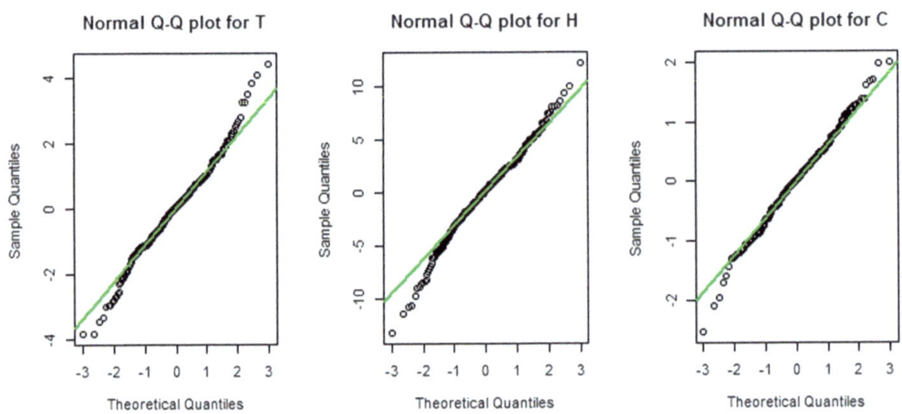

Fig. 2.6 Normal *Q–Q* plot for the residuals of VAR(8) model

Fig. 2.7 Histogram of the residuals for temperature of VAR(8) model

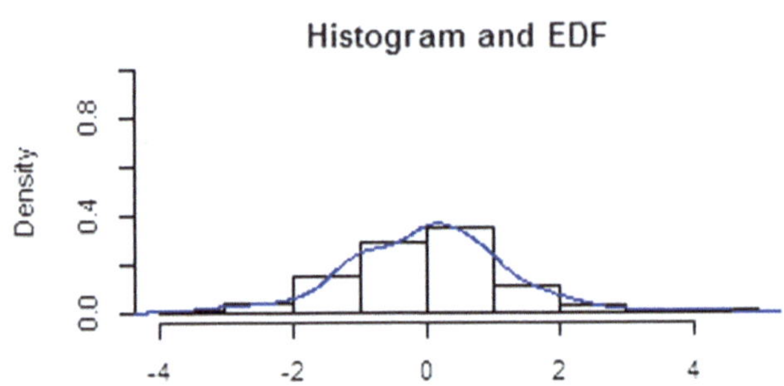

considered outliers. The values of five residuals are January 1998 = −3.118, February 1999 = 3.577, January 2003 = −3.106, June 2005 = 3.104, and February 2006 = 3.327.

Similarly, from Fig. 2.11, we observe that there are three residuals for the series humidity speed falls outside 3.0 with its absolute value. The values of three residuals are March 1986 = −3.648, March 1992 = −3.161, and April 2010 = 3.345.

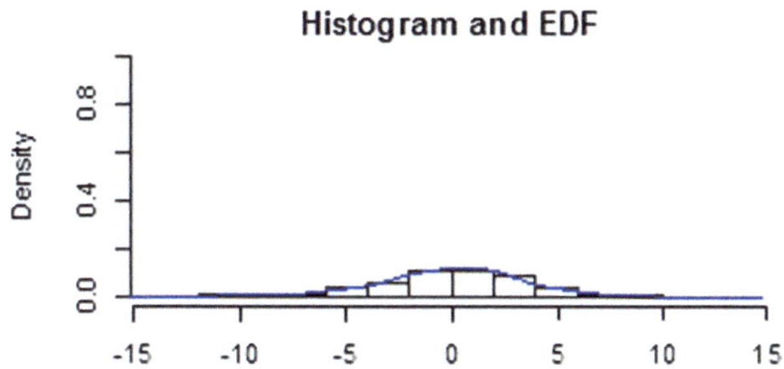

Fig. 2.8 Histogram of the residuals for humidity of VAR(8) model

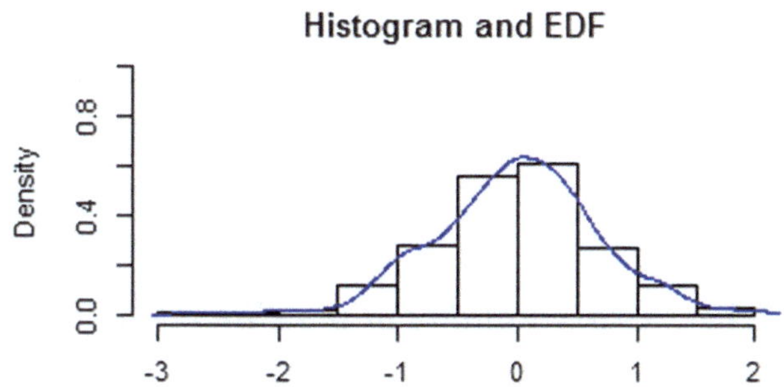

Fig. 2.9 Histogram of the residuals for cloud coverage of VAR(8) model

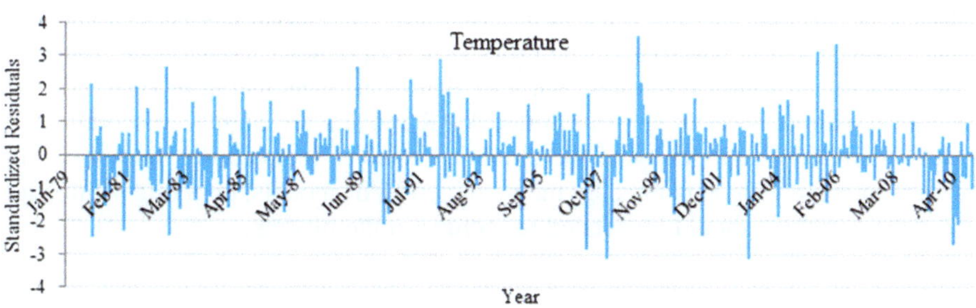

Fig. 2.10 Standardized residuals plot for temperature of VAR(8) model

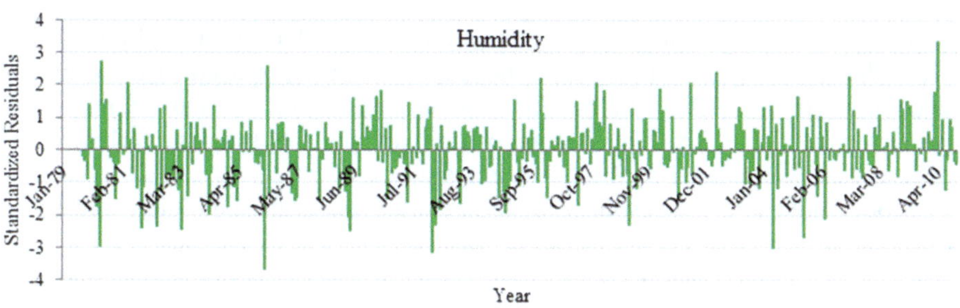

Fig. 2.11 Standardized residuals plot for humidity of VAR(8) model

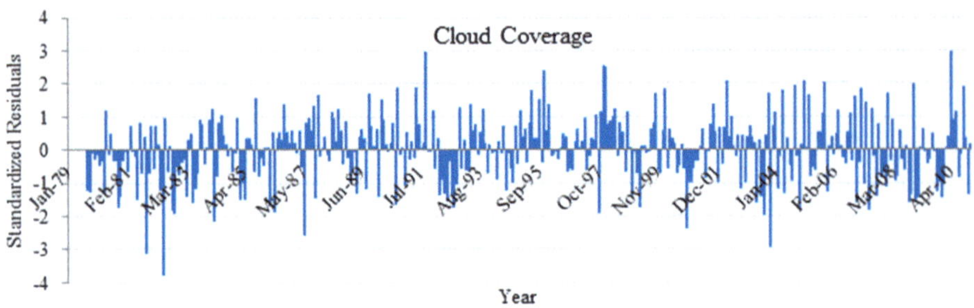

Fig. 2.12 Standardized residuals plot for cloud coverage of VAR(8) model

From Fig. 2.12, we observe that there are two residuals for the series cloud coverage falls outside 3.0 with its absolute value. The values of two residuals are October 1981 = −3.138 and May 1982 = −3.773.

2.3.7 Cross Validity of the Fitted VAR Models

The computed cross validity predicted power (Eq. 2.17) of the estimated equations (Eqs. 2.6, 2.7, and 2.8) for temperature, humidity, and cloud coverage was 0.868, 0.801, and 0.876, respectively. The cross validity predicted power was higher for the equation of cloud coverage (Eq. 2.8) than that of temperature (Eq. 2.6) and humidity (Eq. 2.7). Hence, all the fitted VAR models are highly cross-validated. More than 80 % endogenity of the climatic variables temperature, humidity, and cloud coverage were explained by the predicted VAR models, implying some other climatic variables should be needed to include in the models.

2.3.8 Forecast Error Variance Decomposition

To examine the short-run dynamic interactions between the variables, FEV is used. Here the period is month because the data is monthly spaced. The computed decomposition of FEV with the forecast horizons 1, 6, 12, and 18 months is presented in Table 2.8. The decomposition divides the forecast variance into different parts explained by their own innovations.

From Table 2.8, the response of temperature on itself, humidity, and cloud coverage is observed. The temperature seemed to be less exogenous in the system that explained more than 65 % of its FEV after 1 year (12 periods). The humidity and cloud coverage were accounted for more than 9 % and 19 % of the variation in temperature, respectively. The humidity to itself explained more than 55 % FEV after 1 year. More than 30 % and 11 % FEV were, respectively, explained in temperature and cloud coverage for the variation in humidity. More than 55 %

Table 2.8 Forecast error variance decomposition for VAR(8) model

FEV in	Period (month)	Standard error (SE)	Temperature	Humidity	Cloud coverage
Temperature	01	1.268	100.00	00.000	00.000
	06	1.787	72.364	10.224	17.412
	12	1.913	70.590	09.556	19.854
	18	2.231	66.137	09.460	24.403
Humidity	01	3.738	32.441	67.559	00.000
	06	4.987	30.806	65.968	03.227
	12	5.382	30.356	58.582	11.062
	18	5.726	33.119	54.869	12.012
Cloud coverage	01	0.693	13.490	23.425	63.084
	06	0.743	13.263	25.890	60.847
	12	0.876	22.233	20.845	56.922
	18	0.948	21.664	20.726	57.609

FEV for cloud coverage were explained after 1 year by cloud coverage itself. Also, more than 20 % FEV were explained by temperature and humidity in the variation of cloud coverage. Therefore, the climatic variables are interrelated to each other.

2.3.9 *Impulse Response Function*

Using Monte Carlo simulation with 10,000 repetitions, the IRF with its standard error was estimated and shown in Fig. 2.13. The response in each variable is expressed to 1 SD innovation. The temporary response from temperature to humidity showed a significant effect at 5 % level with its fourth month, implying after 4 months the temperature decreased 0.065 % in response for the 0.076 % increasing shock in humidity. Also, the cloud coverage showed significant effect on temperature at 5 % level and the temperature will be increased by 0.075 % for the 0.071 % increasing shock in cloud coverage after fourth month.

At 5 % level of significance, humidity will be increased by 0.737 % for the 0.227 % increasing shock in temperature after third month. But, after seventh month, humidity will be increased by 0.035 % for the 0.202 % increasing shock in cloud coverage.

After fifth month the cloud coverage will be decreased by 0.045 % for the 0.038 % increasing shock in temperature. Cloud coverage will be decreased by 0.063 % after sixth month for the 0.036 % increasing shock in humidity.

2.3.10 *Forecasting Using VAR(8) Model*

Time series analysis is very important for predicting or forecasting. The forecast of the climatic variables temperature, humidity, and cloud coverage from January

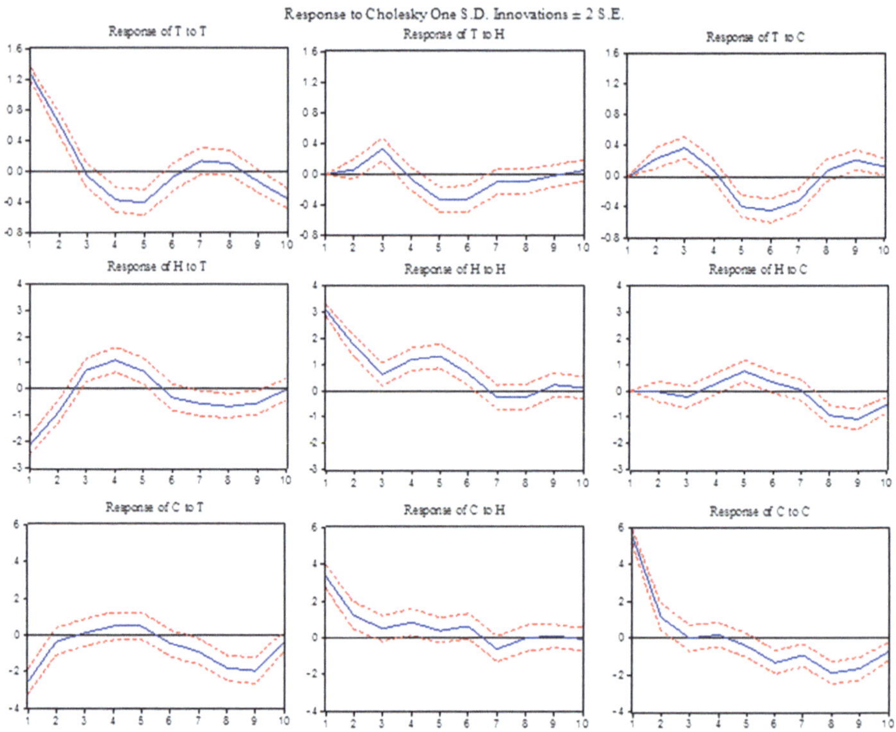

Fig. 2.13 Impulse response function (IRF) from the Cholesky decomposition for VAR(8) model. The scale in the *X*-axis indicates the lag in month

2011 to December 2016 with 95 % confidence intervals (CIs) was generated using fitted VAR(8) model and shown in Fig. 2.14. The forecasted values (blue line in Fig. 2.14) divulged slightly upward trends in temperature and humidity but downward in cloud coverage.

2.4 Conclusion

The climate data analysis is very important in recent time. This study included the analysis of temperature, humidity, and cloud coverage during January 1979 to December 2010. These variables were stationary at level with bidirectional causality among themselves, implying all the variables were interrelated that can be modeled with VAR analysis. Appropriate order (i.e., lag length) of VAR model was ensured using AIC, SC, HQ, FPE, and LRT and it was found that this order is 8. The diagnostic check of VAR(8) model divulged that the residuals were stationary, non-autocorrelated, and approximately normally distributed. All the fitted VAR models are highly cross-validated. More than 80 % endogenity of the climatic

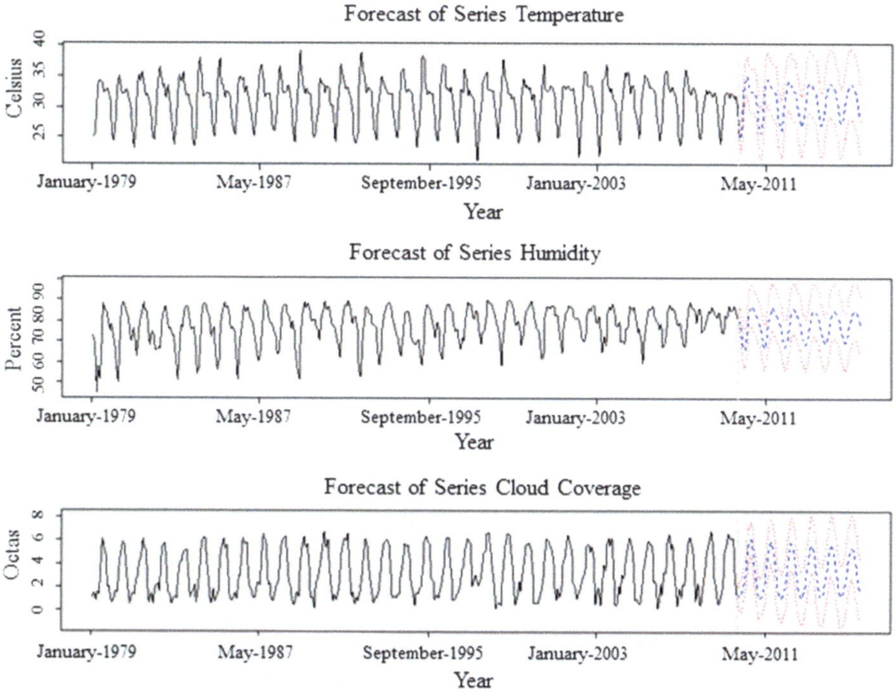

Fig. 2.14 Forecasted value of the climatic variables temperature, humidity, and cloud coverage from the fitted VAR(8) model from January 2011 to December 2016. The *blue* and *red colors lines* indicate forecasted value and 95 % confidence interval, respectively

variables temperature, humidity, and cloud coverage were explained by the predicted VAR models, implying some other climatic variables should be needed to include in the models. The temperature explained more than 65 % of its FEV after 1 year. The humidity and cloud coverage were accounted for more than 9 % and 19 % of the variation in temperature, respectively. The humidity to itself explained more than 55 % FEV after 1 year. For the variation in humidity, more than 30 % and 11 % FEV were explained in temperature and cloud coverage, respectively. More than 55 % FEV for cloud coverage were explained after 1 year by cloud coverage itself. Also, more than 20 % FEV were explained by temperature and humidity in the variation of cloud coverage. After 4 months the temperature decreased 0.065 % in response for the 0.076 % increasing shock in humidity. The temperature will be increased by 0.075 % for the 0.071 % increasing shock in cloud coverage after fourth month. Humidity will be increased by 0.737 % for the 0.227 % increasing shock in temperature after third month. But, after seventh month, humidity will be increased by 0.035 % for the 0.202 % increasing shock in cloud coverage. Cloud coverage will be decreased by 0.045 % for the 0.038 % increasing shock in temperature after fifth month and decreased by 0.063 % after sixth month for the 0.036 % increasing shock in humidity. Finally, forecast value of VAR(8) model showed slightly upward trend in temperature and humidity but downward trend in cloud coverage. The policy makers should consider these results in their policy implication.

Acknowledgments We would like to express our gratitude to Professor Dr. Mohammed Nasser, Chairman, Department of Statistics, University of Rajshahi, Bangladesh, for giving us the opportunity to present this paper in the International Conference and Publish in Conference Proceedings. Very special thanks go to Dr. Tanvir Islam, Department of Civil Engineering, University of Bristol, and his honorable team for publishing this manuscript with a revised version again as a book chapter of Springer Publication.

References

Adenomon MO, Ojehomon VET, Oyejola BA (2013) Modelling the dynamic relationship between rainfall and temperature time series data in Niger State, Nigeria. Math Theory Model 3(4):53–71

Akaike H (1969) Fitting autoregressive models for prediction. Ann Inst Stat Math 21:243–247

Akaike H (1974) A new look at the statistical model identification. IEEE Trans Automat Contr 19:716–723

Altaf MZ, Arshad IA, Ilys MR (2012) Vector autoregression application on macroeconomic variables of Pakistan's economic growth. Sindh Univ Res J 44(2):267–272

Awokuse OT, Bessler DA (2003) Vector autoregressions, policy analysis and directed acyclic graphs: an application to the U.S. economy. J Appl Econ 6(1):1–24

Canova F (1995) VAR models: specification, estimation, inference and forecasting. In: Pesaran H, Wickens M (eds) Handbook of applied econometrics. Blackwell, Oxford, England

Dickey DA, Fuller WA (1979) Distribution of the estimators for autoregressive time series with a unit root. J Am Stat Assoc 74:427–431

Durban M, Glasbey CA (2001) Weather modelling using a multivariate latent Gaussian model. Agr Forest Meteorol 109:187–201

Durbin J, Watson GS (1951) Testing for serial correlation in least squares regression. Biometrika 38:159–171

Ferdous MG, Baten MA (2011) Climatic variables of 50 years and their trends over Rajshahi and Rampur division. J Environ Sci Nat Resour 4(2):147–150

Granger CWJ (1969) Investigating causal relations by econometric models and cross-spectral methods. Econometrica 37:424–438

Gujarati DN (1993) Basic econometrics, 3rd edn. McGraw-Hill, New York

Hamilton DJ (1994) Time series analysis. Princeton University Press, Princeton, NJ

Hannan EJ, Quinn BG (1979) The determination of the order of an autoregression. J R Stat Soc B 41:190–195

Hatanaka M (1996) Time series based econometrics: unit roots and co-integration. Oxford University Press, Oxford

Ivanov V, Kilian L (2001) A practitioner's guide to lag-order selection for vector autoregressions. CEPR discussion paper no. 2685. Centre for Economic Policy Research, London

Janjua PZ, Samad G, Khan NU (2010) Impact of climate change on wheat production: a case study of Pakistan. Pak Dev Rev 49(4):799–821

Johansen S (1995) Likelihood-based inference in cointegrated vector autoregressive models. Oxford University Press, Oxford

Khan MZS, Hossain MI (2010) Democracy and trade balance: a vector autoregressive analysis. Bangladesh Dev Stud 33(4):23–37

Kleiber W, Katz RW, Rajagopalan B (2013) Daily minimum and maximum temperature simulation over complex terrain. Ann Appl Stat 7(1):588–612

Kwiatkowski D, Phillips PCB, Schmidt P, Shin Y (1992) Testing the null hypothesis of stationary against the alternative of a unit root. J Econom 54:159–178

Litterman RB (1986) Forecasting with Bayesian vector autoregression: five years of experience. J Bus Econ Stat 4:25–38

Liu P, Theodoridis K (2012) DSGE model restrictions for structural VAR identification. Int J Cent Bank 8(4):61–95

Liu X, Lindquist E, Vedlitz A (2011) Explaining media and congressional attention to global climate change, 1969–2005: an empirical test of agenda-setting theory. Polit Res Q 64(2):405–419

Lütkepohl H (1991) Introduction to multiple time series analysis. Springer, Berlin

Lütkepohl H (2005) New introduction to multiple time series analysis. Springer, Berlin

Lütkepohl H, Krätzig M (2004) Applied time series econometrics. Cambridge University Press, Cambridge

Moneta A, Chlab N, Entner D, Hoyer P (2011) Causal search in structural vector autoregressive models. Workshop Conf Proc 12:95–118

Mosedale TJ, Stephenson DB, Collins M, Mills TC (2006) Granger causality of coupled climate processes: ocean feedback on the North Atlantic oscillation. J Clim 19:1182–1194

Ni S, Sun D (2005) Bayesian estimates for vector autoregressive models. J Bus Econ Stat 23 (1):105–117

Pankratz A (1991) Forecasting with dynamic regression models. Wiley, New York

Phillips PCB, Perron P (1988) Testing for a unit root in time series regression. Biometrika 75:335–346

Schwarz G (1978) Estimating the dimension of a model. Ann Stat 6:461–464

Shamsnia SA, Shahidi N, Liaghat A, Sarraf A, Vahdat SF (2011) Modeling of weather parameters using stochastic methods (ARIMA model) (case study: Abadeh region, Iran). International conference on environment and industrial innovation IPCBEE. IACSIT Press, Singapore

Sims CA (1972) Money, income and causality. Am Econ Rev 62:540–552

Sims CA (1980) Macroeconomics and reality. Econometrica 48(1):1–48

Stergiou KI, Christou ED, Petrakis G (1997) Modelling and forecasting monthly fisheries catches: comparison of regression, univariate and multivariate time series methods. Fish Res 29(1):55–95

Stevens J (1996) Applied multivariate statistics for the social sciences. Lawrence Erlbaum, Mahwah, NJ

Sun D, Ni S (2004) Bayesian analysis of vector-autoregressive models with non-informative priors. J Stat Plan Inference 121:291–309

Wang W, Niu Z (2009) VAR model of PM2.5, weather and traffic in Los Angeles-long beach area. International conference on environmental science and information application technology, 4–5 July 2009, Wuhan, 3:66–69, ISBN: 978-0-7695-3682-8. DOI: 10.1109/ESIAT.2009.226

Chapter 3
Exploring the Behavior and Changing Trends of Rainfall and Temperature Using Statistical Computing Techniques

Abdus Salam Azad, Md. Kamrul Hasan, M. Arif Imtiazur Rahman, Md. Mustafizur Rahman, and Nashid Shahriar

Abstract The present study aimed at quantifying the change in surface air temperature and monthly total rainfall. The changing trend was detected using Mann–Kendall trend test, seasonal Mann–Kendall trend test, and Sen's slope estimator. *K*-means clustering algorithm was used to identify the rainfall distribution patterns over the years and also their changes with time. A comparative analysis was done among different time series prediction models to find out their suitability for forecasting daily temperature in climatic condition of Bangladesh. The analysis was performed using daily temperature and rainfall data of more than last 40 years (till 2009). The study found an increasing trend in maximum temperature during June to November and in minimum temperature during December to January in Bangladesh. There has been seen no significant change in rainfall over the years. However on the western side of the country, the amount of rain is significantly less than the eastern side. The study found that different prediction models were appropriate for different conditions.

Keywords Climate change • *K*-means clustering algorithm • Mann–Kendall trend test • Sen's slope estimator • Data mining • Pattern recognition • Time series prediction • Statistical analysis

3.1 Introduction

Bangladesh is likely to be one of the countries in the world which is most vulnerable to climate change. In recent times natural hazards are more frequent and intense compared to the similar kind of events occurred in one or two decades ago.

A.S. Azad (✉) • M.K. Hasan • M.A.I. Rahman • M.M. Rahman • N. Shahriar
Department of Computer Science and Engineering, Bangladesh University of Engineering and Technology, Palashi, Dhaka 1000, Bangladesh
e-mail: azadsalam2611@gmail.com; kamrulhasan326@gmail.com; arif.imtiaz216@gmail.com; mustafiz_rahman@cse.buet.ac.bd; nshahriar@cse.buet.ac.bd

T. Islam et al. (eds.), *Computational Intelligence Techniques in Earth and Environmental Sciences*, DOI 10.1007/978-94-017-8642-3_3,

National governments and IPCC (Intergovernmental Panel on Climate Change) scientists accepted that these climate hazards are the result of climate change at the global and regional level. According to the Fourth Assessment Report (AR4) of IPCC in Climate Change 2007: Synthesis Report, during the last hundred years the global temperature increased by 0.74 ± 0.18 °C. The model results of AR4 for Bangladesh are appropriate for global scale. But they did not use the local data of 37 stations in Bangladesh operated by the Bangladesh Meteorological Department (2012). Various researchers have contributed to the study of rainfall and temperature of Bangladesh with local data.

Rana et al. (2007) attempted to construct linear relationship between monthly, seasonal, and annual rainfall over Bangladesh with the southern oscillation index (SOI). Ahasan et al. (2010) analyzed the variability and trends of summer monsoon (Jun–Sep) rainfall over Bangladesh and found that the annual profile of the monthly total country average rainfall shows a unimodal characteristic with highest in July followed by June and August and lowest in January followed by December and February.

Basak et al. (2013) tried to detect trends in the monthly average maximum and minimum temperature and rainfall based on linear regression method. Long-term changes of near surface air temperature over Bangladesh have also been studied by Islam (2009). Various studies like Warrick et al. (1994), Karmakar and Shrestha (2000), Nahrin et al. (1997), Chowdhury and Debsarma (1992), Mia (2003), and Debsarma (2003) also focused on trends of change in rainfall and temperature in the context of Bangladesh.

However, to the best of our knowledge, several effective and advanced methods like Mann–Kendall trend test, clustering, etc. have not been utilized to investigate the climatic conditions to a greater extent.

The Mann–Kendall trend test is a widely known method for finding trends in time series data. Jain et al. (2013) examined trends in monthly, seasonal, annual rainfall, and temperature for the northeast region of India. The magnitude of trend in a time series was determined using Sen's estimator and statistical significance of the trend was analyzed using the Mann–Kendall trend test. Tripathi and Govindaraju (2009) used the Mann–Kendall trend test to study the changes in rainfall and temperature patterns over India. The method was also used to analyze the changing trend of the nonuniformity in rainfall of the upper Yangtze basin by Huang and Wang (2011). A number of researchers like Xi-ting et al. (2011), Gaoliao et al. (2012), Yue and Hashino (2003), Singh et al. (2008a, b), and Kumar and Jain (2010) also employed MK test to find trend in climatic data.

Clustering is a process of partitioning a set of data in a set of meaningful subclasses called clusters, which can effectively be used to analyze the distribution pattern of rainfall. Ramos (2001) used clustering techniques including *K*-means clustering method to analyze the rainfall distribution patterns over the years and their changes over time in a Mediterranean region. The study found that the variations of the mean annual rainfall in the Alt Penedès area throughout a period of 111 years have not followed a consistent trend. Pelczer and Cisneros-Iturbe (2008) also applied the K-means clustering technique to establish intensity classes to identify rainfall patterns over the Valley of Mexico. Major circulation patterns, associated with daily precipitation in Portugal, were classified by Corte-Real et al. (1998) based on the K-means clustering algorithm coupled with principal component analysis.

In our study inspection was done on more than 40 years climatic data of all the stations (37) of Bangladesh. Then five important places—Dhaka, Cox's Bazar, Khulna, Sylhet, and Rajshahi were analyzed with their geographic position in consideration, as these stations give quite a clear picture of the entire country. This study undertook the challenge of finding the trends in daily temperature changes and monthly total rainfall on those stations using the Mann–Kendall trend test and Sen's slope. For finding any trend in monthly total rainfall, seasonal Mann–Kendall trend test was also run to incorporate the seasonality. Monthly total rainfall of these selected regions was investigated to find the distribution of rainfall throughout the year. K-means clustering algorithm was used for this purpose.

As an agricultural country short-term temperature prediction is very important for Bangladesh. The study analyzed some of the well-known time series prediction models for finding their applicability to local data of Bangladesh. The models that were analyzed are autoregressive integrated moving average (ARIMA), Naive, random walk with drift (RWD), and the Theta model.

The later sections provide the details of our study. Section 3.2 provides an overview of the methods and materials that have been used in our study. In Sect. 3.3 the results of our study are discussed and in Sect. 3.4 we conclude the study with our key findings.

3.2 Materials and Methods

In this section we briefly discuss about the different methods and materials used in our study.

3.2.1 K-*Means Clustering*

Clustering is a main task of explorative data mining. It assigns a set of objects into groups (clusters) so that the objects in the same cluster are more similar. For clustering K-means clustering algorithm (MacQueen 1967) was used. It is an algorithm for putting N data points (x_1,x_2, ..., x_n) in an I-dimensional space into K clusters. The mean of each cluster is denoted by $m^{(k)}$. Each vector x has I components x_i. Distances between points in real space is as follows:

$$d(x, y) = \frac{1}{2}\sum_i (x_i - y_i)^2$$

The K-means $m^{(k)}$ is initialized to a random value. Then it follows two steps.

Assignment step: Each data point n is assigned to the nearest mean. For the cluster $k^{(n)}$, the point $x^{(n)}$ belongs to

$$\hat{k}^{(n)} = \arg\min_k d\left(m^{(k)}, x^{(n)}\right)$$

The indicator variable, $r_k^{(n)}$ is set to one if mean k is the closest mean to data point $x^{(n)}$; otherwise, $r_k^{(n)}$ is zero.

$$r_k^{(n)} = \begin{cases} 1 & \text{if } \hat{k}^{(n)} = k \\ 0 & \text{if } \hat{k}^{(n)} \neq k \end{cases}$$

Update step: The model parameters, the means, are adjusted to match the sample means of the data points that they are responsible for.

$$m^{(k)} = \begin{cases} \dfrac{\sum_n r_k^{(n)} x^{(n)}}{R^{(k)}} & \text{if } R^{(k)} > 0 \\ \text{Oldest } m^{(k)} & \text{if } R^{(k)} = 0 \end{cases}$$

where $R^{(k)}$ is the total responsibility of mean k,

$$R^{(k)} = \sum_n r_k^{(n)}$$

The assignment step and update step is repeated until the assignments do not change.

3.2.2 Mann–Kendall Trend Test

The Mann–Kendall test (Mann 1945; Kendall 1975) is a nonparametric test for identifying trends in time series data. The test compares the relative magnitudes of sample data rather than the data values themselves (Gilbert 1987). Here it is assumed that that there exists only one data value per time period. Let $(x_1, x_2, \ldots, x_n)$ represent n data points where x_j represents the data point at time j. Then the Mann–Kendall statistics (S) is given by

$$S = \sum_{k=1}^{n-1} \sum_{j=k+1}^{n} \text{sign}\left(x_j - x_k\right)$$

where

$$\mathrm{sign}\left(x_j - x_k\right) = \begin{cases} +1 & \text{if } x_j - x_k > 0 \\ 0 & \text{if } x_j - x_k = 0 \\ -1 & \text{if } x_j - x_k < 0 \end{cases}$$

A very high positive value of S is an indicator of an increasing trend, and a very low negative value indicates a decreasing trend. However, it is necessary to compute the VAR(S), Sen's slope associated with S and the sample size n, to statistically quantify the significance of the trend. When $n \geq 8$, the S is approximately normally distributed with the mean. The variance of S, VAR(S), by the following equation (Helsel and Hirsch 1992):

$$\mathrm{VAR}(S) = \frac{n*(n-1)*(2n+5) - \sum_{i=1}^{m} t_i(i)(i-1)(2i+5)}{18}$$

where t_i is considered as the number of ties up to sample i. VAR(S) and Sen's slope estimator both are used to estimate the trend in time series data. A positive S value indicates a positive trend and a negative value indicates a negative trend in time series data.

3.2.3 Seasonal Mann–Kendall Trend Test

Hirsch et al. (1982) developed this test that is used to find the monotonic trend in time series data with seasonal variation. Mann–Kendall statistics S is computed separately for each month and then summed to obtain the overall test statistic. Mann–Kendall statistics S_k for each season k ($k = 1, 2, \ldots, p$):

$$S_k = \sum_{i=1}^{n_k - 1} \sum_{j=i+1}^{n_k} \mathrm{sign}\left(x_{jk} - x_{ik}\right)$$

where

x_{jk} = Observation from season k in year j
x_{ik} = Observation from season k in year i

Then these statistics are summed to form overall statistics S_n:

$$S_n = \sum_{k=1}^{p} S_k$$

Here the overall variance of the test statistics $\mathrm{VAR}(S_n)$ is obtained by summing the variances of the Kendall score statistics for each month.

Test interpretation for Mann–Kendall trend test and seasonal Mann–Kendall trend test:

The null and alternative hypotheses for both the Mann–Kendall trend test and seasonal Mann–Kendall trend test:

H_0: There is no trend in the series
H_a: There is a positive/negative trend in the series

Here the significance level alpha (α) is 5 %. If the computed p-value is greater than alpha, then the null hypothesis H_0 cannot be rejected.

3.2.4 Sen's Slope Estimator

In nonparametric statistics, Sen's slope estimator (1968) is a method for robust linear regression that chooses the median slope among all lines through pairs of two-dimensional sample points. The magnitude of linear trend is predicted by the Sen's estimator. The slope (Q) of all data pair is

$$Q = \frac{x_i' - x_i}{i' - i} \quad \text{for } i = 1, 2, 3, \ldots, N$$

where

Q = slope between data points X_i and X_i'
X_i' = data measurement at time i'
X_i = data measurement at time i
i' = time after time i

Sen's estimator of slope is simply given by the median slope (Q'), shown below as:

$$Q' = \begin{cases} Q\left[\dfrac{N+1}{2}\right] & \text{if } N \text{ is odd} \\ \dfrac{Q[N+1] + Q[N+2]}{2} & \text{if } N \text{ is even} \end{cases}$$

where N is the number of calculated slopes.

Then, Q'med is computed by a two-sided test at 100 (1 − α) % confidence interval and then a true slope can be obtained by the nonparametric test. Positive value of Sen's slope indicates increasing trend and a negative value indicates a decreasing trend.

3.2.5 Naive Model

Naive approach is the simplest but an efficient forecasting model of time series data. If the data is stable according to this model, forecast of any periods are equal to the actual value of previous period. When the sequence of observations begins at time $t = 0$, the simplest form of Naive is given by the formulae:

$$y_t = y_{t-1}, \quad t > 0$$

Naive forecast can be used as a benchmark against which other forecasting models can be compared.

3.2.6 Autoregressive Integrated Moving Average Model

ARIMA model is a generalization of an autoregressive moving average (ARMA) model (Box and Jenkins 1970). This model is used to predict future points in the series. ARIMA models aim to describe the autocorrelations in the data. It combines differencing with autoregression and moving average (MA) model.

The full model can be written as:

$$y'_t = c + \sum_{i=1}^{p} \phi_i y'_{t-i} + \sum_{i=1}^{q} \theta_i e_{t-i}$$

ARIMA models are defined for stationary time series. Therefore, if the data shows non-stationary time series, then it is needed to *difference* the time series until a stationary time series is obtained.

Differencing method makes time series stationary by computing the differences between consecutive observations. First-differenced series y'_t can be written as:

$$y'_t = y_t - y_{t-1}$$

When the differenced series is white noise with nonzero mean, the original series can be written as

$$y_t - y_{t-1} = c + e_t$$

where c is the average of the change between consecutive observations.

$$e_t = \text{White noise}$$

Sometimes it is needed to difference the data second time to obtain stationary series:

$$\begin{aligned} y_t'' &= y_t' - y_{t-1}' \\ &= y_t - 2y_{t-1} + y_{t-2} \end{aligned}$$

The backward shift operator B shifts the data y_t back to one period.

$$By_t = y_{t-1} \tag{3.1}$$

So first difference can be written as:

$$y_t' = y_t - y_{t-1} = y_t - By_t = (1 - B)y_t$$

Similarly second difference can be written as:

$$y_t'' = y_t' - y_{t-1}' = (1 - B)^2 y_t$$

In general, a dth order difference can be written as:

$$y_t^d = (1 - B)^d y_t$$

Autoregressive model (*AR*) is a multiple regression model that specifies that the output variable can be computed from linear combination of its own previous values. This model forecasts the value at time t by the weighted average of past few observations. The notation AR(p) indicates an autoregressive model of order p. The AR(p) model is defined as:

$$y_t = c + \sum_{i=1}^{p} \phi_i y_{t-i} + e_t \tag{3.2}$$

where $\phi_1, \phi_2, \ldots, \phi_p$ are the parameters of this model. Using backshift operator B from Eq. 3.1 the AR(p) model can be written as:

$$y_t = c + \sum_{i=1}^{p} \phi_i B^i y_{t-i} + e_t$$

Changing the values of $\phi_1, \phi_2, \ldots, \phi_p$ results in different time series patterns. Autoregressive model can be restricted to stationary data by putting some constraints on the values of the parameters. This equation only gives one step ahead forecast. For n ahead step forecast, past forecast values are used for weighted average on the right side of the equation (Eq. 3.2). There are some problems with forecasting using the AR model. It will not tell very much about future. Forecast will not reach the peaks of the data, because it smoothes out the data by taking the mean of observed values.

The moving average (MA) model is another model to forecast time series data. While the AR model uses past values of forecast variable in regression, the MA model uses past forecast errors. The MA(q) model can be written as:

$$y_t = c + \sum_{i=1}^{q} \theta_i e_{t-q} + e_t$$

where e_t = white noise and $\theta_1, \theta_2, \ldots, \theta_q$ are the parameters of the model.

Any value at time t is the weighted average of past few forecast errors. Some constraints are needed to put on parameters for restricting the model to stationary data. Time series pattern can be changed by changing the parameters. But changing the value of e_t will just change the scale not the pattern.

ARIMA model:

ARIMA model is the combination of all these methods. First the data must be *differenced d* times to obtain stationary series if needed. Then full ARIMA model is applied on stationary data to forecast. The full model is:

$$y_t^{'} = c + \sum_{i=1}^{p} \phi_i y_{t-i}^{'} + \sum_{i=1}^{q} \theta_i e_{t-i} \tag{3.3}$$

This is called ARIMA(p, d, q) model, where

p = order of the autoregressive part
d = degree of first differencing involved
q = order of the moving average part

Equation 3.3 can be written using backshift notation:

$$\left(1 - \sum_{i=1}^{p} \phi_i B^i\right)(1 - B)^d y_t = c + \left(1 + \sum_{i=1}^{q} \theta_i B^i\right) e_t$$

Determining the appropriate value of p, d, and q for the data is difficult. The appropriate value of p, d, and q cannot be found from a time plot. From autocorrelations and partial autocorrelations plot, the value of q and p can be determined. ARIMA model selection becomes problematic with missing observations and other data irregularities.

3.2.7 Random Walk with Drift

When a time series shows irregular pattern, it is better to try to predict the change that occurs from one period to the next. Random walk model can be written as:

$$Y(t) = Y(t-1) + \alpha$$

where

$Y(t)$ = Predicted value at time t
$Y(t-1)$ = Previous value
α = Drift: Average change between periods

Time series value at any period will be equal to the last period's value plus the average change between periods. Average change between two periods is called drift (α) which acts like a trend. This model is known as "random walk" model: it assumes that from one period to the next, the original time series merely takes a random step away from its last position. If the constant term (α) in the random walk model is zero, then it is called random walk without drift. This is similar to the Naive model.

If the series being fitted by a random walk model has an average upward trend ($\alpha > 0$) or downward trend ($\alpha < 0$), then a nonzero constant (α) drift term must be added. This drift along with previous frequency determines new forecast (Pesaran and Pick 2009). It is known as random walk with drift.

3.2.8 *Theta Model*

Theta method is used for obtaining forecast from series of data. Forecasts obtained by theta method are equivalent to simple exponential smoothing (SES) with drift. Theta method is simply a special case of SES with drift, where the drift parameter is half the slope of the linear trend fitted to the data. The method performed well, particularly for monthly series and for microeconomic data. The detail of the Theta model is described by Assimakopoulos and Nikolopoulos (2000).

3.2.9 *Wilcoxon Test*

The Wilcoxon signed rank test is a nonparametric method of testing whether two populations X and Y have the same continuous distribution without assuming them to follow the normal distribution (Wilcoxon 1945). Two data samples are matched if they come from repeated observations of the same subject. The null and alternative hypotheses for the Wilcoxon signed rank test are:

H_0: median difference between the pairs is zero
H_a: median difference is not zero

The output of Wilcoxon signed rank test is p-value. It is used to determine which hypothesis will be accepted. Ninety-five percent confidence level is used. If the value is less than 0.05 significance level, null hypothesis can be rejected which yields that two samples are not identical. Otherwise data samples follow the same distribution.

3.2.10 RMS Error

Root mean square (RMS) error is a measure of the average error, weighted according to the square of the error. It is a good measure of accuracy, but only to compare forecast errors of different models. RMSE can be defined as:

$$\mathrm{RMSE} = \sqrt{\frac{1}{N}\sum_{i=1}^{N}(F_i - O_i)^2}$$

where

F_i = the forecast values
O_t = the corresponding verifying value
N = the number of observation points

3.2.11 SMAPE

Symmetric mean absolute percentage error (SMAPE) is an accuracy measure based on percentage error. For n forecast, the SMAPE can be defined as:

$$\mathrm{SMAPE} = \frac{1}{n}\sum_{i=1}^{n}\frac{|y_i - f_i|}{(y_i + f_i)/2}$$

where

y_i = actual value
f_i = forecast value

The absolute value of y_i and f_i is divided by their average.

3.2.12 Materials Used During the Study

The Mann–Kendall trend test and seasonal Mann–Kendall trend test are implemented in Java. K-means clustering algorithm is implemented using WEKA (Hall et al. 2009). The time series prediction models are implemented using the forecast package of R (R Core Team 2013).

The analysis was done during 2012–2013 in BUET, Bangladesh.

3.2.13 Data Used in the Study

The daily minimum temperature, maximum temperature, and rainfall data in the period 1947–2009 from 37 weather stations of Bangladesh were collected from Bangladesh Meteorological Department (BMD). Daily minimum, maximum temperature and rainfall data were cleaned by filling in missing values with mean values of adjacent days.

With geographic position in consideration, we selected Dhaka, Rajshahi, Khulna, Sylhet, and Cox's Bazar stations for our study.

3.3 Results and Discussion

In this section results of different analysis are presented and investigated. In Sects. 3.3.1 and 3.3.2 the changing trends of temperature and rainfall are inspected by means of the Mann–Kendall trend test and Sen's slope estimator. The distribution of rainfall in Bangladesh is discussed in Sect. 3.3.3. In Sect. 3.3.4 a comparative study is presented on applicability of some of the well-known time series prediction models for providing temperature forecast.

3.3.1 Change in Temperature

Regression analysis and then the Mann–Kendall trend test were run on average maximum and minimum monthly temperature on the selected stations. The results show that the maximum temperature ($T_{\max}$) of the months June to November has increased and the minimum temperature ($T_{\min}$) of winter has increased in Dhaka, Cox's Bazar, and Sylhet. On the contrary $T_{\min}$ shows negative trend in Khulna and Rajshahi.

The result of the Mann–Kendall trend test and Sen's slope estimator on average maximum and minimum temperature are shown in Tables 3.1 and 3.2, respectively. The change in temperature per decade is shown in Fig. 3.1.

In Dhaka an increasing trend in maximum temperature is observed during the months June to November (Table 3.1). On average the maximum temperature has increased 0.19 °C/decade. Figure 3.1c clearly shows the increasing trend during this time span. On the other hand the $T_{\min}$ has significantly increased during November to March (Table 3.2). A very high increase, more than 0.55 °C/decade is observed (Fig. 3.1c).

At Cox's Bazar positive trend is detected throughout the year. Only the month of May has not experienced any trend for $T_{\min}$. Specially from the start of the Monsoon (Jun–Sep) till the month of November $T_{\max}$ showed a significant rise, more than 0.29 °C/decade (Fig. 3.1d). While $T_{\min}$ is increased significantly in

Table 3.1 Mann–Kendall statistics (S) and Sen's slope (SS) of maximum temperature of different regions

Month	Rajshahi	Khulna	Dhaka	Sylhet	Cox's Bazar
January	$S = -227$ $SS = -0.02$	$S = -567$ $SS = -0.026$		$S = +221$ $SS = +0.015$	$S = +427$ $SS = +0.015$
February				$S = +231$ $SS = +0.032$	$S = +639$ $SS = +0.029$
March		$S = -315$ $SS = -0.016$			$S = +679$ $SS = +0.029$
April					$S = +752$ $SS = +0.031$
May				$S = +286$ $SS = +0.026$	$S = +680$ $SS = +0.027$
June	$S = +219$ $SS = +0.03$	$S = +505$ $SS = +0.019$	$S = +625$ $SS = +0.034$	$S = +525$ $SS = +0.036$	$S = +672$ $SS = +0.029$
July	$S = +398$ $SS = +0.03$	$S = +773$ $SS = +0.022$	$S = +732$ $SS = +0.024$	$S = +460$ $SS = +0.033$	$S = +724$ $SS = +0.28$
August	$S = +523$ $SS = +0.037$	$S = +653$ $SS = +0.022$	$S = +836$ $SS = +0.032$	$S = +580$ $SS = +0.039$	$S = +887$ $SS = +0.034$
September		$S = +330$ $SS = +0.012$	$S = +521$ $SS = +0.022$	$S = +431$ $SS = +0.031$	$S = +929$ $SS = +0.032$
October		$S = +435$ $SS = +0.014$	$S = +557$ $SS = +0.028$	$S = +677$ $SS = +0.036$	$S = +955$ $SS = +0.037$
November	$S = +268$ $SS = +0.018$	$S = +438$ $SS = +0.017$	$S = +548$ $SS = +0.03$	$S = +634$ $SS = +0.033$	$S = +892$ $SS = +0.041$
December				$S = +448$ $SS = +0.025$	$S = +905$ $SS = +0.038$

Positive value of SS and S signify positive trend and vice versa
Empty cell denotes no trend

winter (Dec–Feb) by 0.35 °C/decade. These changes are verified by the Mann–Kendall statistics and Sen's slope (Tables 3.1 and 3.2).

Also in Sylhet the temperature has seen a positive trend almost throughout the year. During June to November, the positive trend in T_{max} is about 0.29 °C/decade. In winter T_{min} has increased by about 0.31 °C/decade (Fig. 3.1e). The Mann–Kendall statistics and Sen's slope also suggests that.

In Rajshahi T_{max} has shown a positive trend in June, July, August, and November (Table 3.1). T_{min} increased in July, August (Table 3.2). While January has shown a negative trend both in minimum and maximum temperature. In monsoon T_{max} has shown increasing trend more than 0.2 °C/decade (Fig. 3.1a). In winter (Dec–Feb) T_{min} is found stable (Fig. 3.1a). The Mann–Kendall trend test has shown no trend.

In Khulna T_{max} has shown a positive trend in June to November and negative trend in January and March (Table 3.1). T_{max} has increased about 0.16 °C/decade (Fig. 3.1b). Only in Khulna it is found that in winter T_{min} has shown negative trend. T_{min} has decreased in December to February and May (Table 3.2). A slightly decreasing trend in T_{min}, about 0.11 °C/decade, is found in winter (Fig. 3.1b).

Table 3.2 Mann–Kendall statistics (S) and Sen's slope (SS) of minimum temperature of different regions

Month	Rajshahi	Khulna	Dhaka	Sylhet	Cox's Bazar
January	$S = -266$ $SS = -0.034$	$S = -355$ $SS = -0.04$	$S = +727$ $SS = +0.05$	$S = +339$ $SS = +0.025$	$S = +569$ $SS = +0.025$
February	$S = -226$ $SS = -0.021$	$S = +786$ $SS = +0.055$	$S = +393$ $SS = +0.025$	$S = +759$ $SS = +0.04$	
March			$S = +0.479$ $SS = +0.04$	$S = +358$ $SS = +0.029$	$S = 558$ $SS = +0.033$
April					$S = +347$ $SS = +0.017$
May		$S = -239$ $SS = -0.015$			
June			$S = +319$ $SS = +0.011$		$S = +506$ $SS = +0.016$
July	$S = +169$ $SS = +0.007$		$S = +312$ $SS = +0.008$	$S = +244$ $SS = +0.009$	$S = +581$ $SS = +0.012$
August	$S = +185$ $SS = +0.009$		$S = +431$ $SS = +0.009$	$S = +372$ $SS = +0.012$	$S = +667$ $SS = +0.015$
September				$S = +234$ $SS = +0.011$	$S = +631$ $SS = +0.012$
October			$S = +336$ $SS = +0.013$	$S = +373$ $SS = +0.025$	$S = +660$ $SS = +0.012$
November			$S = +726$ $SS = +0.048$	$S = +495$ $SS = +0.043$	$S = +439$ $SS = +0.034$
December		$S = -263$ $SS = -0.022$	$S = +747$ $SS = +0.056$	$S = +464$ $SS = +0.038$	$S = +559$ $SS = +0.033$

Positive value of SS and S signify positive trend and vice versa
Empty cell denotes no trend

3.3.2 Changes in Monthly Total Rainfall

The Mann–Kendall trend test and seasonal Mann–Kendall trend test were run on monthly total rainfall on the selected stations with intent to find any trend in rainfall over the years. The significance level, α was set to 5 % for both the tests. The period value was set to 12 for the seasonal Mann–Kendall trend test.

Table 3.3 summarizes the tests results. It tabulates the statistics (S), Sen's slope (SS), and p-values for the Mann–Kendall trend test. For the seasonal Mann–Kendall trend test, the statistics (S_k) and p-values are tabulated. p-Values less than 0.05 (significance level, α) indicate the presence of trend in the time series data. If there is a trend, positive values of statistics (S) and Sen's slope (SS) (for the Mann–Kendall trend test) denote positive trend in the series and vice versa.

From the results of the Mann–Kendall trend test, it can be seen that all the p-values are greater than 0.05. So, the null hypothesis cannot be rejected. It indicates that monthly total rainfall has seen no trend over the years in the selected regions, if seasonality is not taken into account.

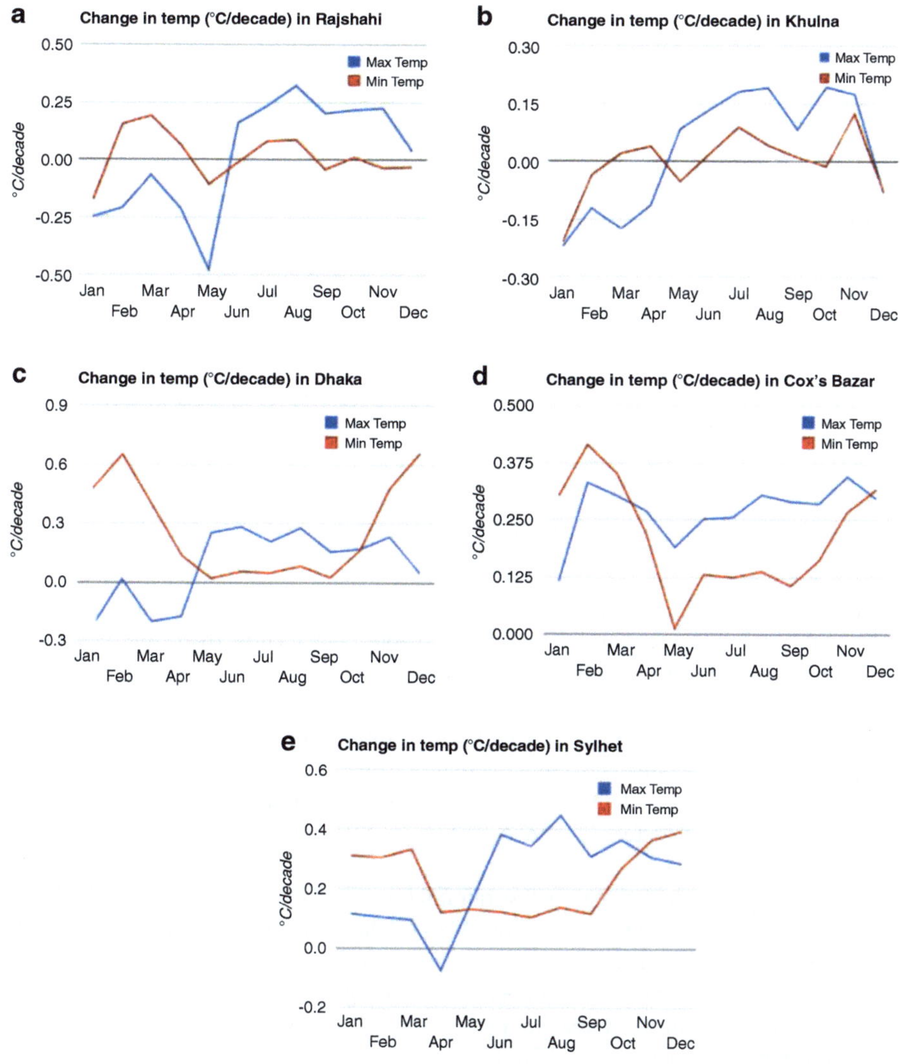

Fig. 3.1 Change in minimum and maximum temperature per decade (°C/decade). (**a**) Rajshahi, (**b**) Khulna, (**c**) Dhaka, (**d**) Cox's Bazar, and (**e**) Sylhet

The seasonal Mann–Kendall trend test shows the same result too, except in Cox's Bazar, where there is seen a positive change over the years.

3.3.3 Clustering of Rainfall

Applying *K*-means clustering algorithm, monthly total rainfall were partitioned into five different clusters. The clusters give five categories of monthly rainfall

Table 3.3 Outcome of Mann–Kendall trend test and seasonal Mann–Kendall trend test on monthly total rainfall data

	Mann–Kendall trend test				Seasonal Mann–Kendall trend test		
Stations	S	SS	p-Value	Trend	S_k	p-Value	Trend
Dhaka	−941	0	0.781	No	−279	0.324	No
Cox's Bazar	4,221	1,464	0.114	No	714	0.007	Positive
Sylhet	−529	0	0.438	No	−364	0.1	No
Khulna	−1,806	1,159	0.303	No	−13	0.484	No
Rajshahi	−1,649	0	0.313	No	−372	0.094	No

Table 3.4 Range of rainfall corresponding to different cluster number

Cluster	Monthly rainfall (mm)	Class
1	0–120	Negligible
2	121–322	Less
3	323–578	Moderate
4	579–966	High
5	967–3,017	Very high

based on its amount. Cluster 1 depicts the months with negligible or no rainfall where Cluster 5 represents the highest (Table 3.4). Thus over the years, the distribution of months in different clusters depicts a clear picture of rainfall of that region on the time period (Fig. 3.2).

In Cox's Bazar there is seen a negligible amount of rain during December to March, falling in Cluster 1. Then it increases from April, reaching its peak rainfall during June to August. During this period, huge amount of rainfall is observed, mostly falling in Clusters 4 and 5. Then it starts decreasing from September (Fig. 3.2a–c).

Study for Sylhet shows slightly less rainfall than Cox's Bazar, negligible rain during November to February, an increase from March. The peak rainfall (Cluster 3–5) is seen during May to August. Then starts to decrease from September (Fig. 3.2m–o).

In Dhaka negligible rainfall is experienced during November to March (Cluster 1). Then it increases from April, reaches peak during May to September (Clusters 2 and 3) and then starts decreasing (Fig. 3.2d–f).

In Khulna during November to February very less rainfall is seen (Cluster 1). Then it increases from March, remains constant in its peak rainfall during May to September (Clusters 2–4) and starts decreasing from October (Fig. 3.2j–l).

The study shows the least amount of rainfall in Rajshahi. Here, the rainfall constantly remains in Cluster 1 from November to April spanning half of the year. The peak rainfall is seen during June to September mostly residing in Clusters 2 and 3 (Fig. 3.2g–i).

Higher cluster number denotes higher amount of rainfall, the actual range can be found in Table 3.4.

The study also shows that over the years the distribution of rainfall has no significant change. Only the month of September shows some increase in Khulna from the year 1996.

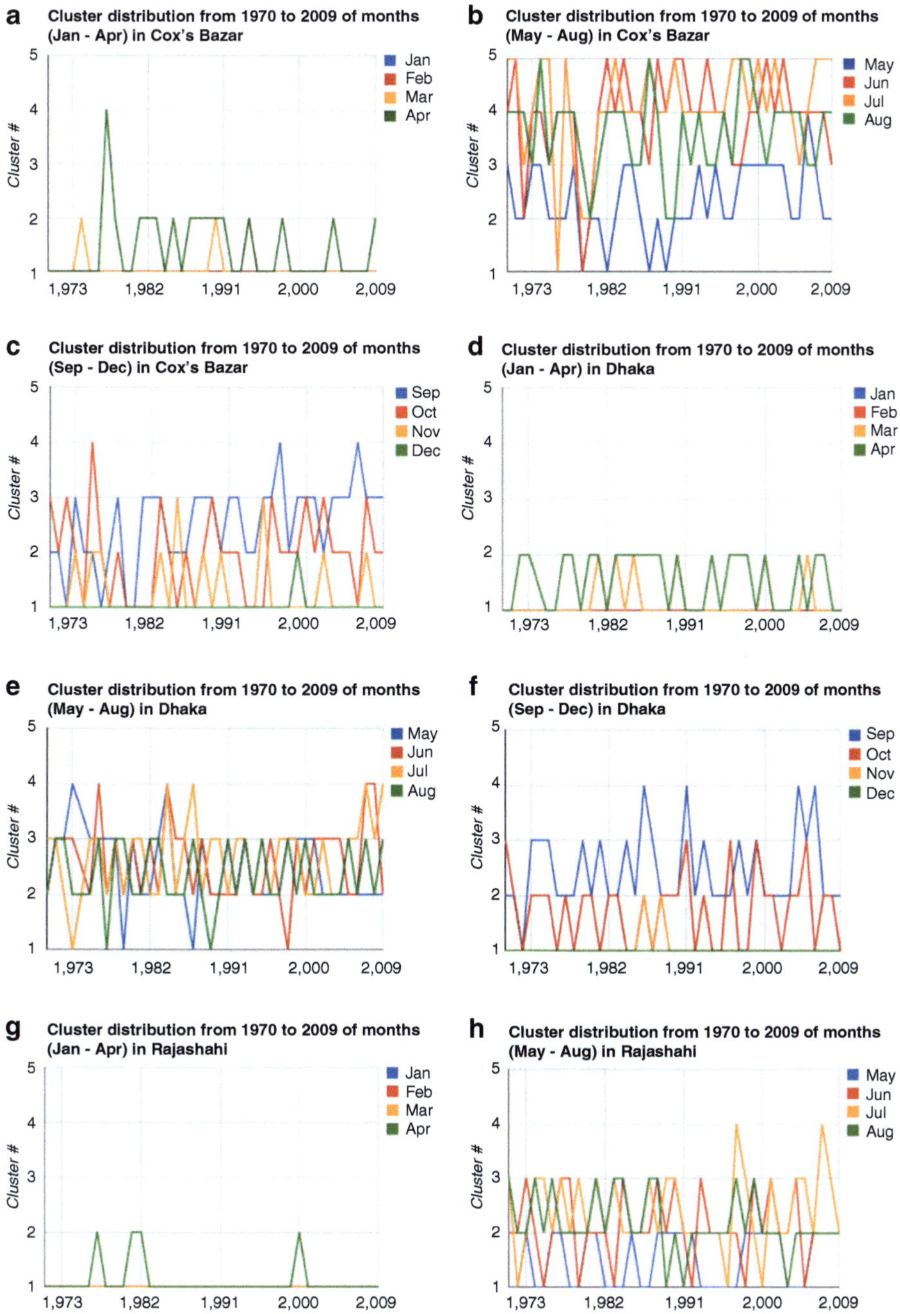

Fig. 3.2 Cluster distribution of rainfall of different regions throughout the year. (**a**) Cox's Bazar (Jan–Apr), (**b**) Cox's Bazar (May–Aug), (**c**) Cox's Bazar (Sep–Dec), (**d**) Dhaka (Jan–Apr), (**e**) Dhaka (May–Aug), (**f**) Dhaka (Sep–Dec), (**g**) Rajshahi (Jan–Apr), (**h**) Rajshahi (May–Aug), (**i**) Rajshahi (Sep–Dec), (**j**) Khulna (Jan–Apr), (**k**) Khulna (May–Aug), (**l**) Khulna (Sep–Dec), (**m**) Sylhet (Jan–Apr), (**n**) Sylhet (May–Aug), and (**o**) Sylhet (Sep–Dec)

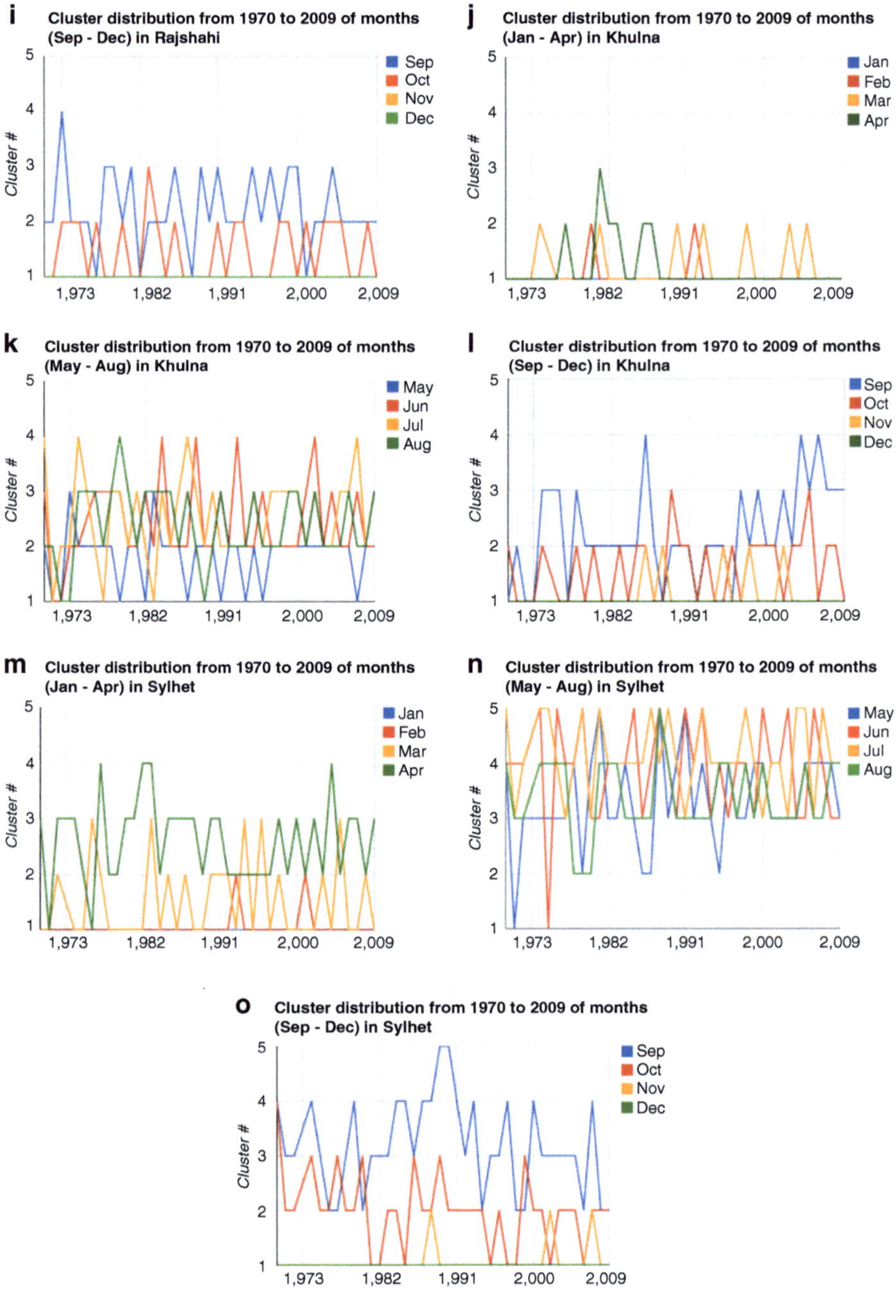

Fig. 3.2 (continued)

It is also evident from the distributions of clusters that in the eastern regions (Cox's Bazar and Sylhet) it rains remarkably more than the western regions (Rajshahi and Khulna).

3.3.4 Comparison of Different Time Series Forecasting Models

Different time series forecasting models were used to predict daily temperature. The models used are the Naive model, RWD model, Theta model, and ARIMA (1,1,0) model. For a particular month, the forecasting models were fed with the first 20 days' data (minimum and maximum temperature) and prediction of next 7 days were obtained. Then the RMS error and SMAPE were computed using the original values. This process was repeated for all the months from 1953 to 2009 for different stations.[1] Then the mean and standard deviation of the RMS error and SMAPE were computed to compare the accuracy of different models. In the cases, where the SMAPE and RMS error provides different ordering of the models, the ordering obtained from the SMAPE is considered for the discussion.

For ARIMA(p, d, q) model, the data is transformed to stationary time series by differencing with degree 1 ($d = 1$). For best fitting, the AR order (p) and MA order (q) were set to 1 and 0, respectively. ARIMA(1,1,0) is called autoregressive model with first difference.

Tables 3.5, 3.6, 3.7, 3.8, 3.9, 3.10, 3.11, 3.12, 3.13, and 3.14 tabulate the mean and standard deviation of the RMS error and SMAPE of minimum and maximum temperature for the selected stations. The models are sorted in descending order according to their mean of the SMAPE.

From the SMAPE and RMS error, it is evident that for all the selected regions, the RWD model shows the greatest SMAPE and RMS error for both minimum and maximum temperature. So it is the least applicable one among the selected models.

No simple ordering among the other three models can be found. Firstly, the scenario for minimum temperature is discussed. The Theta model provides best prediction for Dhaka, Sylhet, and Cox's Bazar while ARIMA and Naive model suits most for Rajshahi and Khulna region, respectively.

On the other hand, for prediction of maximum temperature, ARIMA model is most accurate for Rajshahi, Khulna, and Cox's Bazar. But for Dhaka and Sylhet, the Naive model, beating all the other models gives the best prediction.

The best models for different regions, for both maximum and minimum temperature, are listed in Table 3.15. In most of the cases the best model has the least SMAPE and RMS error simultaneously. In some cases where the SMAPE

[1] For time series prediction, missing data were replaced by the data of the previous day. Months, where more than 50 % data were missing, were not included in calculation.

Table 3.5 Mean RMS error and SMAPE of different models for Dhaka region (minimum temperature)

	RMS error		SMAPE	
Model name	Mean	SD	Mean	SD
Theta model	1.997	1.167	8.929	6.904
ARIMA	2.019	1.25	9.064	7.499
Naive	2.041	1.22	9.134	7.291
Random walk with drift	2.299	1.429	10.429	8.869

Table 3.6 Mean RMS error and SMAPE of different models for Rajshahi region (minimum temperature)

	RMS		SMAPE	
Model name	Mean	SD	Mean	SD
ARIMA	2.005	1.204	9.936	8.04
Naive	2.016	1.167	9.946	7.841
Theta model	1.991	1.182	10.017	8.298
Random walk with drift	2.228	1.378	11.357	9.938

Table 3.7 Mean RMS error and SMAPE of different models for Sylhet region (minimum temperature)

	RMS		SMAPE	
Model name	Mean	SD	Mean	SD
Theta model	1.624	0.96	7.533	5.572
Naive	1.675	0.945	7.705	5.411
ARIMA	1.667	0.982	7.718	5.776
Random walk with drift	1.882	1.147	8.806	6.634

Table 3.8 Mean RMS error and SMAPE of different models for Khulna region (minimum temperature)

	RMS		SMAPE	
Model name	Mean	SD	Mean	SD
Naive	2.004	1.199	8.822	6.857
ARIMA	1.993	1.234	8.862	7.283
Theta model	1.982	1.227	8.929	7.357
Random walk with drift	2.226	1.381	9.906	8.071

Table 3.9 Mean RMS error and SMAPE of different models for Cox's Bazar region (minimum temperature)

	RMS		SMAPE	
Model name	Mean	SD	Mean	SD
Theta model	1.577	0.974	6.541	4.74
ARIMA	1.595	0.984	6.636	4.863
Naive	1.629	1.032	6.714	4.97
Random walk with drift	1.792	1.21	7.462	5.732

Table 3.10 Mean RMS error and SMAPE of different models for Dhaka region (maximum temperature)

	RMS error		SMAPE	
Model name	Mean	SD	Mean	SD
Theta model	2.163	1.331	6.17	4.114
ARIMA	2.145	1.426	6.144	4.879
Naive	2.126	1.329	6.05	4.18
Random walk with drift	2.391	1.56	6.871	4.949

Table 3.11 Mean RMS error and SMAPE of different models for Rajshahi region (maximum temperature)

	RMS		SMAPE	
Model name	Mean	SD	Mean	SD
ARIMA	2.398	1.63	6.656	4.669
Naive	2.406	1.637	6.68	4.743
Theta model	2.41	1.621	6.674	4.542
Random walk with drift	2.713	1.884	7.585	5.49

Table 3.12 Mean RMS error and SMAPE of different models for Sylhet region (maximum temperature)

	RMS		SMAPE	
Model name	Mean	SD	Mean	SD
Theta model	2.655	1.535	7.809	4.924
Naive	2.617	1.542	7.699	5.053
ARIMA	2.618	1.527	7.711	5.114
Random walk with drift	2.934	1.884	8.721	6.32

Table 3.13 Mean RMS error and SMAPE of different models for Khulna region (maximum temperature)

	RMS		SMAPE	
Model name	Mean	SD	Mean	SD
Naive	2.229	1.348	6.162	4.135
ARIMA	2.22	1.378	6.14	4.267
Theta model	2.246	1.353	6.195	4.106
Random walk with drift	2.493	1.592	6.955	4.957

Table 3.14 Mean RMS error and SMAPE of different models for Cox's Bazar region (maximum temperature)

	RMS		SMAPE	
Model name	Mean	SD	Mean	SD
Theta model	1.861	1.054	5.387	3.38
ARIMA	1.807	1.052	5.192	3.396
Naive	1.821	1.093	5.222	3.572
Random walk with drift	2.041	1.289	5.946	4.304

Table 3.15 Listing of the most accurate models of different regions according to SMAPE

Region	Min temp	Max temp
Dhaka	Theta model	Naive
Rajshahi	ARIMA (Theta model)	ARIMA
Sylhet	ARIMA	Naive
Khulna	Theta model (Naive)	ARIMA
Cox's Bazar	Theta model	ARIMA

In cases where orderings obtained from SMAPE and RMS error differ, the best model according to RMS error is written in parentheses

and RMS error disagree about the best model, the SMAPE is given higher priority. The best model according to the RMS error is also listed in parentheses.

The distributions of prediction obtained from the models are then compared by means of their RMS error and SMAPE against the Naive model. The Naive model is an efficient forecasting model and often gives better results than other more

Table 3.16 Wilcoxon test results (p-value) on minimum temperature of different models against Naive model

Region		Arima	Theta	Random Walk
Dhaka	RMS Error	0.002	0.512	2.22E-23
	SMAPE	0.016	0.364	2.56E-22
Cox's Bazar	RMS Error	0.188	0.986	2.22E-23
	SMAPE	0.394	0.627	2.56E-22
Sylhet	RMS Error	0.037	0.578	4.66E-16
	SMAPE	0.051	0.771	4.69E-15
Rajshahi	RMS Error	0.177	0.858	4.07E-09
	SMAPE	0.22	0.539	6.02E-09
Khulna	RMS Error	0.677	0.624	2.79E-16
	SMAPE	0.613	0.804	6.24E-14

Table 3.17 Wilcoxon test results (p-value) on maximum temperature of different models against Naive model

Region		Arima	Theta	Random Walk
Dhaka	RMS Error	0.186	0.232	9.07E-17
	SMAPE	0.423	0.232	1.17E-17
Cox's Bazar	RMS Error	0.154	0.003	7.09E-19
	SMAPE	0.6	0	1.58E-19
Sylhet	RMS Error	0.555	0.01	2.13E-16
	SMAPE	0.215	0.005	2.61E-15
Rajshahi	RMS Error	0.4	0.429	4.05E-13
	SMAPE	0.726	0.325	7.80E-13
Khulna	RMS Error	0.021	0.519	1.03E-14
	SMAPE	0.083	0.611	6.21E-15

sophisticated models, providing a benchmark to compare. Wilcoxon signed rank test was used to determine whether a model follows the same distribution as Naive forecast or not. If the p-value obtained from Wilcoxon test is less than the significance level (0.05), it indicates that the model is different from Naive forecast and vice versa.

Wilcoxon tests were run for both minimum and maximum temperature of the selected stations with the SMAPE vector of different models against the SMAPE vector of the Naive model. The same tests were also run for the RMS errors. Tables 3.16 and 3.17 shows the p-values resulted from the tests. Cells, having p-values greater than 0.05, are colored red, indicating same distribution as the Naive model.

From the results it can be seen that, for time series prediction of temperature value in Bangladesh, the RWD model is significantly different from the Naive model. On the contrary, forecasts given by the Theta model and ARIMA model follow quite the same distribution as the Naive model as expected.

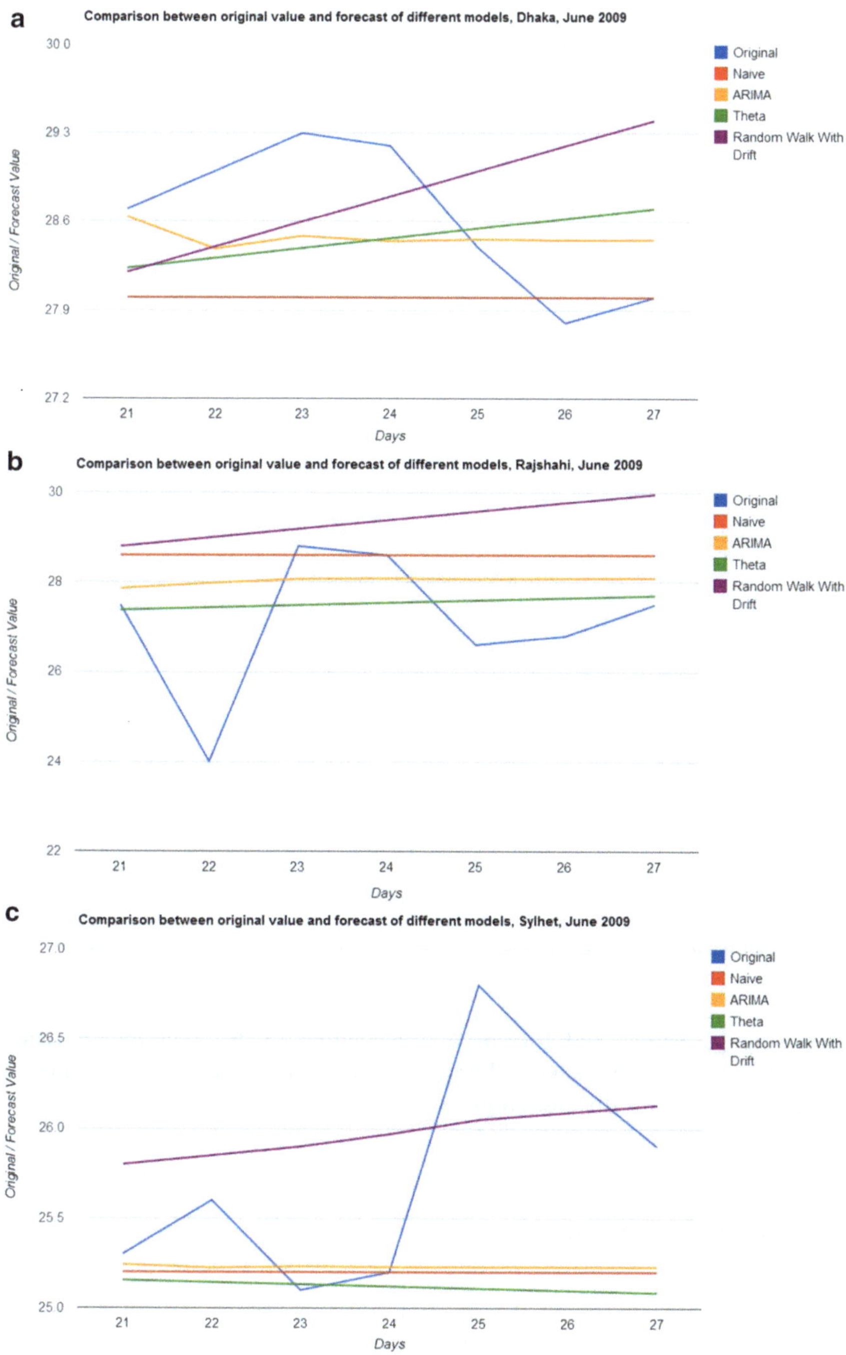

Fig. 3.3 Comparison between original value of temperature and forecast given by different models. (**a**) Dhaka, (**b**) Rajshahi, (**c**) Sylhet, (**d**) Khulna, and (**e**) Cox's Bazar

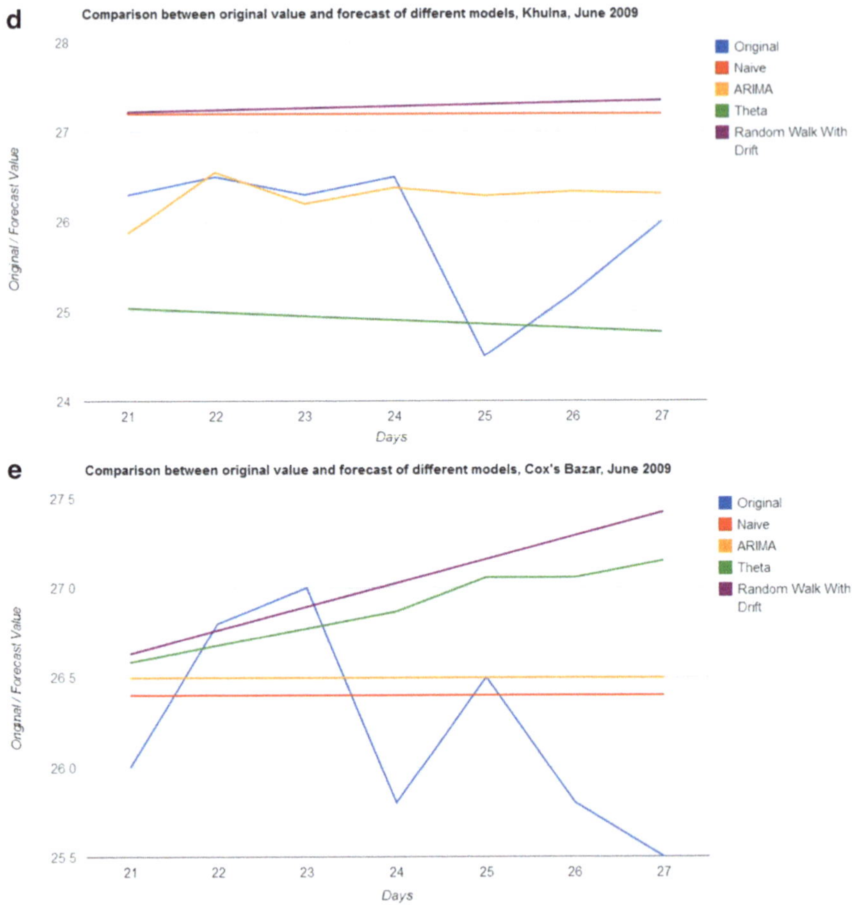

Fig. 3.3 (continued)

The original values and forecast corresponding to different models can be plotted to compare them visually. Figure 3.3 shows the comparison for June 2009 (arbitrarily chosen) for different regions.

3.4 Concluding Remarks

In this study the Mann–Kendall trend test and Sen's slope estimator were used to find the trends in temperature in Bangladesh. The study found that maximum temperature has shown remarkable positive trend during June to November in Bangladesh. On the other hand minimum temperature has increased during

December to January. The results also indicated that the eastern side has faced more change in temperature than the western side. Cox's Bazar and Sylhet exhibit an increasing trend almost throughout the year.

For analyzing the behavior of rainfall, the Mann–Kendall trend test, Sen's slope estimator, and seasonal Mann–Kendall trend test were used. *K*-means clustering algorithm was also employed to identify the rainfall distribution patterns over the years and their changes with time. The peak rainfall throughout the country is experienced during June to August and there has been no significant change in rainfall over the years. The study also reflects that over the years the western side of the country has experienced significantly less rainfall than the eastern side.

Performance of four time series prediction models (ARIMA(1,1,0), Theta, RWD, and Naive) were analyzed with respect to the climate condition of Bangladesh for forecasting daily minimum and maximum temperature. No obvious ordering could be found among ARIMA, Naive, and Theta models for prediction of daily minimum and maximum temperature, each one providing best prediction for different conditions. On the contrary, the RWD model is the least applicable one among the employed models.

References

Ahasan MN, Chowdhary Md AM, Quadir DA (2010) Variability and trends of summer monsoon rainfall over Bangladesh. J Hydrol Meteorol 7(1):1–17

Assimakopoulos V, Nikolopoulos K (2000) The theta model: a decomposition approach to forecasting. Int J Forecast 16:521–530

Bangladesh Meteorological Department (2012) http://www.bmd.gov.bd/

Basak JK, Titumir RAM, Dey NC (2013) Climate change in Bangladesh: a historical analysis of temperature and rainfall data. J Environ 2(2):41–46

Box G, Jenkins G (1970) Time series analysis: forecasting and control. Holden-Day, San Francisco

Chowdhury MHK, Debsarma SK (1992) Climate change in Bangladesh—a statistical review. In: Report of IOC-UNEP workshop on impacts of sea level rise due to global warming, NOAMI, Intergovernmental Oceanographic Commission, Dhaka, 16–19 Nov 1992

Corte-Real J, Qian B, Xu H (1998) Regional climate change in Portugal: precipitation variability associated with large-scale atmospheric circulation. Int J Climatol 18:619–635

Debsarma SK (2003) Intra-annual and inter-annual variation of rainfall over different regions of Bangladesh. In: Proceedings of SAARC seminar on climate variability in the south Asian region and its impacts, SAARC Meteorological Research Centre, Dhaka, 20–24 Dec 2002

Gaoliao J, Zhiwei Z, Hailin Z (2012) Trend analysis of air temperature between 1979–2000 in Hubei Province. Paper presented at the World Automation Congress (WAC), Puerto Vallarta, 24–28 June 2012

Gilbert RO (1987) Statistical methods for environmental pollution monitoring. Van Nostrand Reinhold, New York

Hall M, Frank E, Holmes G, Pfahringer B, Reutemann P, Witten IH (2009) The WEKA data mining software: an update. SIGKDD Explor 11(1):10–18

Helsel DR, Hirsch RM (1992) Statistical methods in water resources. Elsevier, New York

Hirsch RM, Slack JR, Smith RA (1982) Techniques of trend analysis for monthly water quality data. Water Resour Res 18(1):107–121

Huang F, Wang X (2011) Spatial and temporal variation of monthly rainfall nonuniformity of the upper Yangtze basin. Paper presented at international symposium on water resource and environmental protection, Xi'an, Shaanxi Province, 20–22 May 2011
Islam AS (2009) Analyzing changes of temperature over Bangladesh due to global warming using historic data. In: Proceedings of the young scientists of Asia conclave: pressing problems of humankind: energy & climate, Bangalore, 15–17 Jan 2009
Jain SK, Kumar V, Saharia M (2013) Analysis of rainfall and temperature trends in North East India. Int J Climatol 33(4):968–978
Karmakar S, Shrestha ML (2000) Recent climate changes in Bangladesh. In: SAARC Meteorological Research Centre (SMRC), SMRC-No. 4, SMRC Publication, Dhaka
Kendall M (1975) Rank correlation methods. Charles Griffin & Company, London, England
Kumar V, Jain SK (2010) Trends in seasonal and annual rainfall and rainy days in Kashmir valley in the last century. Quatern Int 212:64–69
MacQueen JB (1967) Some methods for classification and analysis of multivariate observations. In: Proceedings of fifth Berkeley symposium on mathematical statistics and probability, University of California Press, Berkeley, pp 81–297
Mann HB (1945) Nonparametric tests against trend. Econometrica 13:245–259
Mia NM (2003) Variations of temperature in Bangladesh. In: Proceedings of SAARC seminar on climate variability in the south Asian region and its impacts, SAARC Meteorological Research Centre, Dhaka, 20–24 Dec 2002
Nahrin Z, Munim AA, Begum QN, Quadir DA (1997) Studies of periodicities of rainfall over Bangladesh. J Remote Sens Environ 1:43–54
Pelczer IJ, Cisneros-Iturbe HL (2008) Identification of rainfall patterns over the Valley of Mexico. Paper presented at the 11th international conference on urban drainage, Edinburgh, 31 Aug–5 Sept 2008
Pesaran MH, Pick A (2009) Forecasting random walks under drift instability. Cambridge working papers in economics, Faculty of Economics, University of Cambridge, Cambridge
R Core Team (2013) R: a language and environment for statistical computing. R Foundation for Statistical Computing, Vienna. ISBN 3-900051-07-0, http://www.R-project.org/
Ramos MC (2001) Divisive and hierarchical clustering techniques to analyse variability of rainfall distribution patterns in a Mediterranean region. Atmos Res 57(2):123–138
Rana M, Shafee S, Karmakar S (2007) Estimation of rainfall in Bangladesh by using southern oscillation index. Pak J Meteorol 4(7):7–23
Sen PK (1968) Estimates of the regression coefficient based on Kendall's tau. J Am Stat Assoc 63:1379–1389
Singh P, Kumar V, Thomas T, Arora M (2008a) Basinwise assessment of temperature variability and trends in the northwest and central India. Hydrol Sci J 53:421–433
Singh P, Kumar V, Thomas T, Arora M (2008b) Changes in rainfall and relative humidity in different river basins in the northwest and central India. Hydrol Process 22:2982–2992
Tripathi S, Govindaraju RS (2009) Change detection in rainfall and temperature patterns over India. In: Proceedings of the third international workshop on knowledge discovery from sensor data, ACM, New York, pp 133–141
Warrick RA, Bhuiya AH, Mirza MQ (1994) The greenhouse effect and climate change: briefing document no. 1. Dhaka, Bangladesh Unnayan Parishad, pp 17–20
Wilcoxon F (1945) Individual comparisons by ranking methods. Biometrics Bull 1(6):80–83
Xi-ting L, Chun-qing G, Xiao Y (2011) Evolvement analysis about rainfall-runoff in the upper stream of Li River under the changeable environment. Paper presented at the international conference on remote sensing, environment and transportation engineering, Nanjing, 24–26 June 2011
Yue S, Hashino M (2003) Temperature trends in Japan: 1900–1990. Theor Appl Climatol 75: 15–27

Chapter 4
Time Series Model Building and Forecasting on Maximum Temperature Data

Amrina Ferdous, Md. Abu Shahin, and Md. Ayub Ali

Abstract Temperature is one of the factors of climate variables and understanding its nature is very important because the effect of temperature on climate change is higher than that of other variables. The purpose of the present study was to build an appropriate model to forecast the monthly maximum temperature of Rajshahi district in Bangladesh. The Box–Jenkins modeling strategy was performed using EViews software. This strategy was performed using the Augmented Dickey–Fuller, Phillip–Perron, Kwiatkowski–Phillips–Schmidt–Shin, autocorrelation function, partial autocorrelation function, ordinary least square method, normal *P–P* plot, Chow's breakdown test, Chow's forecast test, and standardized residuals plot. Seasonal variation, cyclical variation, and a slightly upward trend over time were found in the temperature. The temperature was found to be stationary at level after removing the cyclical variation using the resistant smoothing method, 4253H-twice. The SARMA(2, 1)(1, $2)_{12}$ model was found to be the most appropriate model for forecasting. The fitted model is also stable with no structural change and thus applicable for forecasting and policy purposes. Finally, this model was used for forecasting maximum temperature from January 2010 to December 2015. The forecasted value divulged that the maximum temperature will be increased by 3 °C during 2010–2015. This is an alarming situation for the environment and should take initiative to control and save our environment of Rajshahi district in Bangladesh.

Keywords Box–Jenkins modeling strategy • Correlogram • Normal *P–P* plot • Stability test • Unit root test

---

A. Ferdous (✉)
Department of Mathematics, Washington State University, Pullman, WA 99164, USA

Department of Statistics, University of Rajshahi, Rajshahi 6205, Bangladesh
e-mail: amrina.rafi@gmail.com

M.A. Shahin • M.A. Ali
Department of Statistics, University of Rajshahi, Rajshahi 6205, Bangladesh
e-mail: shahin.hera10@gmail.com; ayubali67@yahoo.com

T. Islam et al. (eds.), *Computational Intelligence Techniques in Earth and Environmental Sciences*, DOI 10.1007/978-94-017-8642-3_4,

4.1 Introduction

Bangladesh is the most vulnerable nation due to global climate change in the world according to German Watch's Global Climate Risk Index (CRI) of 2011 (Harmeling 2011). This is based on the analysis of impacts of major climate events that occurred around the world in the 20-year period since 1990. The temperature of Rajshahi is very high in summer season and very low in winter season among the districts of Bangladesh. Rajshahi weather station is 1 of the 32 weather stations in Bangladesh. Understanding the variability and trend of temperature may help for the proper planning and management of agricultural crop sectors. Investigations on climate change have been documented by many researchers. Eischeid et al. (1995) examined the surface temperature for the validation of climate models as well as for a variety of other climatology uses. Kunkel et al. (1999) studied recent work on trends during this century in societal impacts (direct economic losses and fatalities) in the United States from extreme weather conditions and compared those with trends of associated atmospheric phenomena. Stein and Lloret (2001) worked for forecasting of air and water temperatures for fishery purposes to show that ARIMA models yield better forecasts for highly variable time series than simple models based on the maximum water and air temperatures of 1984–1996. Hughes et al. (2006) showed that the increment is not the same for all the stations in the Antarctic region, and the increment is very significant at the Faraday/Vernadsky station. Only at this station the minimum/maximum monthly temperatures, for the period 1951–2004, are separately available, and they believed that the increment in mean surface temperature at this station is mainly due to the increment in minimum temperatures. They studied the variations in the minimum/maximum temperatures using a multiple regression model with non-Gaussian correlated errors. Mitra and Nath (2007) introduced nearest neighbor-based fuzzy model (NNFM) based on membership values for forecasting the daily maximum temperature at Delhi. Fuzzy membership values had been used to make single point forecasts into the future on the basis of past nearest neighbors. Kaushik and Singh (2008) analyzed monthly temperature and rainfall data and showed that the seasonal autoregressive integrated moving average (SARIMA) model were adequate based on correlation coefficient and root mean square error (RMSE). Tularam and Ilahee (2010) used 50 years data of rainfall and temperature for investigating climatic trends and interactions using the spectral analysis and ARIMA methodology, and successfully concluded that there exists inverse relationship in trend between rainfall and temperature using linear regression, and the ARIMA model for rainfall and temperature data were adequate but residuals of the ARIMA didn't appear to be white noise. Kumar et al. (2011) used to study mathematical properties of singular value decomposition (SVD) and its data exploring capacity and to apply them to make exploratory type clustering for ten climatic variables and 30 weather stations in Bangladesh using a newly developed graphical technique. Shamsnia et al. (2011) analyzed monthly relative humidity, average temperature, and precipitation of Abadeh Station and found that the SARIMA$(2, 1, 1)(1, 1, 0)_{12}$, SARIMA(2, 1, 0)

$(2, 1, 0)_{12}$, and SARIMA(0, 0, 1)$(1, 1, 1)_{12}$ models for respective weather parameters were adequate. Awal and Siddique (2011) showed that the ARIMA (4, 1, 4), ARIMA(2, 1, 1), and ARIMA(2, 2, 3) for Aus, Aman, and Boro rice production, respectively, in Bangladesh are more adequate models. Mahsin et al. (2012) analyzed monthly rainfall in Dhaka Division of Bangladesh and showed that it is followed by the SARIMA(0, 0, 1)$(0, 1, 1)_{12}$ model. Modeling on temperature, especially on the data of Rajshahi district, in Bangladesh for forecasting purpose is not well documented. Rajshahi district is known as Barind area where draught is a serious problem because of increasing maximum temperature (Alam et al. 2012). Sometimes draught is more defenseless than flood and cyclone (Brammer 1987, Shahid and Behrawan 2008, Shahid 2008). A long-term trend of maximum temperature will give us the awareness of appearing and intensity of this draught. For this trend, proper time series modeling on maximum temperature of this Barind area is necessary. Thus, the purpose of the present study is to establish an appropriate time series model for forecasting maximum temperature of Rajshahi district in Bangladesh.

4.2 Materials and Methods

4.2.1 Time Series Data on Temperature

The temperature data of Rajshahi district was collected from Bangladesh Agricultural Research Council. The temperature data was monthly speed over the range from January 1970 to December 2010 (41 years) with 492 realizations. The measurement unit and instrument of maximum temperature are Celsius (°C) and thermometer respectively. The mean is 31.05 and standard deviation is 3.51; these imply high fluctuation of maximum temperature, and positively skewed (skewness $= -0.56$) with platykurtic (kurtosis $= 2.58$) curve. For analyzing purpose, we denote monthly maximum temperature as a temperature.

The temperature data over the range from January 1970 to December 2010 divides the whole sample into two parts for training and test data set. The sample range from January 1970 to December 2009 is considered as training data set for modeling and that from January 2010 to December 2010 is considered as test data set for understanding the performance of model.

The popular software MS Excel, MS Word, EViews, and Minitab were used.

4.2.2 Testing Stationarity of the Time Series

A time series is called stationary if the mean, variance, and autocovariance cannot change over time, i.e., time invariant. Although there are several tests for stationarity, we consider only three tests: (1) graphical analysis (line graph), (2) correlogram test, and (3) unit root (Augmented Dickey–Fuller [ADF], Phillip–Perron [PP], and Kwiatkowski–Phillips–Schmidt–Shin tests) test. The time series plot (line graph) gives an initial clue about the nature of the data. If the line graph shows any trend (upward or downward), indicator of non-stationary of the time series but sometimes the time series is non-stationary without showing any trend. The correlogram, plot of autocorrelation function (ACF), or partial autocorrelation function (PACF) against lags is also useful for testing time series stationarity as well as choosing order of general ARIMA models. If the mean of a series is stationary, then the spikes of correlogram will tend to decay quickly toward 0. The ADF (Dickey and Fuller 1979) and Phillip–Perron (Phillips and Perron 1988) tests are used for testing unit root. In econometrics, a time series that has unit root, called random walk model that are non-stationary. The null hypothesis of the ADF and Phillip–Perron (PP) both tests are same as the given time series is stationary but the first one is parametric and last one is nonparametric tests. The Kwiatkowski–Phillips–Schmidt–Shin (Kwiatkowski et al. 1992) test is also used for testing stationarity but the null hypothesis is different from the ADF and PP tests. The null hypothesis of Kwiatkowski–Phillips–Schmidt–Shin test is the stationary time series. Practically there are some opposite results obtained from these (described above) tests but the comment always done on the basis of majority tests.

4.2.3 Box–Jenkins Modeling Strategy

The Box–Jenkins (BJ) approach for modeling the SARIMA processes was described in a highly influential book by statisticians George Box and Gwilym Jenkins in 1970. The original Box–Jenkins modeling procedure involved an iterative three-stage process of model selection, parameter estimation, and model checking. Recent explanations of the process (Makridakis et al. 1998) often add a preliminary stage of data preparation and a final stage of model application (or forecasting). The complete modeling procedure involved four stages described below:

4.2.3.1 Identification of Order for the SARIMA Structure

Identification means to find out the appropriate values of p, d, q, P, D, and Q of the order of general SARIMA model, i.e., the SARIMA$(p, d, q)(P, D, Q)_s$ model, where p, d, and q are integers greater than or equal to 0 and refer to the order of the autoregressive, integrated, and moving average parts for the nonseasonal model, and P, D, and Q refer to the order of that for seasonal case. We will use the ACF and PACF to find out the value of p, d, q, P, D, and Q. The behavior of the ACF and PACF for the identification tools of the SARIMA$(p, d, q)(P, D, Q)_s$ models are presented in Table 4.1.

Table 4.1 Behavior of the ACF and PACF for seasonal AR, MA, and ARMA models

	$AR(P)_s$	$MA(Q)_s$	$ARMA(P, Q)_s$
ACF	Tails off at lag k's, $k = 1, 2, \ldots$	Cuts off after lag Q's	Tails off at lag k's
PACF	Cuts off after lag P's	Tails off at lag k's, $k = 1, 2, \ldots$	Tails off at lag k's

The general ARIMA model is called the SARIMA model when the time series shows any seasonal effect. The general multiplicative SARIMA model is denoted by the SARIMA$(p, d, q)(P, D, Q)_s$ and can be written as Eq. (4.1) as follows:

$$\Phi(B^S)\phi(B)\nabla_S^D\nabla^d y_t = c + \Theta(B^S)\theta(B)\varepsilon_t \qquad (4.1)$$

where,

$\phi(B) = 1 - \phi_1 B - \phi_2 B^2 - \cdots - \phi_P B^p$ (The p order of AR operator)
$\theta(B) = 1 + \theta_1 B + \theta_2 B^2 + \cdots + \theta_q B^q$ (The q order of MA operator)
$\Phi(B^S) = 1 - \Phi_1 B^S - \Phi_2 B^{2S} - \cdots - \Phi_P B^{PS}$ (The P order of seasonal AR operator)
$\Theta(B^S) = 1 + \Theta_1 B^S + \Theta_2 B^{2S} + \cdots + \Theta_Q B^{QS}$ (The Q order of seasonal MA operator)
$\nabla^d = (1 - B)^d$ and $\nabla_S^D = (1 - B)^D$ refers nonseasonal and seasonal difference operator, c is constant, y_t is any time series, s is the number of season, and ε_t is the usual Gaussian white noise process

4.2.3.2 Parameter Estimation of the SARIMA Model

After getting the appropriate values of p, d, q, P, D, and Q, the next stage is to estimate the parameters c, Φ, ϕ, Θ, and θ of Eq. (4.1) of the SARIMA$(p, d, q)(P, D, Q)_s$ model. The ordinary least squares (OLS) method is used to estimate these parameters.

4.2.3.3 Diagnostic Checking of the Fitted Model

Diagnostic test is applied to understand whether the estimated parameters and residuals of the fitted SARIMA model are significant or not. The t statistic is used to test the significance of the individual coefficient. The normal probability plot and standardized residuals plot are used for checking normality and outliers of the residuals. The in-sample and out-sample of mean squared forecast error, Chow's breakdown test, (Chow 1960) and Chow's forecast test (Chow 1960) are used for checking the stability of the model.

4.2.3.4 Forecasting of the Study Variable

One of the reasons for the popularity of the Box–Jenkins modeling strategy is its success in forecasting (Gujarati 1993, p. 841). In many cases, the forecasts obtained by this method are more reliable than those obtained from the traditional econometric modeling, particularly for short-term forecasts (Gujarati 1993, p. 841) that's why it has been used in this study.

4.3 Results and Discussion

4.3.1 Testing Stationarity Status of Temperature

To observe the stationarity status of temperature data, three methods such as line graph, correlogram, and formal test are considered. The time series plot (line graph) of temperature is presented in Fig. 4.1. This figure shows a slightly upward trend, seasonal variation, cyclical variation, and random variation over time. The resistant smoothing, namely 4253H-twice, is used to remove the cyclical variation and the resulted time series is plotted in Fig. 4.2. Figure 4.2 supports the pattern of stationarity. The data, presented in Fig. 4.2, is used for the next analysis of modeling and forecasting.

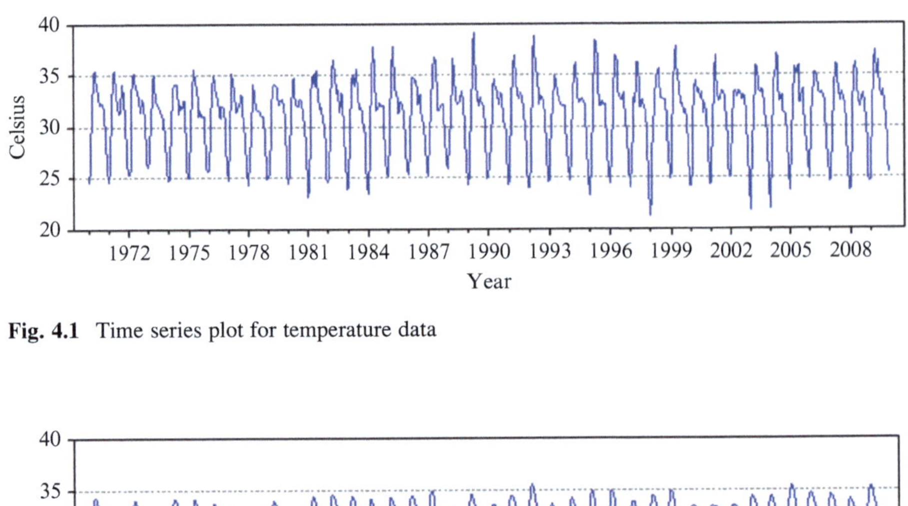

Fig. 4.1 Time series plot for temperature data

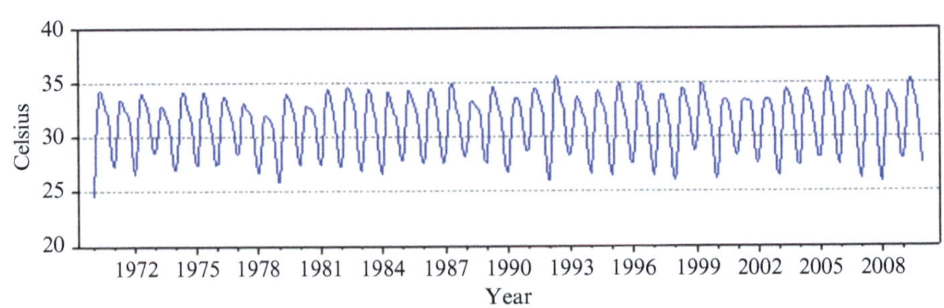

Fig. 4.2 Time series plot for smoothing temperature data

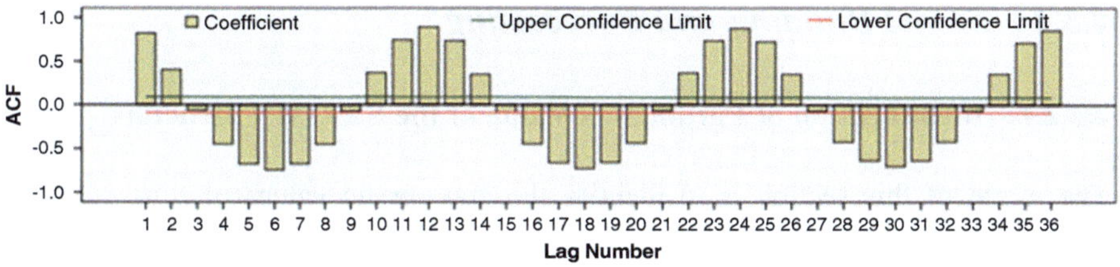

Fig. 4.3 Autocorrelation function of the temperature

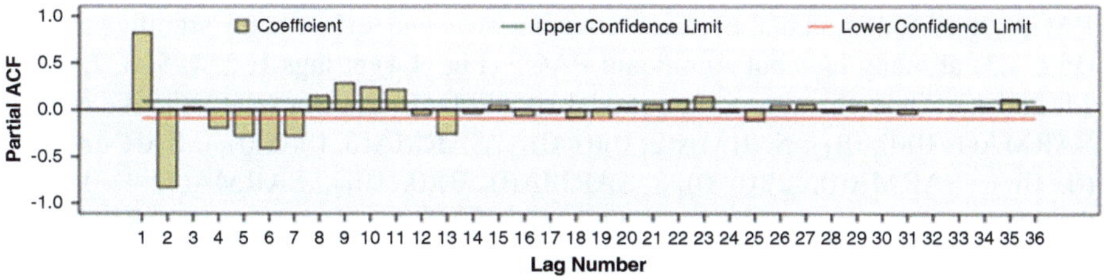

Fig. 4.4 Partial autocorrelation function of the temperature

Table 4.2 The ADF, PP, and KPSS tests for temperature

Test	Test statistics	Probability (pr-value)	Asymptotic critical value at 1 %	Asymptotic critical value at 5 %
ADF	−6.4186	0.0000	−3.4443	−2.8676
PP	−4.3538	0.0004	−3.4438	−2.8674
KPSS	0.4182		0.7390	0.4630

The spikes of correlogram against the ACF or PACF converge to 0 very quickly, implying the time series data is stationary. The stationarity test based on correlogram is presented in Figs. 4.3 and 4.4. Here, the spikes of the ACF (Fig. 4.3) converge to 0 very quickly with seasonal effect. Similarly, the spikes of the PACF (Fig. 4.4) converge to 0 very quickly. Therefore, the temperature can be considered as stationary and modeled considering with seasonal effect.

The ADF test, Phillip–Perron test, and Kwiatkowski–Phillips–Schmidt–Shin test for temperature data are presented in Table 4.2. The calculated values of the test statistics for the ADF test and Phillip–Perron test are −6.4186 and −4.3538 associated with the probabilities 0.0000 and 0.0004, respectively. This implies that the null hypothesis of non-stationary may not be accepted, that is, the temperature may be considered as stationary at level. Similarly, the Kwiatkowski–Phillips–Schmidt–Shin test divulged that the temperature series is stationary at 1 and 5 % at level. All the above tests indicate that the temperature series is stationary at level.

4.3.2 Model Building and Forecasting

4.3.2.1 Identification of Parameters Value of the SARIMA Structure

The motto of this section is to identify the appropriate values of p, d, q, P, D, and Q for the general SARIMA model, SARIMA$(p, d, q)(P, D, Q)_s$. As the temperature is found to be stationary at level, the value of d and D are equal to 0, while the data is monthly implying the value of s is 12. Therefore, the general SARIMA$(p, d, q)(P, D, Q)_s$ model is equivalent to the SARMA$(p, q)(P, Q)_{12}$ model. Now, the value of p, q, P, and Q can be estimated from the ACF and PACF. It is found that the series contains seasonal effect with significant ACF (Fig. 4.3) at many lags but significant PACF (Fig. 4.4) at lags 1, 2, 4, 5, 6, 7, 9, 10, 11, and 13. Thus, the appropriate model should be found from the tentative models SARMA$(1, 0)(0, 0)_{12}$, SARMA$(2, 0)(0, 0)_{12}$, SARMA$(3, 0)(0, 0)_{12}$, SARMA$(0, 1)(0, 0)_{12}$, SARMA$(0, 2)(0, 0)_{12}$, SARMA$(0, 3)(0, 0)_{12}$, SARMA$(1, 1)(0, 0)_{12}$, SARMA$(2, 1)(0, 1)_{12}$, SARMA$(1, 1)(0, 1)_{12}$, SARMA$(1, 1)(1, 1)_{12}$, SARMA$(2, 1)(1, 1)_{12}$, SARMA$(2, 1)(1, 2)_{12}$, and SARMA$(2, 2)(2, 2)_{12}$.

4.3.2.2 Estimation of Parameters for Selected SARMA Models

The OLS method is used to estimate the parameters of the tentative models. The models SARMA$(2, 0)(0, 0)_{12}$, SARMA$(3, 0)(0, 0)_{12}$, SARMA$(0, 2)(0, 0)_{12}$, SARMA$(0, 3)(0, 0)_{12}$, and SARMA$(2, 2)(2, 2)_{12}$ are omitted from the further analysis because of the insignificance of their parameters. The other models are significant and their results are presented in Tables 4.3 and 4.4. The model with highest R^2, lowest AIC, lowest BIC, and DW ≈ 2.0 is considered as the best model. From Table 4.4, the SARMA$(2, 1)(1, 2)_{12}$ model is found to be the best. The DW value is 1.51 for the model SARMA$(2, 1)(1, 2)_{12}$ that implies more autoregressive terms are needed but if we increase the lag of autoregressive terms in this model then it is appeared with insignificant parameters. Hence, finally selected model is SARMA$(2, 1)(1, 2)_{12}$.

4.3.2.3 Diagnostic Checking for Estimated SARMA Models

The diagnostic checking of the model SARMA$(2, 1)(1, 2)_{12}$ has been completed by actual-fitted-residual plot, normal *P–P* plot, Chow's breakdown test, Chow's forecast test, standardized residuals plot, and actual-forecast plot. The actual, fitted, and residual plot (Fig. 4.5) shows that the fitted values are very close to actual value and the residuals are very small implying the model is well fitted. The standardized residuals (Fig. 4.6) lie within the range [−3, +3] indicating that the model is not affected with the problem of outlier or intervention. The normal probability plot (Fig. 4.7) shows that the residuals are approximately normal. To investigate the

Table 4.3 Estimated parameters with probability (in parenthesis) for selected models

Model	$\hat{c}$	$\hat{\phi}_1$	$\hat{\phi}_2$	$\hat{\Phi}_1$	$\hat{\theta}_1$	$\hat{\Theta}_1$	$\hat{\Theta}_2$
SARMA (1, 0) $(0, 0)_{12}$	31.23 (0.00)	0.83 (0.00)					
SARMA (0, 1) $(0, 0)_{12}$	31.19 (0.00)				0.99 (0.00)		
SARMA (1, 1) $(0, 0)_{12}$	31.20 (0.00)	0.82 (0.00)			0.99 (0.00)		
SARMA (2, 1) $(0, 1)_{12}$	31.17 (0.00)	0.84 (0.00)	−0.41 (0.00)		0.93 (0.00)	−0.39 (0.00)	
SARMA (1, 1) $(0, 1)_{12}$	31.18 (0.00)	0.73 (0.00)			0.99 (0.00)	0.65 (0.00)	
SARMA (1, 1) $(1, 1)_{12}$	31.29 (0.00)	0.96 (0.00)		−0.40 (0.00)	0.98 (0.00)	−0.97 (0.00)	
SARMA (2, 1) $(1, 1)_{12}$	31.20 (0.00)	0.84 (0.00)	−0.46 (0.00)	0.30 (0.00)	0.99 (0.00)	−0.33 (0.00)	
SARMA (2, 1) $(1, 2)_{12}$	31.20 (0.00)	0.84 (0.00)	−0.47 (0.00)	0.31 (0.00)	0.99 (0.00)	−0.26 (0.00)	0.19 (0.00)

Table 4.4 Summary of the various tentative models

Model	R^2	AIC	BIC	DW
SARMA(1, 0)$(0, 0)_{12}$	0.69	3.51	3.53	0.49
SARMA(0, 1)$(0, 0)_{12}$	0.72	3.40	3.42	0.37
SARMA(1, 1)$(0, 0)_{12}$	0.91	2.31	2.34	0.64
SARMA(2, 1)$(0, 1)_{12}$	0.96	1.52	1.56	1.12
SARMA(1, 1)$(0, 1)_{12}$	0.92	2.12	2.16	0.97
SARMA(1, 1)$(1, 1)_{12}$	0.94	1.80	1.84	0.65
SARMA(2, 1)$(1, 1)_{12}$	0.97	1.33	1.39	1.47
SARMA(2, 1)$(1, 2)_{12}$	0.97	1.29	1.35	1.51

structural change (test for stability) of the selected model, the Chow's breakdown test, Chow's forecast test, root mean squared forecast error (RMSFE) of in-sample and out-sample are considered. The calculated values of F-statistic and likelihood ratio statistic are 0.652161 and 4.681666 associated with probabilities are 0.712603 and 0.698745, respectively, indicating that the null hypothesis of no structural change can be accepted. Similar result is obtain from Chow's forecast test (F-statistic = 0.8175079 with probability = 0.910119 and likelihood ratio statistic = 132.5569 with probability = 0.470004). Finally, the RMSFE of in-sample and out-sample are 0.459251 and 0.457730, respectively, implying that the selected

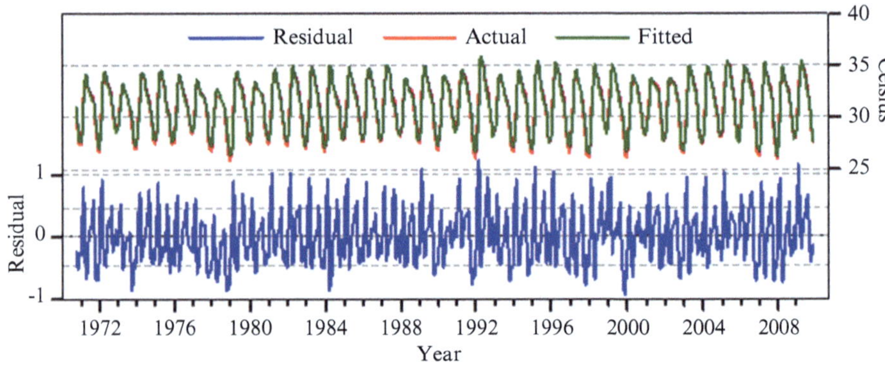

Fig. 4.5 Actual, fitted, and residual graph of the SARMA(2, 1)(1, 2)$_{12}$ model

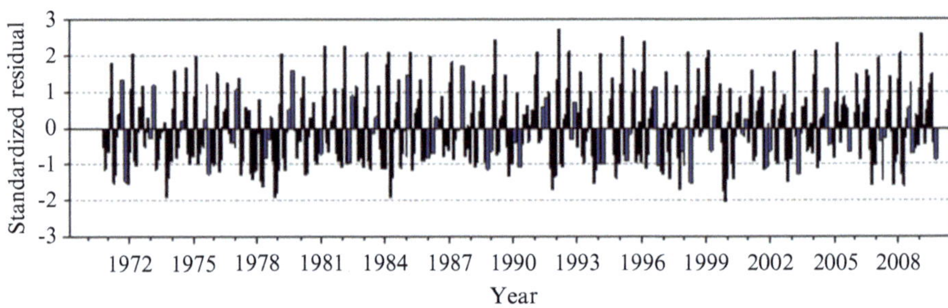

Fig. 4.6 Bar diagram of the standardized residuals of the SARMA(2, 1)(1, 2)$_{12}$ model

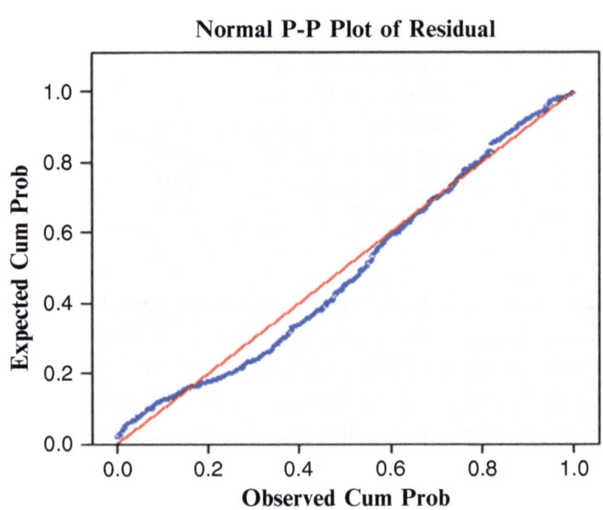

Fig. 4.7 Normal *Q–Q* plot of the residuals of the model SARMA(2, 1)(1, 2)$_{12}$

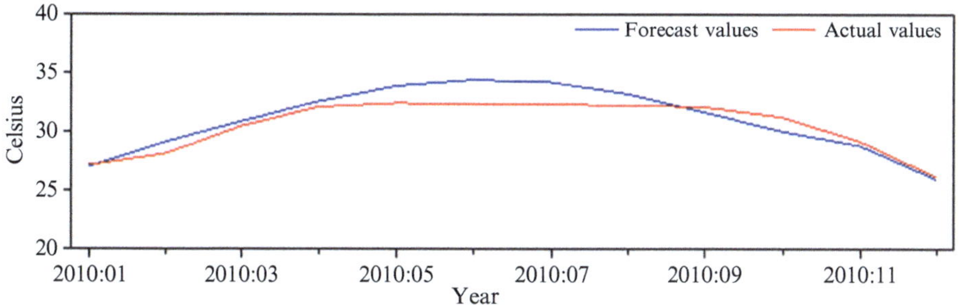

Fig. 4.8 Graph for forecast and actual values of SARMA(2, 1)(1, 2)$_{12}$ model

Table 4.5 Summary of overall forecasting performance of SARMA(2, 1)(1, 2)$_{12}$ model

Criteria	Forecast sample from January 1970 to December 2009	Forecast sample from January 2009 to December 2015
Root mean squared error	2.389621	1.321173
Mean absolute error	2.042642	1.068788
Mean absolute percent error	6.713163	3.247742
Theil inequality coefficient	0.038247	0.020872
Bias proportion	0.000019	0.354952
Variance proportion	0.752367	0.360041
Covariance proportion	0.247614	0.285007

model is adequate because the RMSFE for in-sample is greater than that for out-sample. The test sample and forecast value (Fig. 4.8) imply that the actual value is very close to its forecasted value. Thus, the selected SARMA(2, 1)(1, 2)$_{12}$ model is found to be an approximately normal, stable, and hence adequate model that can be used for forecasting purpose.

4.3.2.4 Forecasting of Temperature Using Selected SARMA Models

The selected SARMA(2, 1)(1, 2)$_{12}$ model has been successfully passed the three stages: identification, parameter estimation, and diagnostic checking. Thus, the model should be used for forecasting purpose. Overall forecasting performance of selected model is shown in Table 4.5 and the forecasted values are shown in Fig. 4.9. The RMSE, mean absolute error, and mean absolute percentage error for the training data set range from January 1970 to December 2009 are 2.389621, 2.042642, and 6.713163, respectively, which are always greater than that for the forecasted data range from January 2010 to December 2015 which are equal to 1.321173, 1.068788, and 3.247742, respectively. This implies that the forecasted values are reliable.

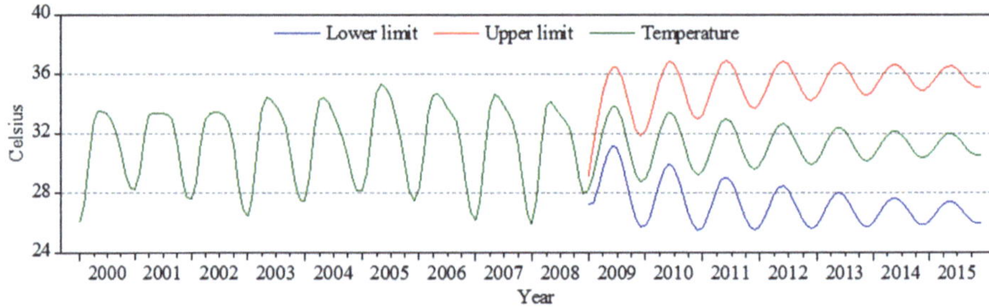

Fig. 4.9 Forecasting value of SARMA(2, 1)(1, 2)$_{12}$ model. The lines separated by *blue*, *green*, and *red* colors represent the 95 % confidence lower limit, 95 % confidence upper limit and temperature respectively

4.4 Conclusion

Time series analysis and forecasting on maximum temperature are very important in recent time. The common practice of time series analysis has the underlying assumption that the series must be stationary. The temperature series is found to stationary at level. The model SARMA(2, 1)(1, 2)$_{12}$ is found to be as the best model comparing with other possible models. Also, the model is found to be free from the problem of outliers and intervention and the estimated residuals follow the usual assumptions of principle of least square. The fitted model is also stable with no structural change and thus applicable for forecasting and policy purposes. The forecasted value divulged that the maximum temperature will be increased by 3 °C during 2010–2015. This is an alarming situation for the environment and should take initiative to control and save our environment of Rajshahi district in Bangladesh.

Acknowledgements We would like to express our gratitude to Professor Dr. Mohammed Nasser, Chairman, Department of Statistics, University of Rajshahi, Bangladesh for giving us the opportunity to present this chapter in the International Conference and publish in Conference Proceedings. A very special thank goes to Dr. Tanvir Islam, Department of Civil Engineering, University of Bristol, and to his honorable team for publishing this manuscript with a revised version again as a Book Chapter of Springer Publication.

References

Alam ATMJ, Rahman MS, Sadaat AHM (2012) Markov Chain analysis of agricultural drought at Barind, Bangladesh. In: Proceedings of international conference on statistical data mining for bioinformatics health, agriculture and environment, Rajshahi, 21–24 Dec 2012, pp 398–404. ISBN: 978-984-33-5876-9

Awal MA, Siddique MAB (2011) Rice production in Bangladesh employing by ARIMA model. Bangladesh J Agric Res 36(1):51–62

Brammer H (1987) Drought in Bangladesh, lessons for planners and administrators. Disasters 11:21–29
Chow G (1960) Tests of equality between sets of coefficients in two linear regressions. Econometrica 28(3):591–605. doi:10.2307/1910133
Dickey DA, Fuller WA (1979) Distribution of the estimators for autoregressive time series with a unit root. J Am Stat Assoc 74:427–431
Eischeid JK, Baker CB, Diaz HF, Karl TR (1995) The quality control of long-term climatological data using objective data analysis. J Appl Meteorol 34:2787–2795
Gujarati DN (1993) Basic econometrics, 3rd edn. McGraw-Hill, New York
Harmeling S (2011) Global Climate Risk Index 2011. Germanwatch e.V, Bonn. ISBN 978-3-939846-74-1. Available at: www.germanwatch.org/cri
Hughes GL, Rao SS, Rao TS (2006) Statistical analysis and time-series models for minimum/maximum temperatures in the Antarctic Peninsula. Proc R Soc A 463:241–259. doi:10.1098/rspa.2006.1766
Kaushik I, Singh SM (2008) SARIMA model for forecasting of monthly rainfall and temperature. J Environ Res Dev 3(2):506–514
Kumar N, Nasser M, Sarker SC (2011) A new singular value decomposition based robust graphical clustering technique and its application in climatic data. J Geogr Geol 3(1):227–238
Kunkel KE, Changnon SA, Pielke RA (1999) Temporal fluctuations in weather and climate extremes that cause economic and human health impacts: a review. Bull Am Meteorol Soc 80(6):1077–1098
Kwiatkowski D, Phillips PCB, Schmidt P, Shin Y (1992) Testing the null hypothesis of stationary against the alternative of a unit root. J Econom 54:159–178
Mahsin M, Akhter Y, Begum M (2012) Modeling rainfall in Dhaka division of Bangladesh using time series analysis. J Math Model Appl 1(5):67–73
Makridakis S, Wheelwright SC, Hyndman RJ (1998) Forecasting: methods and applications, 3rd edn. Wiley, New York
Mitra AK, Nath S (2007) Forecasting maximum temperatures via fuzzy nearest neighbour model over Delhi. Appl Comput Math 6(2):288–294
Phillips PCB, Perron P (1988) Testing for a unit root in time series regression. Biometrika 75:335–346
Shahid S (2008) Spatial and temporal characteristics of droughts in the western part of Bangladesh. Hydrol Process 22:2235–2247. doi:10.1002/hyp.6820
Shahid S, Behrawan H (2008) Drought risk assessment in the western part of Bangladesh. Nat Hazards 46:391–413. doi:10.1007/s11069-007-9191-5
Shamsnia SA, Shahidi N, Liaghat A, Sarraf A, Vahdat SF (2011) Modeling of weather parameters using stochastic methods (ARIMA model) (case study: Abadeh Region, Iran). In: International conference on environment and industrial innovation, 2011 IPCBEE, vol 12. IACSIT Press, Singapore
Stein M, Lloret J (2001) Forecasting of air and water temperatures for fishery purposes with selected examples from Northwest Atlantic. J Northwest Atl Fish Sci 29:23–30
Tularam GA, Ilahee M (2010) Time series analysis of rainfall and temperature interactions in coastal catchments. J Math Stat 6(3):372–380

Chapter 5
GIS Visualization of Climate Change and Prediction of Human Responses

P.K. Nag, Priya Dutta, Varsha Chorsiya, and Anjali Nag

Abstract Estimation of heat stress based on WBGT (wet bulb globe temperature) index is widely accepted as international standard. The purpose of the present study was to provide tolerance limit for people and interventions required to protect individuals from the dangerous consequences of heat. The meteorological data collected from Indian Meteorological Department of Ahmedabad (2001–2011) was used for estimating the WBGT. Multiple regression analysis was used to explore relationship between variables dry bulb temperature (T_a), wet bulb temperature (T_{wb}), and globe temperature (T_g) across the districts varied widely in two different seasons, i.e., summer and winter months. The linear regression analysis was applied for the purpose of future prediction, with respect to the WBGT index, and heat tolerance limit and visualized using GIS tool. The average tolerance time for 2001–2011 arrived at 82 ± 16 and 159 ± 36 min for the months of summer and winter, respectively. Thus, the WBGT and tolerance limit maps might prevail working population from heat stress fury.

Keywords WBGT • Tolerance time • Heat stress • GIS • Prediction

5.1 Introduction

Human being has a unique thermoregulatory mechanism to cope up against extreme climatic exposures and variations. But, frequent heat events indicate a risk to health in human. The strategies that minimize the physiological stress and strain depend

P.K. Nag • V. Chorsiya • A. Nag
National Institute of Occupational Health, Ahmedabad 380 016, India

P. Dutta (✉)
Indian Institute of Public Health Gandhinagar, Public Health Foundation of India, Sardar Patel Institute Campus, Drive In Road, Thaltej, Ahmedabad 380 054, India
e-mail: priya9909@yahoo.co.in; priyadutta@iiphg.org

T. Islam et al. (eds.), *Computational Intelligence Techniques in Earth and Environmental Sciences*, DOI 10.1007/978-94-017-8642-3_5,

on suitable adjustment to thermal, cardiovascular, and respiratory response characteristics that govern ones susceptibility limit to work in extreme hot climate. However, when the heat stress exceeds beyond the upper limit of acceptable strain, often called the limit of tolerance, heat strain may lead to heat casualty (Davies 1993). A state of dehydration may occur with the loss of body water as sweat, including loss of fluid through respiration, gastrointestinal tract as well as kidney (Gisolfi et al. 1995). This in turn disturbs the homeostasis of the body (Maughan et al. 1996), leading to decreased skin blood flow, elevated body core temperature (T_{cr}), decreased sweating rate, tolerance to work, and increased risks of heat injuries (Nag et al. 2013).

To understand past and future climate by using observations and theoretical models, GIS technology has been used in obtaining, storing, managing, analyzing and visualizing geographical, enviro-climatic, socioeconomic, and health data for effective decision making (Brooker and Utzinger 2007; Simoonga et al. 2008). Once the data are brought into the GIS database, the user can display large database by mapping with different tools and techniques. Researchers have also used GIS and remote sensing techniques in monitoring the groundwater quality relation to the land use/land cover (Srivastava et al. 2012). Research on satellite pictures has revealed that the glacier Siachen Glacier reduced to 5.9 km in its longitudinal extent from the time period of 1989–2009 (Searle 2013).

To develop a simple and authentic method to calculate increasing heat stress among population and working groups as a result of global warming. WBGT (wet bulb globe temperature) index is universally accepted and the method can be used to estimate daily weather station data. These estimates from weather station data can further be used to analyze past events and visualize the present scenario of climate change. But, when it comes to predicting change in the future, becomes a real challenge. Though the projections are exhaustive, extensive endeavor is needed to describe the impact of such dynamics on human health, in order to devise preventive and mitigating measures including warning alert for early preparedness. Therefore, to compute weather station data for future estimation is useful to prevent heat exposure at any places of the world with nearby authentic weather station data.

The Indian Meteorological Department (IMD) provides useful meteorological data from well-distributed weather stations across the country. These data include high resolution gridded daily rainfall data, temperature, and cloud. These data can be used for research work and projection. This chapter made a modest attempt to depict the climate change in the past 10 years (2001–2011) in 25 districts of Gujarat using the Indian meteorological data collected from 14 weather stations and to predict the probable future scenario of 2021 in light of biophysical variables of human. This study was also attempted to determine the tolerance limit of people and interventions required to protect individuals from the dangerous consequences of heat.

5.2 Method and Materials

The secondary data from the IMD from 2001 to 2011 were compiled and compared to the past and present environmental variables and further treated for regression analysis for expected future prediction. The data included minimum and maximum temperature (°C), dry bulb temperature (°C), dew point temperature (°C), relative humidity (%), bright hours of sunshine (hour and min), and wind direction (36 point of compass) of 14 weather stations across 25 districts of Gujarat. A single weather station could be shared by more than one district and vice versa. For calculation and predictions of heat stress and strain, these data were compiled, expressing into the WBGT index. The calculation was based on the average 10 days/season maximum temperature for each weather stations of Gujarat from 2001 to 2011 for summer (May and June) and winter (December and January). An entire day data was omitted if any hourly observations within that day were missing. The linear regression analysis was applied for the purpose of future prediction, with respect to the WBGT index, and heat tolerance limit and visualized using GIS tool.

The WBGT index is compared with the experiments of climatic chamber that includes three environmental conditions based on the average of seasonal parameters. These three conditions form the climatic load in which the subject had to work for three 3 days, i.e., ambient temperature 39 °C, 42 °C, and 45 °C with relative humidity 65–70 %, 50–55 %, and 40–45 %, respectively. However, the total period of the study program for each subject lasted about 1 week because consecutive days of exposure to heat may influence the individual state of acclimatization. The ergometric workload and air velocity kept constant irrespective of the heat exposure and maintained at 75 W and 0.4 m/s, respectively. The subjects selected for the study were usually habituated to moderate physical work in the occupational field. Prior to starting the experiment, each subject was familiarized with the standard test procedure and consent was taken according to the Indian Council Medical Research (ICMR) ethical guidelines (2000) was taken. The safe exposure period achieved by predetermined end point criteria for T_{cr} (40 °C) or heart rate (95 % maximal heart rate), or physiological symptoms forced the subject to discontinue was taken as tolerance time in the present study (Nag et al. 2007; Montain et al. 1994).

5.3 Wet Bulb (T_{wb}) and Globe (T_g) Temperature as Physical Model to Predict WBGT

Estimation of the heat stress on human, based on the "WBGT (wet bulb globe temperature) index," is the accepted international standard that provides a simple method for the assessment and control of hot environments (ISO 7423 2003; ISO 7933 1989). The WBGT is calculated from dry bulb temperature (T_a), wet bulb temperature (T_{wb}), and globe temperature (T_g) using the following equation:

$$\text{WBGT outdoor} = 0.7T_{wb} + 0.2T_g + 0.1T_a \quad (5.1)$$

$$\text{WBGT indoor} = 0.7T_{\text{wb}} + 0.3T_{\text{g}} \quad (5.2)$$

The T_a, T_{wb}, relative humidity, and wind velocity were direct observation from meteorological data. Whereas the globe temperature (T_g) was determined from the experimental data of OHM thermal monitor (32.1, Italy), ambient temperature (T_a), and globe temperature (T_g) measurements, the following linear regression equation was derived:

$$T_{\text{g}}(^{\circ}\text{C}) = 1.27T_{\text{a}} - 5.25 \quad (r = 0.912, p < 0.001, \text{df} : 2,993) \quad (5.3)$$

The prediction equation was obtained from T_a and T_g observations in the range from 27–40 °C to 28–45 °C, respectively.

The upper limit of acceptable strain, referred as tolerance time depends on the environment, work performed, and personal characteristics. Tolerance time can become a good biophysical indicator to predict heat load on a particular occupational group. It depends upon anthropometric dimensions of individuals, magnitude, and rate of climate variation to which a person is exposed to and their adaptive capacity. There is an obvious need to determine heat tolerance of an individual to avoid heat illness and disorders. Based on the relationships between the WBGT and tolerance time through simulated experimentation were utilized for calculation of tolerance time of people in different districts. Tolerance time shows moderate strong relation with the WBGT index.

$$\text{TT} = 235.13 - 4.9\text{WBGT} \quad (r = 0.596, p < 0.05, \text{df} = 30) \quad (5.4)$$

Table 5.1 illustrates the secondary data collected from 14 weather stations, covering 25 districts of Gujarat and the spread of T_a, T_{wb}, T_g, WBGT, and tolerance time across the districts varied widely in two different seasons. The mean and SD values presented in these tables were based on the compiled secondary data from the IMD sources (2001–2011). The direct observed values obtained from the IMD source were T_a, T_{wb} which is utilized for generating their respective predicted value of T_g, WBGT, and TT with the use of abovementioned equations. The obtained data were further treated for prediction as presented in Table 5.2. Since, the data required to make projection for the future prediction equations are derived, taking "year" as an independent variable for the WBGT and TT is finally used to predict the values for the year 2016 and 2021.

The regression coefficients as well as intercepts of the equations for the WBGT and heat tolerance time varied across the weather stations, due to variability in the meteorological data and also possible influence of the number of data points used in each equation. However, taking into account the degrees of freedom, all the regression equations and coefficient of determination (R^2) were found to be statistically highly significant.

In view of the fact that population characteristics such as age group, gender, bodily status, ethnic diversity, as well as pre-diagnosed health ailment have

Table 5.1 Descriptive statistics for T_a, T_{wb}, T_g, WBGT, and tolerance time

District name	Weather station	T_a (°C)	T_{wb} (°C)	T_g (°C)	WBGT (°C)	Tolerance time (min)
Summer						
Ahmedabad ($N = 832$)	Ahmedabad	36.3 ± 3.4	26.9 ± 2.9	39.3 ± 2.8	29.5 ± 2	119 ± 26
Amreli ($N = 322$)	Bhavnagar, Rajkot	33.5 ± 2.4	26.5 ± 1.6	37.1 ± 1.9	29.4 ± 1.0	118 ± 12
Anand ($N = 645$)	Ahmedabad, Anand	36.4 ± 3.4	26.4 ± 2.4	39.5 ± 2.8	29.9 ± 1.4	113 ± 16
Banaskantha ($N = 808$)	Banaskantha, Mahesana	35.5 ± 3.1	26.6 ± 2.4	38.8 ± 2.5	29.3 ± 1.9	121 ± 25
Bharuch ($N = 635$)	Surat, Vadodara	32.4 ± 2.6	27.3 ± 1.7	36.3 ± 2.1	29.1 ± 1.1	118 ± 14
Bhavnagar ($N = 709$)	Bhavnagar	35.2 ± 3.3	26.7 ± 2.1	38.5 ± 2.7	29.5 ± 1.6	118 ± 19
Dahod ($N = 1,446$)	Vadodara, Sabarkantha	35.8 ± 3.1	26.7 ± 1.5	39 ± 2.5	30.3 ± 1.4	110 ± 16
The Dangs ($N = 637$)	Surat	32.3 ± 2.6	26.8 ± 1.2	36.2 ± 2.1	29.4 ± 1.4	119 ± 18
Gandhinagar ($N = 696$)	Ahmedabad, Kheda, Mahesana	35.6 ± 3.8	26.5 ± 2.4	38.8 ± 3.1	29.7 ± 1.7	117 ± 21
Jamnagar ($N = 673$)	Porbandar, Rajkot	31.1 ± 1.1	27.6 ± 0.8	35.2 ± 0.9	29.8 ± 0.7	113 ± 8
Junagadh ($N = 951$)	Veraval, Rajkot	32.2 ± 1.6	28.2 ± 1.4	36 ± 1.3	30 ± 1.4	112 ± 16
Kheda ($N = 1,461$)	Kheda, Sabarkantha	35.7 ± 3.3	26.7 ± 1.5	38.9 ± 2.6	30.3 ± 1.3	109 ± 14
Kutch ($N = 549$)	Kutch, Rann of Kutch	35.6 ± 2.6	27.1 ± 2.4	38.8 ± 2.1	29.5 ± 2.4	120 ± 35
Mahesana ($N = 744$)	Mahesana, Ahmedabad	36 ± 3.3	26.1 ± 2	39.1 ± 2.7	29.9 ± 1.4	114 ± 17
Narmada ($N = 654$)	Surat, Vadodara	35.8 ± 3	26.7 ± 1.4	39 ± 2.4	30.4 ± 1.2	108 ± 12
Navsari ($N = 620$)	Surat	32.8 ± 2.2	26.9 ± 1.3	36.6 ± 1.8	29.7 ± 0.9	116 ± 11
Panchmahals ($N = 739$)	Vadodara, Kheda	35.5 ± 3.1	27.1 ± 2.1	38.8 ± 2.5	30.5 ± 1.5	107 ± 14
Patan ($N = 894$)	Mahesana, Banaskantha	34.9 ± 3.1	26.7 ± 2.2	38.3 ± 2.5	29.9 ± 1.4	113 ± 16
Porbandar ($N = 525$)	Porbandar, Veraval	32.4 ± 2.0	27.3 ± 1.5	36.2 ± 1.6	29.6 ± 1.1	116 ± 12
Rajkot ($N = 667$)	Rajkot, Rann of Kutch	36 ± 3.5	26.2 ± 2.4	39.1 ± 2.9	29.2 ± 2.3	123 ± 34
Sabarkantha ($N = 750$)	Sabarkantha, Mahesana	35.5 ± 3	26.4 ± 2.1	38.8 ± 2.4	29.6 ± 1.6	117 ± 21
Surat ($N = 626$)	Surat	32.6 ± 2.5	26.9 ± 1.3	36.3 ± 2.1	29.6 ± 1.0	116 ± 11
Surendranagar ($N = 611$)	Ahmedabad, Rann of Kutch	32 ± 1.2	27.5 ± 0.8	35.9 ± 0.9	30 ± 0.7	111 ± 8
Valsad ($N = 650$)	Surat	32.4 ± 2.5	27 ± 1.4	36.2 ± 2	29.4 ± 1.2	118 ± 14
Vadodara ($N = 580$)	Vadodara	35.7 ± 3.4	27.1 ± 2.0	38.9 ± 2.7	30.1 ± 1.7	112 ± 20

(continued)

Table 5.1 (continued)

District name	Weather station	T_a (°C)	T_{wb} (°C)	T_g (°C)	WBGT (°C)	Tolerance time (min)
Winter						
Ahmedabad ($N = 767$)	Ahmedabad	26.4 ± 2.4	16.8 ± 2.4	31.4 ± 2.3	20.9 ± 2.3	266 ± 61
Amreli ($N = 304$)	Bhavnagar, Rajkot	26 ± 2.5	18 ± 3.4	31 ± 2	21.3 ± 2.3	256 ± 54
Anand ($N = 685$)	Ahmedabad, Anand	26.3 ± 2.8	16.8 ± 2.2	31.3 ± 2.3	21.1 ± 2.0	255 ± 46
Banaskantha ($N = 786$)	Banaskantha, Mahesana	26.1 ± 3.2	16.6 ± 2.3	31.1 ± 2.6	20.8 ± 2.2	267 ± 62
Bharuch ($N = 656$)	Surat, Vadodara	28.3 ± 2.7	19.0 ± 2.6	32.9 ± 2.2	22.6 ± 2.4	228 ± 58
Bhavnagar ($N = 630$)	Bhavnagar	26.4 ± 2.7	17.3 ± 2.4	31.3 ± 2.2	21.3 ± 2.0	255 ± 48
Dahod ($N = 811$)	Vadodara, Sabarkantha	28.2 ± 2.8	18.7 ± 2.3	32.8 ± 2.2	22.6 ± 2.5	229 ± 64
The Dangs ($N = 661$)	Surat	28 ± 2.8	19 ± 2.5	32.7 ± 2.2	22.4 ± 2.7	235 ± 71
Gandhinagar ($N = 760$)	Ahmedabad, Kheda, Mahesana	27.4 ± 2.8	17.5 ± 2.4	32.1 ± 2.3	21.7 ± 2.3	271 ± 71
Jamnagar ($N = 743$)	Porbandar, Rajkot	26 ± 2.3	17.8 ± 2.7	31 ± 1.8	21.8 ± 8	244 ± 50
Junagadh ($N = 922$)	Veraval, Rajkot	28.6 ± 2.2	19.7 ± 3.6	33.1 ± 1.8	23.2 ± 3.1	218 ± 70
Kheda ($N = 1{,}600$)	Kheda, Sabarkantha	28.4 ± 2.8	18.9 ± 2.3	33 ± 2.3	22.8 ± 2.4	227 ± 63
Kutch ($N = 656$)	Kutch, Rann of Kutch	25.2 ± 3.2	17.2 ± 4	30.4 ± 2.6	20.3 ± 3.4	287 ± 88
Mahesana ($N = 764$)	Mahesana, Ahmedabad	26.4 ± 2.8	16.8 ± 2.4	31.4 ± 2.3	20.9 ± 2.1	264 ± 54
Narmada ($N = 855$)	Surat, Vadodara	28.4 ± 2.8	19 ± 2.6	32.9 ± 2.2	22.5 ± 2.6	230 ± 64
Navsari ($N = 622$)	Surat	28.2 ± 2.7	19 ± 2.7	32.8 ± 2.2	22.5 ± 2.4	230 ± 60
Panchmahals ($N = 853$)	Vadodara, Kheda	28.7 ± 2.8	19 ± 2.6	32.9 ± 2.3	22.9 ± 2.3	220 ± 48
Patan ($N = 753$)	Mahesana, Banaskantha	26.5 ± 2.8	16.5 ± 3.4	31.4 ± 2.3	20.6 ± 2.8	275 ± 71
Porbandar ($N = 514$)	Porbandar, Veraval	28.7 ± 2.8	17.8 ± 3.2	33.2 ± 2.3	22 ± 2.7	242 ± 63
Rajkot ($N = 633$)	Rajkot, Rann of Kutch	26.9 ± 3.1	16.6 ± 2.7	31.8 ± 2.5	20.9 ± 2.3	266 ± 60
Sabarkantha ($N = 792$)	Sabarkantha, Mahesana	26.1 ± 3.2	16.7 ± 2.4	31.1 ± 2.6	20.8 ± 2.2	268 ± 58
Surat ($N = 670$)	Surat	28.4 ± 2.8	19 ± 2.5	33 ± 2.2	22.8 ± 2.3	223 ± 54
Surendranagar ($N = 620$)	Ahmedabad, Rann of Kutch	27.4 ± 1.8	18.3 ± 2.5	30.1 ± 1.6	21.8 ± 2.1	243 ± 48
Valsad ($N = 663$)	Surat	28.2 ± 2.7	19.2 ± 2.9	32.8 ± 2.2	22.5 ± 2.3	228 ± 54
Vadodara ($N = 613$)	Vadodara	28.4 ± 2.9	18.8 ± 2.2	33 ± 2.3	22.9 ± 2.3	221 ± 51

Values are mean ± SD (2001–2011)

Table 5.2 Weather station-wise estimation WBGT and tolerance time future projection of Gujarat

	Linear regression analysis			
Weather station	WBGT (°C)	R^2	Tolerance time (min)	R^2
Summer month				
Ahmedabad	0.69 (year)–1,351.2	0.78	8,797.5–4.35 (year)	0.76
Mahesana	0.78 (year)–1,525.7	0.77	9,352.8–4.62 (year)	0.7
Bhavnagar	0.24 (year)–438.3	0.51	3,601.8–1.75 (year)	0.5
Kheda	0.08 (year)–133.3	0.12	1,443.9–0.67 (year)	0.11
Veraval	0.18 (year)–321.5	0.41	3,021.8–1.46 (year)	0.41
Kutch	0.37 (year)–701.9	0.48	5,176.7–2.54 (year)	0.43
Porbandar	0.49 (year)–941.3	0.73	7,433.6–3.67 (year)	0.72
Rajkot	0.68 (year)–1,330.0	0.72	8,577.8–4.23 (year)	0.54
Surat	0.4 (year)–765.4	0.51	6,059.2–2.98 (year)	0.52
Vadodara	0.56 (year)–1,086.4	0.58	6,966.1–3.44 (year)	0.53
Banaskantha	0.19 (year)–355.8	0.62	6,966.1–3.44 (year)	0.53
Sabarkantha	0.64 (year)–1,243.2	0.4	12,005–5.94 (year)	0.44
Anand	0.07 (year)–110.2	0.45	13,590–6.72 (year)	0.72
Rann of Kutch	0.03 (year)–19.5	0.16	879.7–0.42 (year)	0.27
Winter month				
Ahmedabad	0.45 (year)–868.6	0.37	12,259–6.03 (year)	0.32
Mahesana	0.3 (year)–583.9	0.35	9,598.4–4.70 (year)	0.31
Bhavnagar	0.37 (year)–707.7	0.21	8,957.1–4.39 (year)	0.15
Kheda	0.278 (year)–531.1	0.2	7,628–3.72 (year)	0.17
Veraval	0.479 (year)–932.5	0.61	11,345–5.59 (year)	0.57
Kutch	2.22 (year)–4,443.4	0.47	52,487–25.98 (year)	0.48
Porbandar	0.55 (year)–1,082.6	0.49	15,380–7.59 (year)	0.51
Rajkot	0.22 (year)–409.5	0.11	6,260.3–3.04 (year)	0.1
Surat	0.25 (year)–481.3	0.19	6,056.2–2.95 (year)	0.17
Vadodara	0.29 (year)–559.97	0.15	6,600.2–3.22 (year)	0.11
Banaskantha	0.22 (year)–417.2	0.16	6,734.2–3.28 (year)	0.15
Sabarkantha	0.87 (year)–1,718.9	0.35	2,884.6–14.24 (year)	0.33
Anand	0.22 (year)–417.2	0.16	6,734.2–3.28 (year)	0.15
Rann of Kutch	0.75 (year)–1,480.2	0.52	14,989–7.38 (year)	0.42

All R^2 values were statistically significant

influence on environmental warmth perception and one's tolerance time to work in hot environment, the present prediction is a general direction of environmental warmth and tolerance in the respective districts. These predictions can be further refined by longitudinal responses of people from different regions. As the number of data points across the districts and seasons were unequal, there was variation in the coefficients of determinations.

5.4 WBGT and Tolerance Time a Tool for Climate Change Assessment

Climate modeling with respect to human health as well as biophysiological indicators is important to determine the vulnerability limit of population. The IMD possesses a vast weather observational network and is involved in regular data collection, data bank management, research, and weather forecasting for national policy needs. With the trend of climatic change recorded for the decade 2001–2011, it was evident that the state like Gujarat faced increased length and intensity of heat exposure, with consequent effects on human physiological and pathological processes. For example, the meteorological data presented in Table 5.1 were further treated for GIS visualization. Further extrapolation and application may be very much possible to the block or ward level, depending on the input data.

Figure 5.1 shows the spatial pattern of warming of the districts of Gujarat. During the summer of 2001, the districts like Dahod, Panchmahal, Kheda,

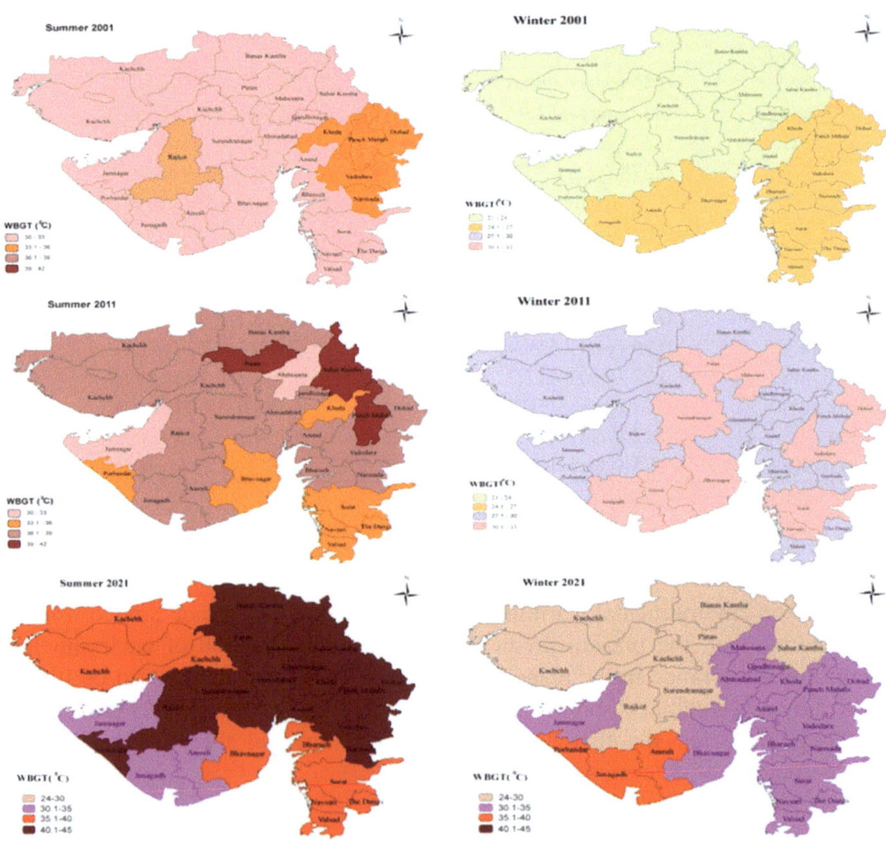

Fig. 5.1 Environmental warmth in terms of WBGT index in different districts of Gujarat

Vadodara, Narmada, and Rajkot showed a higher range of the WBGT temperature (33.1–36 °C), as compared to other districts of the state (30–33 °C). But in 2011, all districts showed the WBGT in the range of 36.1–42 °C, only with the exception of Jamnagar and Mahesana where the WBGT values were 31.2 °C and 33.7 °C, respectively. The overall environmental warmth (WBGT) elevated in the range of 2–3 °C from 2001 to 2011.

In the winter months of 2001, the WBGT levels in southern districts varied from 21 to 24 °C and in northern part 24.1–27 °C. The situation was very clear in 2011 showing that around 15 districts such as Banaskantha, Kutch, Sabarkantha, Gandhinagar, Jamnagar, Porbandar, Rajkot, Ahmedabad, Kheda, Panchmahal, Anand, Bharuch, Narmada, Valsad, and The Dang recorded the WBGT in between 27.1 and 30 °C and the remaining 10 districts such as Patan, Mahesana, Surendranagar, Junagadh, Amreli, Bhavnagar, Dahod, Vadodara, Surat, and Navsari ranged between 30.1 and 33 °C.

Apart from the variables included in the WBGT assessment (i.e., humidity and mean radiant field), other environmental factors, such as rainfall, cloud, and wind velocity, may also influence the environmental condition and, consequently, environmental warmth, as perceived by humans. The exposure pattern of the decade may repeat itself with the changing climate. In the present analysis, the internationally recognized the WBGT index was applied to indicate that the environmental warmth had distinctive changing pattern 2001–2011, during summer, as well as in winter months. In the year 2001, the WBGT ranged between 21 and 24 °C in winter, and 30 and 33 °C in summer. But, in 2011, the WBGT ranged from 31.2 to 42.5 °C. From 2001 to 2011, the WBGT increased in summer months by 2.08 °C and about 1.5 °C in winter, respectively. It was evident that the working population in different districts of Gujarat are at high level of exposure of environmental warmth as per standardize recommended the WBGT temperature in all different seasons.

The linear regression analysis yielded for the year 2021 shows that the temperatures build-up might be at a rate of 2 °C, since the environmental warmth assessment of the WBGT is based on humidity and mean radiant temperature.

5.4.1 Tolerance Time

The calculated WBGT can be compared with critical WBGT for calculating tolerance time. Human tolerance time for exposure to high heat was calculated based on experimental observations, the meteorological data recorded, and presented in GIS map (Fig. 5.2). The tolerance time is arrived at based on the rate of body core temperature (T_{cr}) build up from the basal level to the critical level of ~39 °C. Beyond this level of T_{cr}, a person may be at risk of hyperthermia and at critical state of thermoregulatory adjustment.

The average tolerance time for 2001–2011 arrived at 82 ± 16 and 159 ± 36 min for the months of summer and winter, respectively. During the summer of 2001, the tolerance time predicted for the district like Rajkot, Bhavnagar, Kheda,

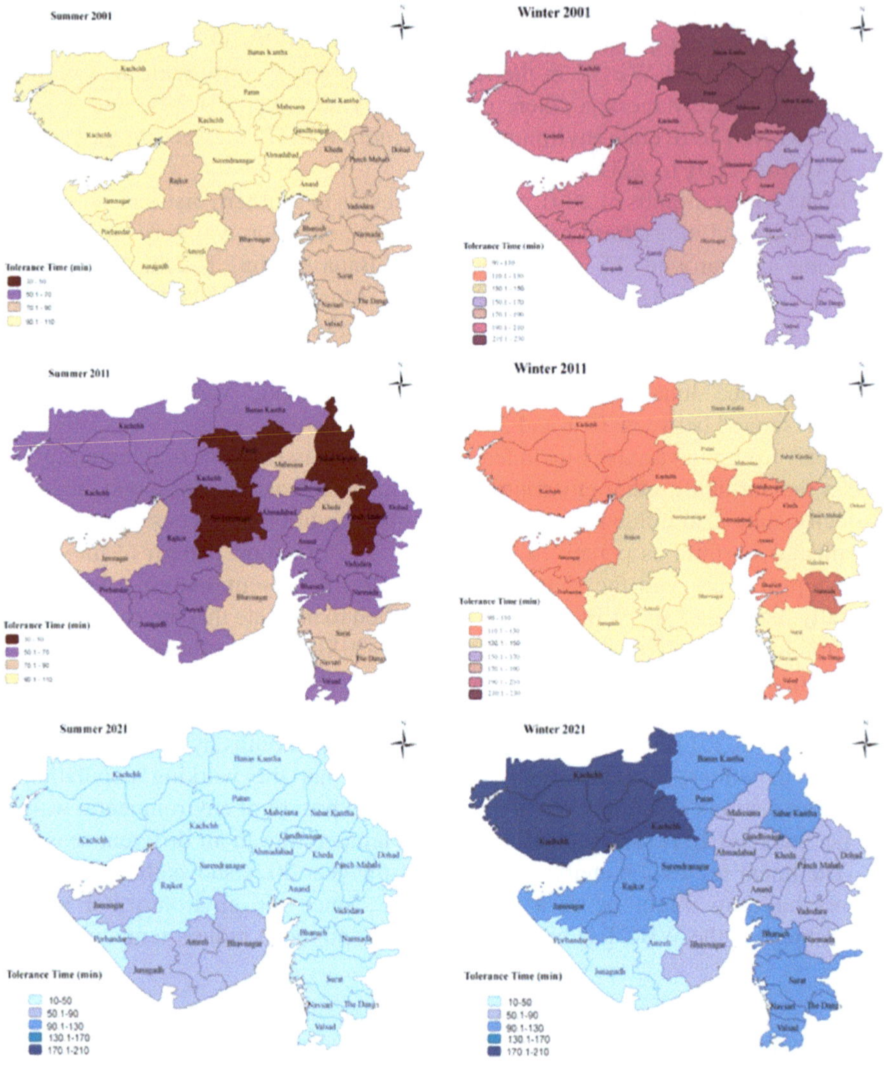

Fig. 5.2 Predicted heat tolerance time for human exposure in different districts of Gujarat

Panchmahals, Dahod, Vadodara, Bharuch, Narmada, Surat, Navsari, The Dangs, and Valsad as 70–90 min. In 2011, the estimated tolerance time decreased to a great extent in these districts. Surendranagar, Patan, Sabarkantha, and Panchmahal were recorded estimated lowest tolerance time in 2011 as 30–50 min.

Regions or places encounter extreme heat wave situation in the summer months further a little deviation in climatic condition from winter to summer, the humans pushed over threshold that manifest at places with higher morbidity and mortality among people in certain geographical regions.

Due to a variety of modifying factors, the estimated increase of ambient temperature may not be linear for a whole century. Despite uncertainty to degree of climate change and temperature build up, any increase in environmental warmth will cause significant impacts on physiological parameters for adjustment and tolerability. When the human ability to tolerate heat is hampered, the work dimensions like productivity, quality of work, and performance are also hampered. Fatigue increases with potential risk to workplace accidents and injuries (Parsons 2009). The prediction suggests that 1 °C increase in environmental warmth as the WBGT might cause 8 min loss in tolerance time.

5.5 Discussion

The WBGT map depicts that the climate change is taking place and if these scenarios prevailed then the working population may be more affected due to interruption of heat dissipating mechanism. The changing climate and warming may become worse in coming decades. There is a perceived realization that the climatic variability with respect to extreme heat potentially causes direct and indirect health effects and loss of productivity of people in different occupational settings (WMO 2011; Ladochy et al. 2007).

Davis and Kalkstein (1990) established threshold temperatures, which represent the temperature beyond which human mortality significantly increases, for cities throughout the United States. He further explains that the significant excess in mortality rate at New York occurs when the apparent temperatures exceed 32 °C during the summer. Nakai et al. (1996) revealed that the death from heat stroke in the occupational field occurred at 27 °C or more WBGT. Kalkstein (1991) reported heat-related disorder mainly in summer at 41 °C.

As the Western India being a rapidly industrializing region, the local climate change in the districts of Gujarat though depend on geographic and meteorological conditions are influenced by urbanization, industrialization, power plants, and burning of fossil fuels, and also on concerted actions to limit green gas emissions. And the inevitable health problem ranges from mild to even serious life-threatening scenarios. Early action by municipalities, health officials, and governmental institute are imperative to help face heat health vulnerability of people in the region and to build resilience in protecting human health. Heat acclimatization is specific to environmental conditions and depends on the body composition profile of population involved. The likely heat-intolerant population living in an area of low environmental heat load may be less at risk than a group living in an area of high heat load.

Therefore, information of the WBGT and the type of work being performed are required in determining how long a person can safely work or remain in a particular hot environment. The ambient temperature is a dominant factor shaping the distribution of the human habitat with overarching effect that result from the influence of temperature on physiological and thermal responses. At high temperatures, there is an increased secretion of sweat, with simultaneous loss of water and salts from the body, which lead to exhaustion and impairment of body functions (Bates and Miller 2008). In the process of thermoregulation, the physiological changes take place to prevent muscles from producing heat. These changes are associated with tiredness, tendency to sleep and fatigue, leading to performance decrement, errors, and accidents. It is suggested that the population in the susceptible regions to be fully aware of the necessity to avoid dehydration through systematic fluid supplement.

The biophysical derivations to arrive at the limit of tolerance time of human exposure to heat might be taken to ascertain vulnerability of a population group. During the months of summer, people in most districts were limited by the prevailing climatic conditions and the tolerance time might be in the range of 40–100 min, or less for habitual exposures. Researchers have found that human physical comfort not only depends on temperature but also is related to the interaction of physical characteristics and thermoregulatory adjustment (Kenney and Havenith 1993).

Vast populations in indoor and outdoor environment are engaged in manual labor. These workers are face-to-face with the fury of extreme heat in summer and post-monsoon months. Calosi et al. (2008) have shown the physical strain of humans in the tropics and its direct effect on the limits of tolerance. Better understanding of physiological mechanism and the limits of exposure of different population groups, including young workers, pregnant mothers, and elderly, would be useful to predict the direct impact of climatic warming on the vulnerable population. Environmental warmth depends on the characteristics of environment and anthropogenic activities, which reflect on the physiological and biophysical criteria of heat stress and strain.

The marked extra circulatory and thermal strain observed in the present study of heat stress indicates that continuous work at the WBGT value of about 40.8 °C (beyond tolerance limit) should be avoided and rest pauses in shade shelters should be incorporated into the work-shift. This may help to improve preparedness from heat extremities for detrimental health effects and give some idea for preparedness for future.

5.6 Conclusion

To summarize, the study proves that Gujarat exhibits extreme heat-related physiological stress in summer months. Given the scenarios of regional warming during the next decades, the expected change in tolerance time due to direct climate effects in Gujarat districts is alarming. Gujarat may be less prone to heat-related stress in winter. Projections of future climate change can be used as inputs into models that assess the physiological impact of climate change on public health. With the known ambient temperature, establishing relationships between spatial, demographic, biophysical, and environmental factors may yield a robust approach to provide working personnel a practical tool to better prepare for heat-related eventuality and tailor intervention measures for spatial examination of vulnerability. Preparing for the health consequences of climate change requires skill professionals, public health, and other physiological and biophysical surveillance data to provide effective health communication to vulnerable population.

References

Bates GP, Miller VS (2008) Sweat rate and sodium loss during work in the heat. J Occup Med Toxicol 3:1–6
Brooker S, Utzinger J (2007) Integrated disease mapping in a polyparasitic world. Geospat Health 1(2):141–146
Calosi P, Bilton DT, Spicer JI (2008) Thermal tolerance, acclimatory capacity and vulnerability to global climate change. Biol Lett 4(1):99–102
Davies CTM (1993) Thermal limits to severe prolonged exercise in man. J Therm Biol 18:605–607
Davis RE, Kalkstein LS (1990) Development of an automated spatial synoptic climatological classification. Int J Climatol 10:769–794
Gisolfi CV et al (1995) Effect of sodium concentration in a carbohydrate-electrolyte solution on intestinal absorption. Med Sci Sports Exerc 27:1414–1420
Indian Council of Medical Research (2000) Ethical guidelines for biomedical research on human subject. ICMR, New Delhi, pp 1–77
ISO 7243 (2003) Hot environments—estimation of the heat stress on working man, based on the WBGT-index (wet bulb globe temperature). International Standards Organization, Geneva
ISO 7933 (1989) Hot environments—analytical determination and interpretation of thermal stress using calculation of required sweat rate index. International Standards Organization, Geneva
Kalkstein LS (1991) A new approach to evaluate the impact of climate on human mortality. Environ Health Perspect 96:145–150
Kenney WL, Havenith G (1993) Thermal physiology of the elderly and handicapped, heat stress and age: skin blood flow and body temperature. J Therm Biol 18:341–344
Ladochy S, Medina R, Patzert W (2007) Recent California climate variability: spatial and temporal patterns in temperature trends. Climate Res 33:159–169
Maughan RJ, Leiper JB, Shirreffs SM (1996) Restoration of fluid balance after exercise-induced dehydration effects of food and fluid intake. Eur J Appl Physiol 73:317–325
Montain SJ et al (1994) Physiological tolerance to uncompensable heat stress: effects of exercise intensity, protective clothing and climate. J Appl Physiol 77:216–219
Nag PK, Nag A, Ashtekar SP (2007) Thermal limits of men in moderate to heavy work in tropical farming. Ind Health 45(1):107–117
Nag PK, Dutta P, Nag A (2013) Critical body temperature profile as indicator of heat stress vulnerability. Ind Health 51:113–122
Nakai S, Shinzato K, Morimoto T (1996) Epidemiological analysis of heat disorders in Japan—an analysis of gleaned cases from newspaper report between 1990 and 1994. Jpn J Biometeor 33:71–77
Parsons K (2009) Maintaining health, comfort and productivity in heat waves. Glob Health Action 2:39–45
Searle M (2013) Colliding continents: a geological exploration of Himalaya, Karakoram and Tibet, Kindleth edn. Oxford University Press, Oxford
Simoonga C et al (2008) The epidemiology and small-scale spatial heterogeneity of urinary schistosomiasis in Lusaka province, Zambia. Geospat Health 3(1):57–67
Srivastava PK, Gupta M, Mukherjee S (2012) Mapping spatial distribution of pollutants in groundwater of a tropical area of India using remote sensing and GIS. Appl Geomatics 4(1):21–32
World Meteorological Organization (2011) WMO statement on the status of the global climate in 2011. WMO-No-1085

Part III
Probabilistic and Transforms Intelligence Techniques in Earth and Environmental Sciences

Chapter 6
Markov Chain Analysis of Weekly Rainfall Data for Predicting Agricultural Drought

A.T.M. Jahangir Alam, M. Sayedur Rahman, and A.H.M. Sadaat

Abstract In the semiarid Barind region, episodes of agricultural droughts of varying severity have occurred. The occurrence of these agricultural droughts is associated with rainfall variability and can be reflected by soil moisture deficit that significantly affects crop performance and yield. In the present study, an analysis of long-term (1971–2010) rainfall data of 12 rain monitoring stations in the Barind region was carried out using a Markov chain model which provides a drought index for predicting the spatial and temporal extent of agricultural droughts. Inverse distance weighted interpolation was used to map the spatial extent of drought in a GIS environment. The results indicated that in the Pre-Kharif season drought occurs almost every year in different parts of the study area. Though occurrence of drought is less frequent in the Kharif season the minimum probability of wet weeks leads to reduction in crop yields. Meanwhile, the calculation of 12 months drought suggests that severe to moderate drought is a common phenomenon in this area. Drought index is also found to vary depending on the length of period. The return period analysis suggests that chronic drought is more frequent in the Pre-Kharif season and the frequency of moderate droughts is higher in the Kharif season. On the contrary severe drought is more frequent for a 12-month period.

Keywords Markov chain model • GIS • Agricultural droughts • Barind region

A.T.M.J. Alam (✉) • A.H.M. Sadaat
Department of Environmental Sciences, Jahangirnagar University,
Savar, Dhaka 1342, Bangladesh
e-mail: kallal1022@gmail.com

M.S. Rahman
Department of Statistics, University of Rajshahi, Rajshahi 6205, Bangladesh

T. Islam et al. (eds.), *Computational Intelligence Techniques in Earth and Environmental Sciences*, DOI 10.1007/978-94-017-8642-3_6,

6.1 Introduction

Drought is a complex natural phenomenon that has affected civilization throughout history. It occurs with varying frequency in all types of economic systems and in both developed and less developed countries. Almost all year around, drought occurs in one part of the globe. The wide variety of sectors affected by drought, its diverse geographical and temporal distribution, and the human demand placed on water supply make it difficult to develop a single definition of drought (Richard and Heim 2002). According to Al-Salihi (2003), Ben-Zvi (1987), and Darcup et al. (1980), drought is mainly caused by deficiency of rainfall which leads to severe agricultural and hydrological hazard. They described drought as a lack of rainfall, which cannot be prevented by social forces. In the context of Bangladesh, Brammer (1987) defined drought as a condition when soil moisture is less than the requirement of satisfactory crop growth during the seasons when crops are normally grown. Similarly, Heathcote (1974) defined drought as a situation when the shortfall of water is harmful for man's agricultural activities.

Drought is a recurring phenomenon in the northwestern part of Bangladesh. Though the drought has attracted less attention from the science community than flood or cyclone, several authors found that the impact of drought can be more defenseless than flood and cyclone (Brammer 1987; Shahid and Behrawan 2008; Shahid 2008; Alam et al. 2012a, b). This is also evident by statistics. For example, the loss of 1978/1979 drought was greater than the loss of flood in 1974 (Paul 1998). Rice production losses in the drought of 1982 were 50 % more than the losses caused by the flood in the same year (Ramsey et al. 2007). Furthermore, in the 1997 drought the national production of food grains reduced by one million tons, of which 0.6 million tons were T. aman (Ramsey et al. 2007).

On the global level, impacts of natural hazards and disasters are staggering. In Bangladesh, the major natural hazards are also in line with global patterns. In the context of global warming, most of the climate models project a decrease in precipitation in the dry season and an increase during the monsoon season in South Asia (Christensen et al. 2007; Shahid and Behrawan 2008). This will cause a cruel combination of more extreme floods and droughts in the region. Due to the land use changes within the country and its neighboring countries, Bangladesh has already showed an increased frequency of droughts in recent years (Shahid and Behrawan 2008). Concern among scientists has grown on changes of precipitation and frequent occurrence of droughts in Bangladesh. Moreover, agriculture is the largest sector of Bangladesh's economy and has an overwhelming impact on major macroeconomic activities like employment generation, poverty alleviation, human resource development, and food security (Shahid, 2010). Consequently, the impact of agricultural drought ultimately affects the overall social status of the people of Bangladesh. Therefore, it is important to analyze the agricultural drought severity and predict the time, area, and probability of its occurrence.

Markov chain models have been successfully used to derive agricultural drought indices in the Indian subcontinent (Alam 2010; Alam et al. 2011,

Alam et al. 2013; Banik et al. 2002; Barkotulla 2007). They have been used for determining drought proneness (Banik et al. 2002; Barkotulla 2007), analyzing rainfall drought correlation (Alam et al. 2011), analyzing agricultural droughts (Biamah et al. 2005), and predicting critical wet and dry spells (Ochola and Kerkidis 2003). Several studies also suggested that the reliability of meteorological persistence can be best described through Markov chain models of proper order (Rahman 1999a, b). However, the first-order Markov processes are adequate in representing the occurrence of rainfall events (Jimoh and Webster 1999). In addition, Markov chain models are found to be promising in simulating the length of the longest dry and wet spells and largest rainfall during monsoons (Sharma 1996; Biamah et al. 2005). In the present study, the first-order Markov chain model was chosen because of its flexibility to give a variety of useful statistical results. The spatial and temporal patterns of agricultural droughts were analyzed and the probability of wet weeks also estimated because it determines the soil moisture holding capacity, thus influences crop yields.

6.2 Study Area and Method

6.2.1 Climatic Condition of the Barind Region

Barind Tract is the largest Pleistocene physiographic unit of the Bengal Basin which covers an area of about 7,770 km^2. Three distinct seasons can be recognized in this region: (1) the dry winter season from December to February, (2) the pre-monsoon hot summer season from March to May, and (3) the rainy monsoon season which lasts from June to October (Shahid 2011). Climatically, the region is characterized by high fluctuating rainfall and the ratio of dry to rainy months is found to be highest in Bangladesh. It is observed that the annual average rainfall vary between 1,300 and 1,600 mm and the seasonal distribution of rainfall shows that more than 90 % rainfall occurs during May to October (Shahid 2011). Although the whole region has long cool winter, the maximum number of days with temperature above 40 °C is observed in summer (Rahman 1999a). The location of the rainfall stations used in this study is given in Fig. 6.1.

Geologically Barind Tract belongs to an old alluvial formation which is usually composed of massive argillaceous beds of pale reddish brown color that often turns yellowish on weathering. Lime nodules and pisolitic ferruginous concretions occur throughout the soil. Locally the soils are rich in lime. Soils are deficient in nitrogen and phosphorus (Moslehuddin et al. 2008). The top soil of Barind Tract is reddish in color which is mainly in origin of Madhupur clay. Most of the soil is clay though a small amount of silt and fine sands are also encountered (Yasmin 2008). In the dry season the soil becomes very hard and in the wet season it becomes slippery rather than soft. The mean elevation of the area is 25 m above the mean sea level and is quite flat. The digital elevation model of the study area is given in Fig. 6.2.

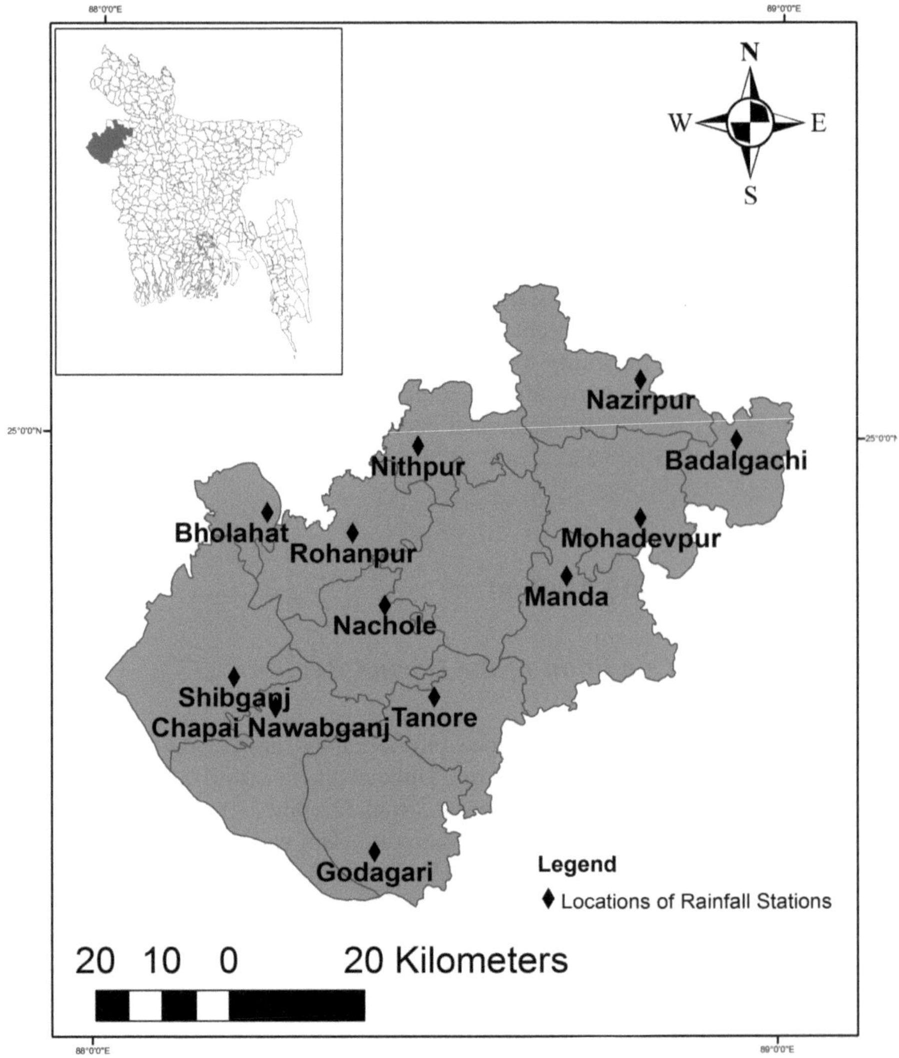

Fig. 6.1 Location of the rainfall stations

6.2.2 *Markov Chain Agricultural Drought Index*

Several authors have found that the sequences in daily rainfall occurrences can be described by a simple Markov chain model (Kazt 1974; Anderson and Goodman 1957; Rahman 1999a, b; Banik et al. 2002; Barkotulla 2007; Alam et al. 2011). The theory of Markov chain is described below by following Banik et al. (2002):

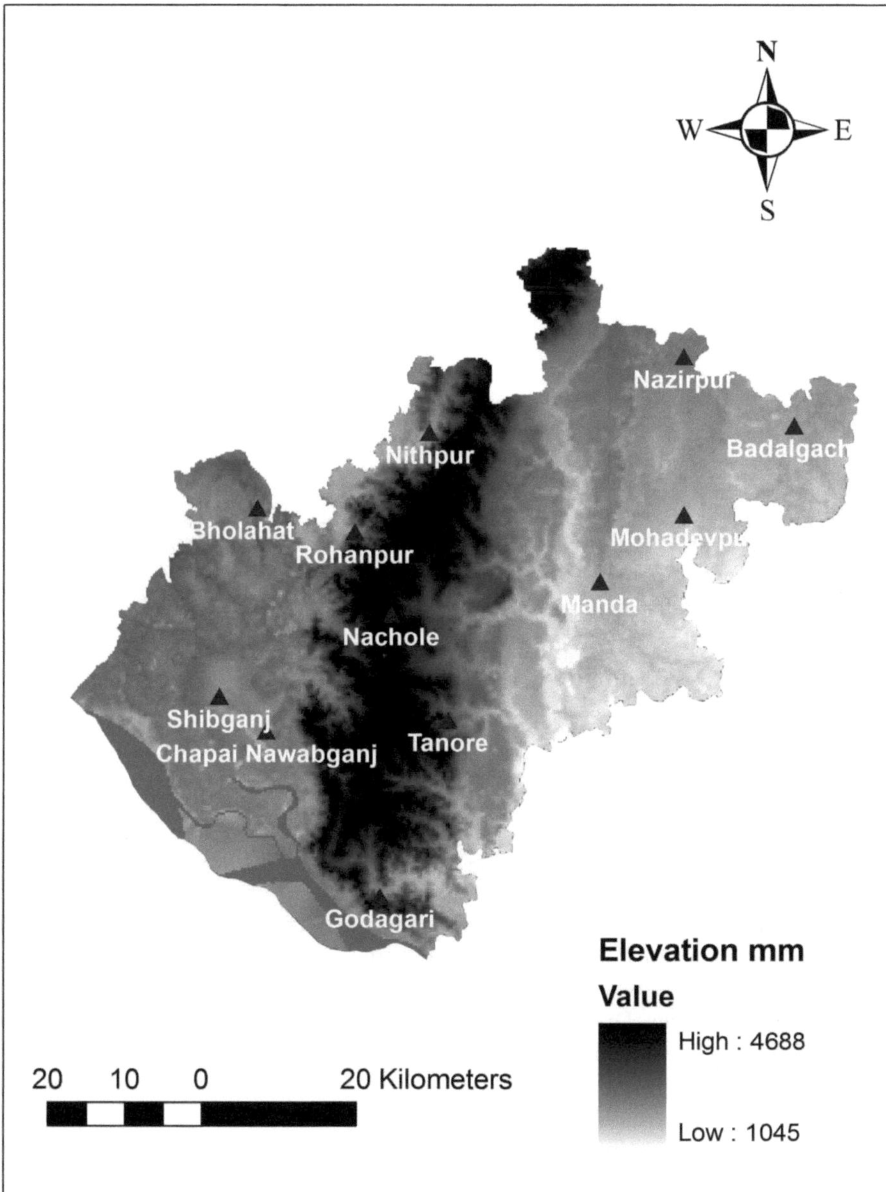

Fig. 6.2 Digital elevation model of the study area

Let $X_0, X_1, X_2, \ldots, X_n$, be random variables distributed identically and taking only two values, namely 0 and 1, with probability one, i.e.,

$$X_n = \begin{cases} 0 & \text{if the } n\text{th week is dry} \\ 1 & \text{if the } n\text{th week is wet} \end{cases}$$

Firstly, it is assumed that,

$$P\left(X_{n+1} = x_{n+1} \middle| X_n = x_n, X_{n-1} = x_{n-1} \cdots X_0 = x_0\right) \\ = P\left(X_{n+1} = x_{n+1} \middle| X_n = x_n\right) \tag{6.1}$$

where $x_0, x_1, \ldots, x_{n+1} \in \{0,1\}$.

In other words, the probability of wetness of any week depends on whether the previous week was wet or dry. Given the event on the previous week, the probability of wetness is assumed independent of further preceding weeks. So, the stochastic process $\{X_n, n = 0, 1, 2, \ldots\}$ is a Markov chain.

Considering the following transition matrix

$$\begin{bmatrix} P_{00} & P_{01} \\ P_{10} & P_{11} \end{bmatrix}$$

where $P_{ij} = P(X_1 = j|X_0 = i), i, j = 0, 1, P_{00} + P_{01} = 1$ and $P_{10} + P_{11} = 1$.

Let $p = P(X_0 = 1)$. Here p is the absolute probability of a week being wet during the monsoon period. Clearly, $P(X_0 = 0) = 1 - p$.

For a stationary distribution,

$$\begin{bmatrix} 1-p & p \end{bmatrix} \begin{bmatrix} P_{00} & P_{01} \\ P_{10} & P_{11} \end{bmatrix} \begin{bmatrix} 1-p & p \end{bmatrix} \tag{6.2}$$

which gives

$$p = \frac{P_{01}}{1 - (P_{11} - P_{01})} \tag{6.3}$$

It is further assumed that P_{ij} remains constant over the years. The maximum likelihood estimates of P_{01} and P_{11} are appropriate relative functions.

A wet spell of length k is defined as a sequence of k wet weeks preceded and followed by dry weeks. Dry spells are defined correspondingly. The probability of a wet spell of length k, given that this week is wet, is calculated as

$$P(W = k) = (1 - P_{11}){P_{11}}^{k-1}$$

The probability of wet sequences with length greater than k is

$$P(W > K) = \sum_{t=k+1}^{\infty} P(W = t) = {P_{11}}^{k} \tag{6.4}$$

Similarly, the probability of a dry spell of length m is

$$P(D = m) = (1 - P_{01}){P_{01}}^{m-1}$$

Table 6.1 Index of drought proneness

Criteria	Degree of drought proneness
$0.000 \leq \text{DI} \leq 0.125$	Chronic
$0.125 < \text{DI} \leq 0.180$	Severe
$0.180 < \text{DI} \leq 0.235$	Moderate
$0.235 < \text{DI} \leq 0.310$	Mild
$0.310 < \text{DI} \leq 1.000$	Occasional

And the probability of dry sequences with length greater than m is

$$P(D > m) = (1 - P_{01})^m \tag{6.5}$$

Let Y be a random variable such that $Y =$ the number of wet weeks among an n weeks period, i.e.,

$$Y = X_0 + X_1 + \cdots + X_{n-1}$$

For a large n, Y follows a normal distribution with

$$\text{Mean} = n \times p \tag{6.6}$$

P_{11} gives the probability of a week to be wet given that the previous week was wet too. When P_{11} is large, the chance of wet weeks is also large. But a small P_{11} may not indicate high drought proneness. In this case, a large P_{01} implies a large number of short wet spells which can prevent the occurrence of drought.

Hence, an index of drought proneness may be defined as

$$\text{DI} = P_{11} \times P_{01} \tag{6.7}$$

This index of drought proneness is bounded by 0 and 1. The higher the value of DI, the lower the degree of drought proneness. The extent of drought proneness is given in Table 6.1 (Banik et al. 2002).

6.2.3 Test of Null Hypothesis

Hypothesis testing arises for Markov chains just as it does for independent processes. The theory of hypothesis testing described below follows Medhi (1994):

Consider a time homogeneous Markov chain with a finite number, m, of states $(1, 2, \ldots, m)$ and having a transitional probability matrix $P = (P_{jk}), j, k = 1, 2, \ldots, m$. Suppose that the number of observed direct transitions from the state j to the state k is n_{jk}, and that the total number of observations is $(N + 1)$, then we can put

$$\sum_{k=1}^{m} n_{jk} = n_j \quad \text{and} \quad \sum_{j=1}^{m} n_{jk} = n_k, \quad j, k = 1, 2, \ldots, m$$

Suppose the observation comes from a Markov chain with a given transition, i.e., matrix $P^0 = (P^0_{jk})$ then t the null hypothesis is

$$H_0 : P = P^0$$

For a large N, the test statistic is given as

$$\sum_{k=1}^{m} \frac{n_j \left(P_{jk} - P^0_{jk}\right)^2}{P^0_{jk}}, \quad j = 1, 2, \ldots, m \tag{6.8}$$

which is distributed as χ^2 with df (degrees of freedom) of $m - 1$. Here, P^0_{jk} is excluded when it equals to 0. Alternatively, a test for all P_{jk} can be obtained by calculating,

$$\sum_{j=1}^{m}\sum_{k=1}^{m} \frac{n_j \left(P_{jk} - P^0_{jk}\right)^2}{P^0_{jk}} \tag{6.9}$$

which has an asymptotic χ^2 distribution with $m(m - 1)$ df (the number of df being reduced by the number of P^0_{jk} which equals to 0, if any for j, $k = 1, 2, 3, \ldots, m$).

The likelihood ratio criterion for H_0 is given by

$$\lambda = \prod_j \prod_k \left(\frac{P_{jk}}{P^0_{jk}}\right)^{n_{jk}} \tag{6.10}$$

Under the null hypothesis, the test statistic

$$-2\log\lambda = 2\sum_j \sum_k n_{jk} \log \frac{n_{jk}}{(n_j) P^0_{jk}} \tag{6.11}$$

has an asymptotic χ^2 distribution with $m(m - 1)$ df

Here n_{jk} is the number of observed direct transition from state j to state k and P^0_{jk} is the probability of transition from state j to k under null hypothesis.

6.2.4 Mapping the Spatial Extent of Agricultural Drought

Inverse distance weighted (IDW) interpolation was used for mapping the spatial extent of agricultural droughts, which is expressed as

$$\hat{Z}(s_0) = \sum_{i=1}^{N} \lambda_i Z(s_i) \tag{6.12}$$

where

$\hat{Z}(s_0)$ = The drought index value to be predicted for location s_0

N = the number of the measured sample points (rainfall stations in this case) surrounding the prediction location

λ_i = the weight assigned to the measured point i which decreases with distance from the prediction location

$Z(s_i)$ = the measured drought index value at location s_i

As the IDW is chosen to predict spatial variations of agricultural drought, cross validation is essential to validate critical parameters that could affect the interpolation accuracy of drought index. In this case, a value was evaluated for optimal parameters. This ensures the overall utility of the IDW models and enables optimal data prediction that is comparable to the observed data.

The IDW surface calculation depends on the selection of a power parameter (p) and neighborhood search strategy. In this study, the root mean squared prediction error (RMSPE) was adopted to assess the IDW models performances. The best IDW model has the lowest RMSPE with mean error (ME) nearest 0 (Johnston et al. 2001). In the present study Geostatistical Analyst tool of Arc Map 9.3 was used to find the optimal powers of the IDW which produce minimum RMSPE and mean prediction error.

6.3 Results and Discussion

The daily rainfall data from 1971 to 2010 of the 12 rainfall stations were obtained from Bangladesh Water Development Board. The daily data were converted to weekly data because week is close to the optimum period for agriculture (Banik et al. 2002). As a Markov chain model is a threshold-based model, 20 mm weekly rainfall is considered as the threshold value on the basis of the findings of Alam et al. (2012a). Moreover, 13 and 21 standard weeks were considered for the Pre-Kharif and Kharif seasons, respectively. In order to find out the yearly drought scenario of Barind Tract, 12 months drought has also been calculated using 52 standard weeks.

6.3.1 Temporal and Spatial Characteristics of Agricultural Drought

The time series of agricultural drought occurrence is given in Fig. 6.3. Figure 6.3a indicated that during the Pre-Kharif growing season almost all years were affected by chronic to moderate agricultural drought. Although the severity of drought varies from year to year, the occurrence of chronic drought was more frequent

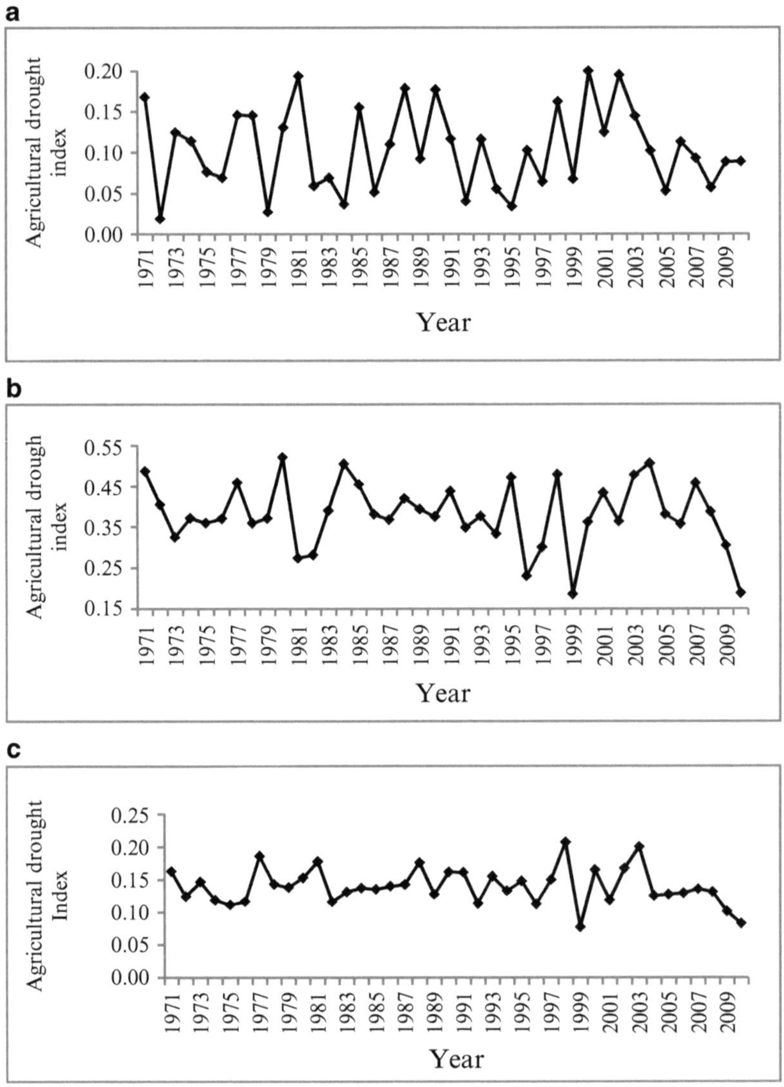

Fig. 6.3 Time series of average agricultural drought index (**a**) Pre-Kharif season (**b**) Kharif season (**c**) over a 12-month period

during the period of 1991–1999. On contrast, the occurrence of the Kharif agricultural drought was less frequent.

During 1971–2010 around 5–6 Kharif agricultural droughts occurred in the area (Fig. 6.3b). As these drought indexes were calculated on the basis of the average data of the rainfall stations, they are not representative to some areas where chronic droughts occurred almost every year. Indeed the Kharif droughts also occur every year in some places in the Barind region. In addition, the Kharif agricultural droughts were more frequent during the late 1990s, and after 2006 there were four consecutive drought events observed during this season. The 12 months drought indexes are given in Fig. 6.3c. It depicts that severe to moderate droughts occurred every year in the Barind region. Similar to the Kharif season, there were four consecutive chronic drought events observed during 2006–2010. This is because most of the rainfall in the Barind region occurs during the Kharif season. The Kharif season drought index primarily determines the occurrence of drought in a 12-month period.

The study also tried to understand the spatial distribution of the major drought events in the Pre-Kharif and Kharif seasons as well as over 12 months periods. The spatial distributions of the most important Pre-Kharif drought events every year is shown in Fig. 6.4. During this growing season, chronic droughts occur almost every year in some parts of the study area. From Fig. 6.4 it is evident that during 1971–2010 there were at least 5 years (1972, 1979, 1995, 2005, and 2008) when the entire study area was affected by chronic droughts, resulted from very low rainfall (almost 0) and high temperature which influenced the moisture holding capacity of soil (Alam 2010). The chronic agricultural droughts during the Pre-Kharif growing season increased the groundwater irrigation demand in *Boro* rice fields which share almost 70 % of the total rice production of Bangladesh. But in the Barind region, the lack of rain or chronic droughts compelled farmers not to seed their *Boro* rice. Moreover, around 85 % of the land remained fallow during this season in the Barind region (Alam et al. 2011; Saadat et al. 2009).

During the Kharif growing season, nine major drought years were identified during the period of 1971–2010. Among these, four drought years were identified during the last decade of the twentieth century (Fig. 6.5). A large variation is also observed in the spatial extent of these drought events. In 1982, around 32 % of the area was affected by moderate droughts and the drought severity was higher in the central part compared to other parts of the study area (Fig. 6.5). However, in 1994 the percentage of the moderately affected area is reduced to around 4 % and most severe drought events shifted from the central part to the central eastern part (Fig. 6.5). Among the Kharif drought events, in 2010 around 6 and 30 % of the area was affected by chronic and severe droughts, respectively (Fig. 6.5), and the northern part of the area was hardest hit. Along with these, in 1997, 1999, and 2009 the drought severity was higher in the western corner. These imply that the Kharif agricultural droughts may affect any place in the area in any year. The results of these drought events are also in line with the global patterns (Shahid and Behrawan 2008). It is well known that the last decade of the twentieth century was the driest in the last 100 years. The agricultural droughts of 1994 led to a decrease in rice and wheat production by 3.59×10^6 tons (Shahid 2008 and Shahid and Behrawan 2008). Moreover, Dey et al. (2011) marked 1994 as the driest year in the contemporary period which had devastating impacts on the northwestern part of Bangladesh. The drought events of 1997 produced a shortfall of one million tons of food grains, of which 0.6 million tons were T. aman (Ramsey et al. 2007). Ramsey et al. (2007) also confirmed that around 2.32 million hectares of T. aman rice crops were damaged due to agricultural droughts every year in Bangladesh. On the other hand, Alam et al. (2011) and Rahman (2000) indicate this amount is 0.574–1.748

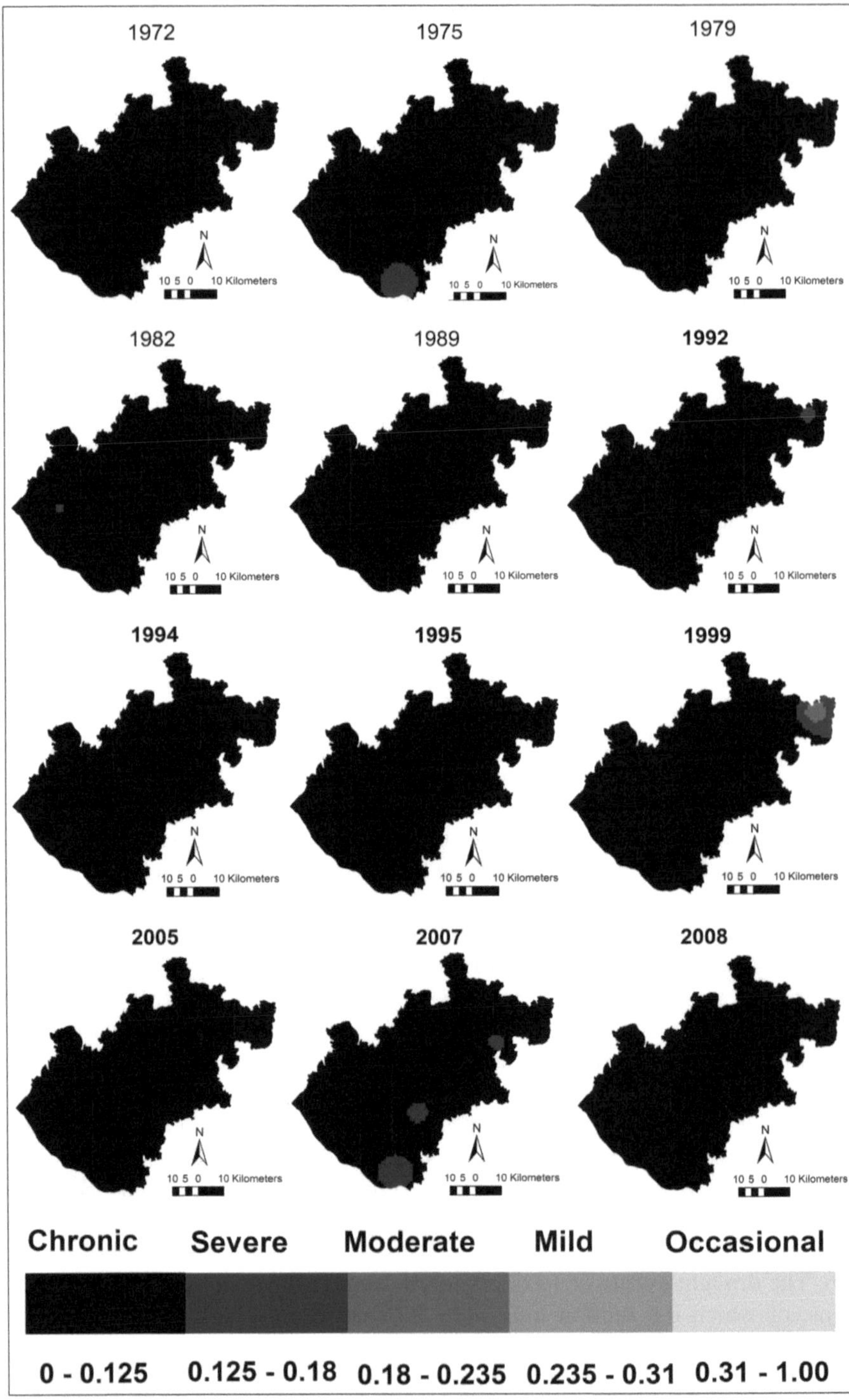

Fig. 6.4 Major agricultural drought events in the Pre-Kharif season

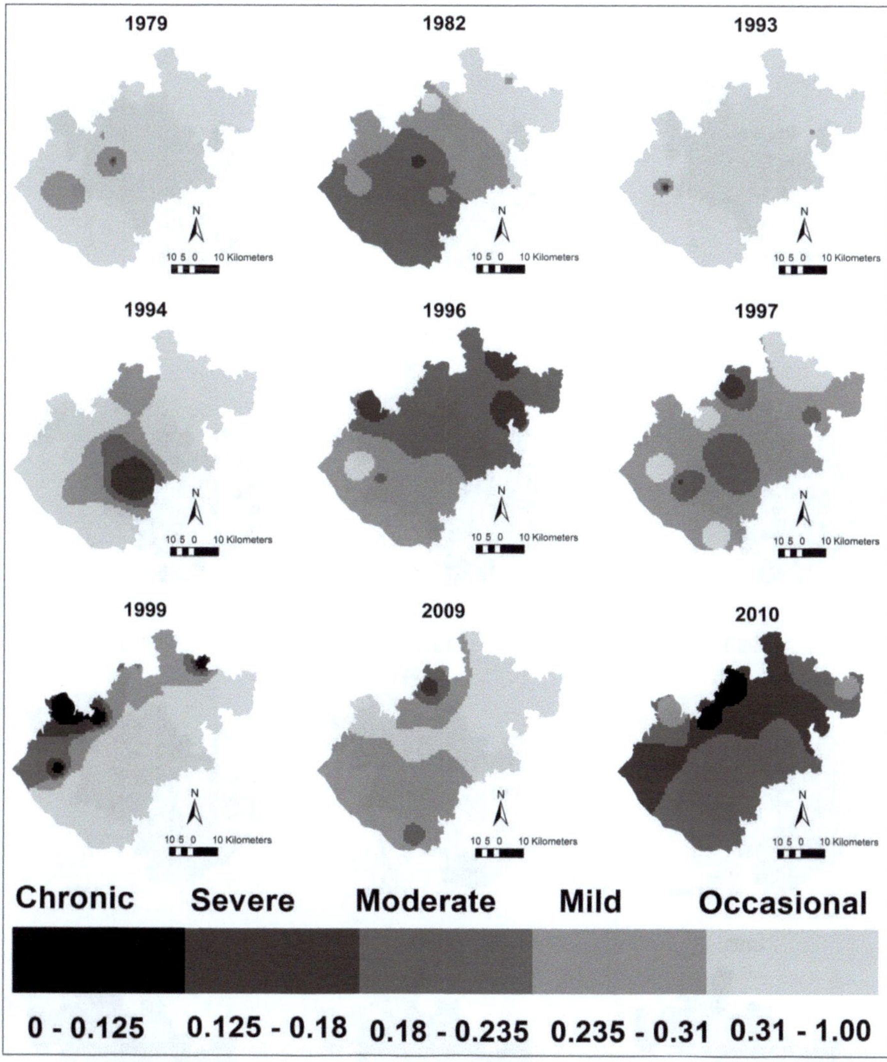

Fig. 6.5 Major agricultural drought events in the Kharif season

millions of hectares. Therefore it can be concluded that the results of these drought indexes are in line with historical drought events of the country.

The 12 months period drought index is calculated for the period from January to December. The major 12 months drought events are given in Fig. 6.6, which shows that drought may occur in any place of the Barind region. During the 12 months period, the agricultural drought does not follow any regular pattern. In 1972 the whole study area was affected by chronic droughts except the north eastern and north western corners. In 1974 the chronic droughts affected only the southern part.

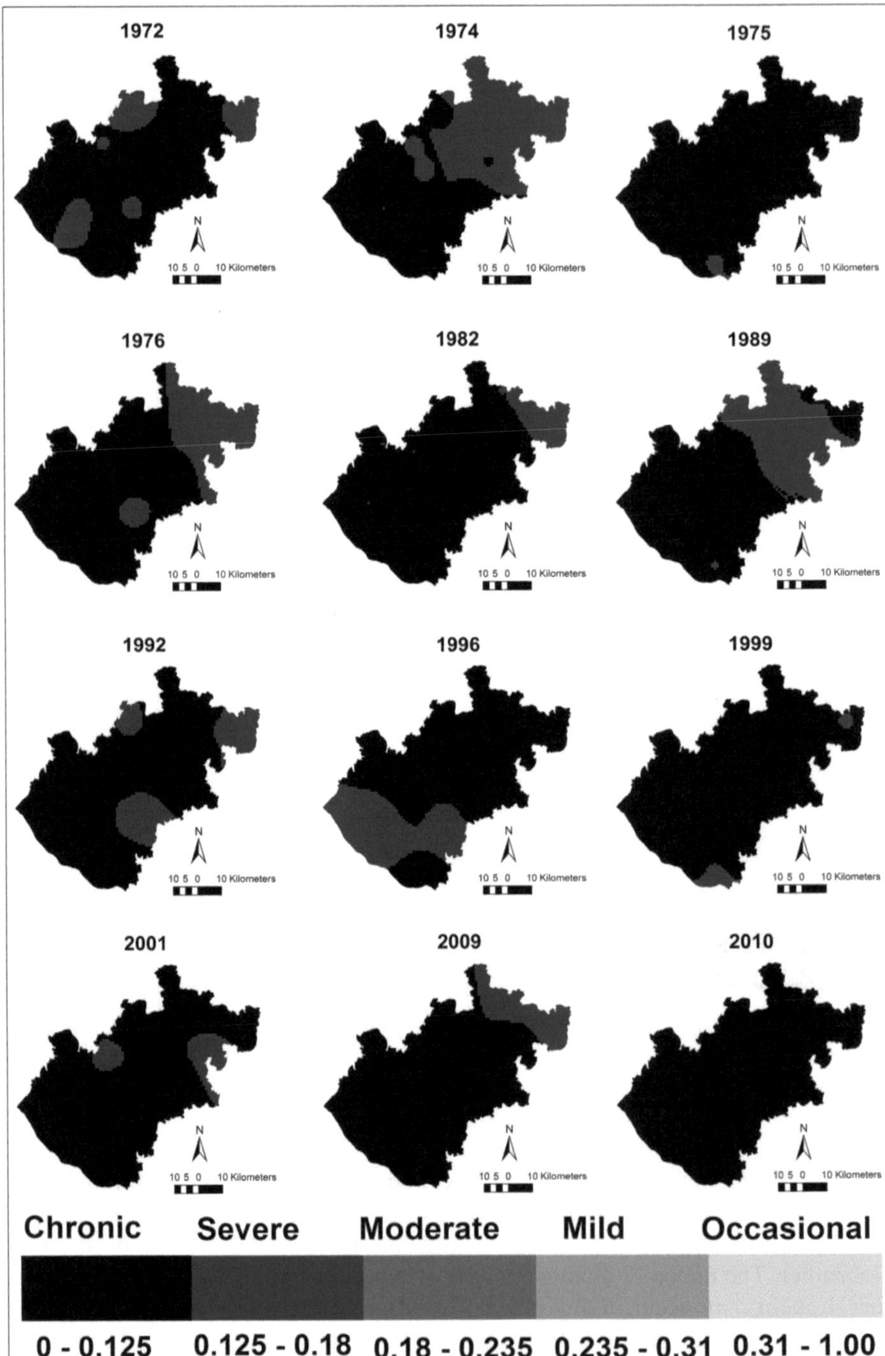

Fig. 6.6 Major agricultural drought events in 12 months period

On the contrary in 1996 the drought severity was higher in the northern Barind region (Fig. 6.6). The most severe consecutive drought years were observed during 2009 and 2010. The spatial distribution of drought in 2009 indicates that almost the whole study area except a small northern portion was affected by chronic droughts and in 2010 the whole study area experienced chronic droughts.

6.3.2 *Probability of Wet Spell*

As the soil moisture storage of the area depends on rainfall, an analysis of the probability of wet spell is carried out to predict the water holding capacity of the soil (Rahman et al. 2012). According to Banik et al. (2002) and Alam (2010), it is necessary to have 10–12 wet weeks for better crop production. Equation (6.4) was used to calculate the probability of wet spell. It is evident from Fig. 6.7a that the average probability of a wet spell of 5 or more weeks during the Pre-Kharif season approaches 0 at all rainfall stations. In Godagari station the probability of consecutive 4 or more wet weeks tends to 0 (Fig. 6.7a). Therefore, cropping is quite impossible without artificial irrigation during this season. In the Kharif season, the probability of a wet spell of longer than 5 weeks is close to 0. This result validates the statement of Saadat et al. (2009) that the most devastating drought occurred during this season because farmers totally depend on rain to sow their seeds. A small amount of rainfall (more than the threshold value) may be hindering to identify severe droughts in the Kharif season, but the very low probability of wet spell indicates that the area may be affected by future chronic to severe droughts.

To find out a suitable time for crop growing, the probability of annual wet spells was calculated. This probability suggests low chances of more than 5 consecutive wet weeks (Fig. 6.7c) and recommends the necessity of irrigation for better cropping in this region.

6.3.3 *Relative Frequency of Occurrence of Agricultural Drought*

The frequency analysis is commonly used in meteorology and hydrology to assess the return period of a particular event. The percentage of drought occurrence is computed by taking the ratio of agricultural drought occurrence in each season to the total agricultural drought occurrence in the same season for different drought categories. From Table 6.2 it can be noticed that in Chapai Nawabganj station the frequencies of chronic drought in the Pre-Kharif and Kharif seasons and over 12 months are 67.5 %, 0 %, and 52.5 %, respectively. This suggests that in the Pre-Kharif season the chronic drought may occur around three times in the 4 years duration, while as a 12-month event, the return period of chronic drought is 2 years.

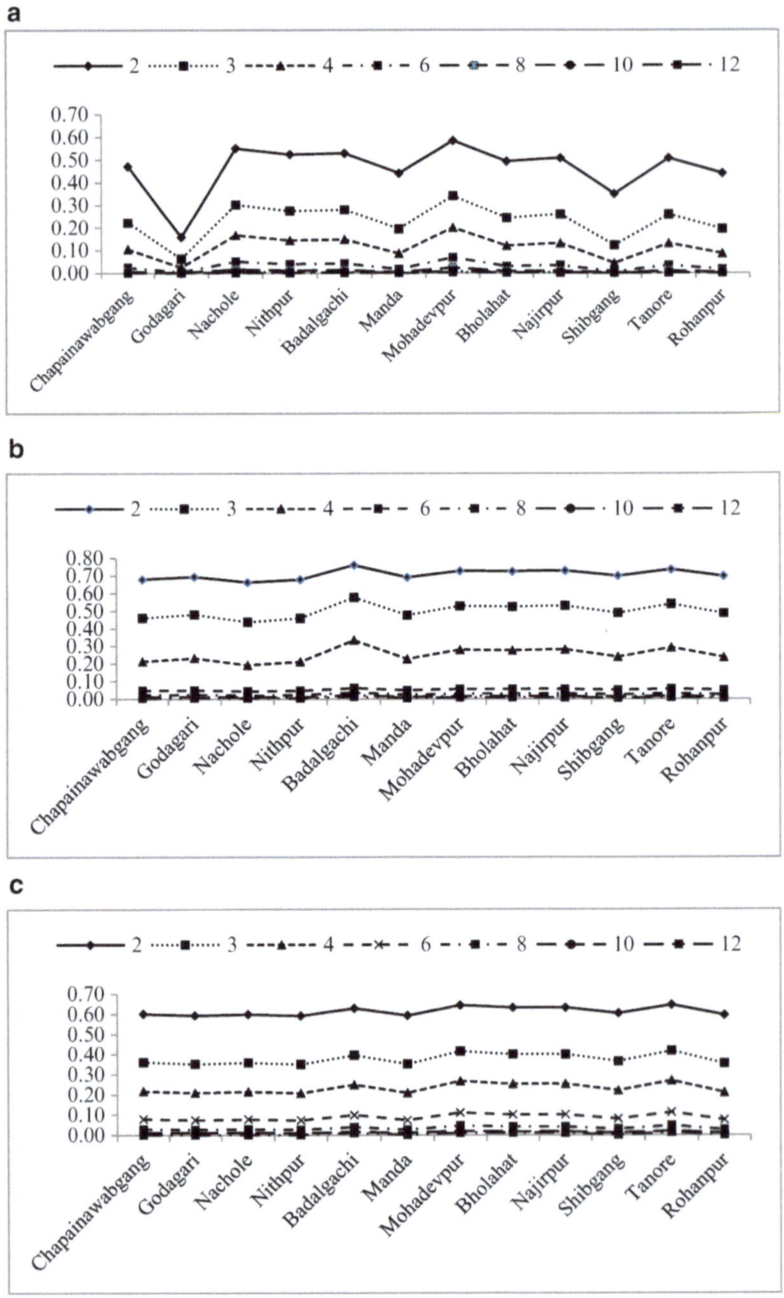

Fig. 6.7 Forty years average probability of wet weeks in (**a**) Pre-Kharif season (**b**) Kharif season (**c**) 12 months period

Table 6.2 Relative frequency of occurrences (%) of agricultural drought at different rainfall stations of Barind in different seasons

	Pre-Kharif			Kharif			12 months period		
Name of the station	*C*	*S*	*M*	*C*	*S*	*M*	*C*	*S*	*M*
Chapai Nawabganj	67.5	17.5	10	0	2.5	17.5	52.5	32.5	10
Godagari	75	10	7.5	0	10	10	35	40	20
Nachole	77.5	5	10	2.5	7.5	12.5	42.5	45	7.5
Nithpur	65	25	2.5	2.5	5	7.5	35	45	12.5
Badalgachi	47.5	27.5	15	0	2.5	12.5	22.5	52.5	17.5
Manda	75	12.5	5	0	5	2.5	35	52.5	5
Mohadevpur	57.5	20	12.5	0	5	5	30	47.5	12.5
Bholahat	80	7.5	5	2.5	5	7.5	42.5	40	10
Nazirpur	72.5	12.5	7.5	2.5	2.5	5	32.5	55	2.5
Rohanpur	82.5	10	2.5	5	0	2.5	25	65	5
Shibganj	82.5	10	5	7.5	2.5	7.5	47.5	42.5	5
Tanore	57.5	22.5	5	0	2.5	7.5	57.5	32.5	2.5

C chronic drought, *S* severe drought, *M* moderate drought

Table 6.3 χ^2 test for reliability of the occurrence and nonoccurrence rainfall recorded at different weather stations of BWDB at Barind region during 1971–2010

	Value of $-2\log \lambda$		
Stations	Pre-Kharif	Kharif	12 months period
Chapai Nawabganj	13.21	7.00	17.77
Godagari	5.51	8.18	5.5
Nachole	24.57	6.7	24.57
Nithpur	19.45	6.19	19.45
Badalgachi	16.23	7.84	16.23
Manda	16.95	6.62	9.45
Mohadevpur	27.96	8.79	27.96
Bholahat	17.18	19.27	17.18
Nazirpur	16.94	23.28	16.94
Shibganj	3.98	13.22	3.98
Tanore	14.85	13.41	14.85
Rohanpur	10.45	9.26	10.46

The frequency analysis of different categories of drought suggests that the occurrence of chronic drought is higher in the Pre-Kharif season, but during the Kharif season, the return period of moderate drought is shorter. Meanwhile, severe droughts more frequently affected the study area when considered as 12 months events.

6.3.4 Result of Hypothesis Testing

Hypothesis testing is an important requirement for running any statistical model. In the present study, a null hypothesis was tested against seasonal and annual wet and dry spells in every year using χ^2 test. The test of null hypothesis that the chain is of order 0 against the alternative hypothesis that is of order 1 is given in the Table 6.3.

The χ value of the transition matrix in all the rainfall stations being very small ($\chi^2 = 0.00393$ at 1 df at 5 % level) indicating the null hypothesis that chain of order 0 is rejected. Therefore, it can be said that the performance of the model is statistically satisfactory.

6.4 Conclusions

In the present study, seasonal and annual drought indexes of the Barind region of Bangladesh were calculated. Chronic droughts were found to occur in the Pre-Kharif season and farmers abstain from seeding. In the Kharif season good crops may grow depending on the blessing of adequate rain in time. Also moderate droughts were found to be more frequent during the Kharif growing season. A long-term probability calculation suggests that the area is potentially prone to future chronic and severe droughts. The modeling method can be replicated by considering crop diversity and various crop growing periods in different seasons under prevailing agroclimatic conditions. The outcome of this study might be helpful for the agricultural planners and irrigation engineers in operational responses to agricultural drought risk reduction in Bangladesh.

References

Alam ATMJ (2010) Spatial and temporal characteristics of agricultural drought at Barind region, Bangladesh: an application of GIS and Markov chain model. Master's thesis, submitted to the Department of Environmental Sciences, Jahangirnagar University, Dhaka, Bangladesh

Alam ATMJ, Saadat AHM, Rahman MS, Barkotulla MAB (2011) Spatial analysis of rainfall distribution and its impact on agricultural drought at Barind region, Bangladesh. Rajshahi Univ J Environ Sci 1(1):40–50

Alam ATMJ, Rahman MS, Saadat AHM (2012a) Comparison of threshold values of Markov chain for determining agricultural drought at Barind, Bangladesh. In: Abstract of North Bengal drought conference (NBDC) 2012 on sharing knowledge combating climate change disaster, University of Rajshahi and Kakonhat, 27–28 March 2012, p 21

Alam ATMJ, Rahman MS, Saadat AHM, Huq MM (2012b) Gamma distribution and its application of spatially monitoring meteorological drought in Barind, Bangladesh. J Environ Sci Nat Resour 5(2):287–293

Alam ATMJ, Saadat AHM, Rahman MS, Rahman S (2013) Spatio-temporal variation of agricultural drought in the Barind region of Bangladesh: An application of Markov chain model, Irrig. and Drain. doi: 10.1002/ird.1800

Al-Salihi AH (2003) Drought identification and characterization in Jordan. J Arid Environ 53:585–606

Anderson TW, Goodman LA (1957) Statistical inference about Markov chains. Ann Math Stat 28:89–110

Banik P, Mandal A, Rahman MS (2002) Markov chain analysis of weekly rainfall data in determining drought-proneness. Discrete Dyn Nat Soc 7:231–239

Barkotulla MAB (2007) Markov Chain analysis of rainfall in agricultural drought of Barind region. Ph. D. thesis, submitted to the Department of Statistics, University of Rajshahi, Bangladesh

Ben-Zvi A (1987) Indices of hydrological drought in Israel. J Hydrol 92:179–191

Biamah EK, Sterk G, Sharma TC (2005) Analysis of agricultural drought in Iiuni Eastern Kenya: application of Markov model. Hydrol Process 19:1307–1322

Brammer H (1987) Drought in Bangladesh, lessons for planners and administrators. Disasters 11:21–29

Christensen JH, Hewitson B, Busuioc A, Chen A, Gao X, Held I, Jones R, Kolli RK, Kwon WT, Laprise R, Magaña Rueda V, Mearns L, Menéndez CG, Räisänen J, Rinke A, Sarr A, Whetton P (2007) Regional climate projections. In: Solomon SD, Qin M, Manning Z, Chen M, Marquis KB, Averyt MT, Miller HL (eds) Climate change 2007: the physical science basis. Contribution of working group I to the fourth assessment report of the intergovernmental panel on climate change. Cambridge University Press, Cambridge

Darcup JA, Lee KS, Paulson EG (1980) On the definition of drought. Water Resour Res 16:297–302

Dey NC, Alam MS, Sajjan AK, Bhuiyan MA, Ghose L, Ibaraki Y, Karim F (2011) Assessing environmental and health impact of drought in the Northwest Bangladesh. J Environ Sci Nat Resour 4(2):89–97

Heathcote RL (1974) Drought in South Australia. In: White GF (ed) Natural hazards: local, national, global. Oxford University Press, New York

Heim RR Jr (2002) A review of twentieth-century drought indices used in the United States. Bull Am Meteorol Soc 83(8):1149–1165

Jimoh OD, Webster P (1999) Optimum order of Markov chain for daily rainfall in Nigeria. J Hydrol 185:45–69

Johnston K, Ver Hoof JM, Krivoruchko K, Lucas N (2001) Using ArcGIS™ geostatistical analyst. ESRI, Redlands

Kazt RW (1974) Computing probabilistic associated with the Markov chain model for precipitation. J Appl Meteorol 53:953–958

Medhi J (1994) Stochastic process. New Age International Publishers, New Delhi

Moslehuddin AZM, Habibullah MM, Egashira K (2008) Mineralogy of soils from different agroecological regions of Bangladesh: region 25—level Barind tract and region 27—north–eastern Barind tract. J Fac Agric Kyushu Univ 53(1):163–169

Ochola WO, Kerkidis P (2003) A Markov chain simulation model for predicting critical wet and dry spells in Kenya: analysing rainfall events in the Kanoplains. Irrig Drain 52:327–342

Paul BK (1998) Coping mechanisms practiced by drought victims (1994/5) in North Bengal, Bangladesh. Appl Geogr 18:355–373

Rahman MS (1999a) A stochastic simulated Markov chain model for daily rainfall at Barind, Bangladesh. J Interdiscip Math 2(1):7–32

Rahman MS (1999b) Logistic regression estimation of a simulated Markov chain model for daily rainfall in Bangladesh. J Interdiscip Math 2(1):33–40

Rahman MS (2000) A rainfall simulation model for agricultural development in Bangladesh. Discrete Dyn Nat Soc 5:1–7

Rahman M, Alam ATMJ, Saadat AHM (2012) Assessment of groundwater potential for irrigation in Barind region Bangladesh: an application of TMWB model. In: Proceedings of the second international conference on environmental technology and construction engineering for sustainable development, 10–12 March 2012, Sylhet, Bangladesh, pp 357–361

Ramsey S, Subbia AR, Bass S, Juergens I (2007) Livelihood adaptation to climate variability and change in drought-prone areas of Bangladesh. Asian Disaster Preparedness Center Food and Agriculture Organization of the United Nations, Rome

Saadat AHM, Alam ATMJ, Alam M, Shovon J, Uzzaman R (2009) Impact of drought on agriculture of Barind tract: a case study of Dharmapur Chapainawabgang. In: Workshop on impacts of climate change on livelihoods, agriculture, aquaculture and fisheries sector of Bangladesh, BAU, Mymensingh, 1 Oct 2009, pp 54–64

Shahid S (2008) Spatial and temporal characteristics of droughts in the western part of Bangladesh. Hydrol Process 22:2235–2247

Shahid S (2010) Spatio-temporal variation of aridity and dry period in term of irrigation demand in Bangladesh, American–Eurasian Journal of Agricultural & Environmental Sciences 7(4): 386–396

Shahid S (2011) Impact of climate change on irrigation water demand of dry season Boro rice in northwest Bangladesh. Clim Change 105:433–453

Shahid S, Behrawan H (2008) Drought risk assessment in the western part of Bangladesh. Nat Hazards 46:391–413. doi:10.1007/s11069-007-9191-5

Sharma TC (1996) Simulation of the Kenyan longest dry and wet spells and the longest rain-sums using a Markov model. J Hydrol 178:55–67

Yasmin R (2008) Ground water modelling of the north-eastern part of Barind tract for its sustainable development and management, Bangladesh. Asian J Inf Technol 7(5):218–225

Chapter 7
Forecasting Tropical Cyclones in Bangladesh: A Markov Renewal Approach

Md. Asaduzzaman and A.H.M. Mahbub Latif

Abstract Bangladesh frequently suffers from tropical cyclones possibly due to its unique location. The funnel-shaped northern part of the Bay of Bengal causes tidal bores when cyclones make landfall. These tropical cyclones can be very devastating and can severely affect the coastline of Bangladesh. In this study we analyzed 135 tropical cyclones occurred in Bangladesh during 1877–2009 considering the physical characteristics of the storm surge process. For analyzing the storm surge process, a Markov renewal model that takes into account both the sojourn times and the transitions between different types of cyclones simultaneously was considered. Exponential distribution for the sojourn times was assumed to derive the probabilities of occurrence of different types of cyclones for various lengths of time intervals. Given the type of the last cyclone occurred probabilities of occurrence of the next cyclone are reported using the fitted Markov renewal model. The mean recurrence times of different type of cyclones were also calculated assuming ergodicity of the Markov renewal process.

Keywords Cyclone prediction • Semi-Markov process • Non-Poisson process • Stationary process

7.1 Introduction

Bangladesh is a low-lying, riverine country located between 20.30° and 26.38° north latitude and 88.04° and 92.44° east longitude. The country is formed by a delta plain at the confluence of the Ganges (Padma), Brahmaputra (Jamuna), and

M. Asaduzzaman (✉) • A.H.M.M. Latif
Institute of Statistical Research and Training (ISRT), University of Dhaka, Dhaka 1000, Bangladesh
e-mail: asad@isrt.ac.bd

T. Islam et al. (eds.), *Computational Intelligence Techniques in Earth and Environmental Sciences*, DOI 10.1007/978-94-017-8642-3_7,

Meghna rivers and their tributaries. Bangladesh has a tropical monsoon climate characterized by heavy seasonal rainfall, high temperatures, and high humidity. Natural disasters such as tropical cyclones affect the country very frequently, and the country is highly vulnerable to tropical cyclones and associated storm surges because of its 440-mile long coastline.

Over the years Bangladesh experienced a number of tropical cyclones and the one that struck on November 11, 1970 is considered as the world's deadliest tropical cyclone ever recorded (equivalent to a strong category 3 hurricane), where more than half a million people lost their lives as a result of the storm surge that flooded much of the low-lying islands of the Ganges Delta. The 1991 cyclone, struck the southeastern part of Bangladesh, is considered as the second deadliest that forced a 6 m storm surge inland over a wide area, killing about 140 thousand people and leaving as many as 10 million people homeless. Recently two devastating tropical cyclones *Sidr* in 2007 and *Aila* in 2009 struck the southern part of Bangladesh that took 4,234 and 190 human lives, and caused about USD 2.3 and 0.27 million property damage, respectively (Centre for Research on the Epidemiology of Disasters).

Although a good number of studies have been carried out on the Bay of Bengal cyclones, a few of them have focused on the Bangladesh coast. Among these studies both the climatological (Rai Sircar 1956; Mooley 1980) and statistical (Raghavendra 1973) analyses of the cyclones have been performed. Emphasizing on the tropical cyclones in Bangladesh, some mitigative measures of storm surges were studied by Khalil (1992), vulnerability by Haque and Blair (1992) and Ali (1996), and warning process by Haque (1995). However, these studies mainly concentrated on frequency, vulnerability, and climatological analyses of the cyclones. Using numerical simulation models of tropical cyclones, Dube et al. (1985, 1986) and Sinha et al. (1986) studied storm surges in Bangladesh coastline while Murty et al. (1986) in the Bay of Bengal. As-Salek and Yasuda (1995) proposed a numerical simulation model for the prediction of the storm surges incorporating the effects of islands and the bathymetrical details, and the effects of island, the sea level rises and the bottom level changes on storm surges were also investigated. Considering the stochastic occurrence pattern, Mooley (1981) proposed a Poisson model for severe cyclones that struck around the Bay of Bengal during 1877–1977. Among recent works, studies by Alam et al. (2003) and Islam and Peterson (2009) are notable, where Alam et al. (2003) analyzed on the frequency of landfalling Bay of Bengal cyclones over a period of data from 1974 to 1999. Islam and Peterson (2009) performed a comprehensive climatological analysis on the landfalling Bay of Bengal cyclones, where the coastal region of Bangladesh is divided into five segments, and landfalling cyclones and vulnerability were compared among the coastal segments. Variations on year-to-year and pre- and post-monsoon occurrence of landfalling tropical cyclones in Bangladesh have also been reported. Although several studies on cyclones of Bangladesh have been carried out but main concentration has been given on the frequency analysis describing the occurrence, types, nature, and damages of the cyclones occurred in Bangladesh.

Stochastic counting process models are common in forecasting natural disasters and have recently been proved as an important tool to analyze occurrence of such events, for instance, earthquake (Ogata 1988, 1998; Alvarez 2005; Garavaglia and Pavani 2011), drought (Gupta and Duckstein 1975), flood (Fiorentino et al. 1984; Lardet and Obled 1994), and storm (Jagger et al. 2002; Lu and Garrido 2005; Rumpf et al. 2007). In this study, a Markov renewal model is developed to forecast tropical landfalling cyclones in Bangladesh considering the inter-occurrence times and types of cyclones. In more details, this chapter is organized as follows: the data sources and description are given in Sect. 7.2, and the mathematical models are described in Sect. 7.3. Section "Results and Discussion" contains results on the fit of the proposed model, and some final remarks are given in section "Conclusion."

7.2 Data Sources and Description

The data for this study have been collected from two separate sources. Data on 115 cyclones that struck the coastal areas of Bangladesh during the period 1877–2003 have been taken from Islam and Peterson (2009), where the data were obtained from (1) the Global Tropical Cyclone Climatic Atlas (GTCCA), validated by the hurricane database of the Unisys Corporation, and (2) Center for Research on the Epidemiology of Disasters (CRED). The data contain information mainly on occurrence date, place, and historical tracks of tropical cyclones that hit Bangladesh during 1877–2003. Data on 10 cyclones of the recent years 2003–2009 are obtained from the CRED, where the date, place, causalities of the cyclones are available. Therefore, date of cyclone occurrence and the place of the landfall are available for a total of 125 cyclones that struck Bangladesh coast during 1877–2009. The cyclones are classified into three categories depending on the wind speed, which are tropical depression (TD), tropical storm (TS), and hurricane (HU). Table 7.1 shows the classification of the 125 cyclones, of which 40 were tropical depressions (TDs), 52 were tropical storms (TSs), and the remaining 33 were hurricanes (HUs).

Table 7.1 Classification and number of land-falling cyclones occurred in Bangladesh during 1877–2009

Type	Wind speed (in knot)	Number of storms
Tropical depression (TD)	0–34	40
Tropical storm (TS)	34–63	52
Hurricane (HU)	64+	33
Total		125

7.3 Methods

Stochastic counting models are popular to describe processes with the random occurrences. To model tropical cyclones in Bangladesh, we consider a Markov renewal process (MRP) which takes into account the inter-occurrence times of cyclones as sojourn times of the process and the types of cyclones as states of the process. In this section, we describe an MRP with its properties, likelihood construction, parameter estimation, and stationary behavior.

7.3.1 Markov Renewal Process and Its Properties

Let us consider a process called $S(t)$ with m possible states, where m is a finite natural number. Let J_0 be the initial state at time 0, i.e., the state from the process $S(t)$ starts, and let X_1 be the length of time the process stays at J_0 before moving to state J_1 and the process stays at state J_1 for a length of time X_2 before moving into state J_2, and so on. The two-dimensional stochastic process in continuous time $\{(J_n, X_n), n \geq 0\}$ is called a positive (J-X) process. Suppose $X_0 = 0$ a.s., where the sequence $(J_n, n \geq 0)$ gives the successive states of S in time, and the sequence $(X_{n;}\ n \geq 0)$ gives the successive sojourn times, i.e., X_n is the time spent by S in state J_{n-1} $(n > 0)$. The $\{(J_n, X_n), n \geq 0\}$ is then called an MRP (Janssen and Manca, 2007).

Let $\mathbf{F} = \{F_{ij}(\cdot), i, j \in E\}$, where $E = \{1, \ldots, m\}$ with $m \in N$, denote a matrix of distribution functions on $\mathfrak{R}+$ and $\mathbf{P} = \{P_{ij}(\cdot), i, j \in E\}$ denote a transition matrix on E, and $\mathbf{a} = \left\{ a_i, i \in E, a_i > 0, \sum_{i=1}^{m} a_i = 1 \right\}$ be a probability distribution on E. Then the two-dimensional stochastic process $\{(J_n, X_n), n \geq 0\}$ defined on a complete probability space $(\Omega, \Im, P)$ satisfies

(i) $X_0 = 0$ a.s.
(ii) $P(J_0 = k) = a_k$ for every $k \in E$
(iii) $P\left(J_n = k, X_n \leq x \middle| J_0, J_1, X_1, \ldots, J_{n-1}, X_{n-1}\right) = p_{J_n-1,k} F_{J_n-1,k}(x)$ a.s. for every $x \in (0, +\infty)$ and $k \in E$

Then, clearly the process $\{(J_n, X_n), n \geq 0\}$ is an MRP. Some of its important properties of the process are

(a) $(J_{n-1}, n > 0)$ is an E-valued Markov chain with transition matrix $\mathbf{P}$ and initial distribution $\mathbf{a}$.
(b) For every $n > 1, X_1, X_2, \ldots, X_n$ are conditionally independent, given J_n $(n > 0)$ and $\left(X_1 \leq x_1, X_2 \leq x_2, \ldots, X_n \leq x_n \middle| J_n, n \geq 0\right) = \prod_{i=1}^{n} F_{J_i-1, J_i}(x_i)$.

7.3.2 Likelihood Construction and Parameter Estimation

The Markov chain (J_n, $n \geq 0$) represents the states that are successively visited by the process and the process (X_n, $n \geq 0$) represents the successive waiting times. In our application the states are the types of cyclone classified with respect to their severity E, and may take three values which are tropical depression (TD), tropical storm (TS), and hurricane (HU). The X_ns are the times between successive cyclones of the type TD, TS, or HU. If a cyclone can be classified of severity $i \in E$ and the next cyclone is of severity $j \in E$, the time between the two cyclones is a positive random variable with distribution function F_{ij} with the corresponding density f_{ij}, for every $i, j = 1, \ldots, m$.

Let (j_0, j_1, x_1, ..., $j_{\tau-1}$, $x_{\tau-1}$, x_τ) be a realization of an MRP on the time window [0, T], where τ represents the number of times the states of E are visited in [0, T] and for the last event J_τ the sojourn time between the last event and T is x_τ considered as censored, i.e., $x_\tau > [T - (x_1 + x_2 + \cdots + x_{\tau-1})]$. Then the conditional likelihood function given $J_0 = j_0$ can be expressed as

$$L(j_0) = \left[\prod_{i=1}^{\tau-1} p_{j_{i-1}j_i} f_{j_{i-1}j_i}(x_i)\right] \times \left[\sum_{k=1}^{m} p_{j_{\tau-1}k}\left(1 - F_{j_{\tau-1}k}(x_\tau)\right)\right].$$

Then the corresponding log-likelihood function is

$$l(j_0) = \sum_{i=1}^{\tau-1} \ln p_{j_{i-1}j_i} + \sum_{i=1}^{\tau-1} \ln f_{j_{i-1}j_i}(x_i) + \ln\left[\sum_{k=1}^{m} p_{j_{\tau-1}k}\left(1 - F_{j_{\tau-1}k}(x_\tau)\right)\right].$$

The log-likelihood function $l(j_0)$ contains two types of parameters involving the transition probability matrix $\mathbf{P}$ and the distribution of sojourn times corresponding to different transitions $\mathbf{F}_{ij}$.

A number of distributions can be assumed for sojourn time between the cyclones and in this study exponential distribution is considered. So the probability that sojourn time between cyclones i and j is less than or equal to x is

$$F_{ij}(x) = \mathrm{e}^{-x/\mu_{ij}}, \ x > 0,$$

where μ_{ij} is the transition rate for the transition from state i to state j. Then the likelihood equation becomes

$$l(j_0) = \sum_{i=0}^{\tau-1} \ln p_{j_i j_{i+1}} - \sum_{i=1}^{\tau-1} \ln \mu_{j_i j_{i+1}} - \sum_{i=0}^{\tau-1} \left(x_i/\mu_{ij}\right) + \ln\left[\sum_{k=1}^{m} p_{j_\tau k} \exp\left(-x_\tau/\mu_{j_\tau,k}\right)\right]. \quad (7.1)$$

The maximum likelihood estimate (MLE) of p_{ij} and μ_{ij} are obtained in two stages; in the first stage, the elements of the transition probability matrix are estimated using the following expression

$$\hat{p}_{ij} = \frac{\#\,\text{of transition from state } i \text{ to state } j}{\#\,\text{of transition from state } i}, \tag{7.2}$$

which can be derived by considering the data on different transitions over the period [0, T] as a sample from a multinomial distribution. In the second stage, the estimates $\hat{p}_{ij}$ are plugged into the conditional log-likelihood functions, which becomes only a function of μ_{ij} and then maximizing the likelihood function MLE of μ_{ij}s are obtained.

7.3.3 Cross-State Prediction

One of the objectives of this work is to predict of the state of the next event being known the state of the last event and the time passed by the last occurred event. The probability that the cyclone k will occur after t^* time of the last event i occurred at t_0 can be calculated as

$$P\left(t^*, k \middle| t_0, i\right) = \frac{\left[F_{ik}(t_0 + t^*) - F_{ik}(t_0)\right] p_{ik}}{\sum_{i,j \in E} \left[1 - F_{ij}(t_0)\right] p_{ij}}. \tag{7.3}$$

7.3.4 Asymptotic Behavior: Mean Recurrence Time

If we assume that the Markov chain is ergodic, i.e., the states are non-null recurrent and aperiodic, then it will have a unique stationary distribution. For stationary Markov chains, the mean recurrence time exists and it can be defined as the expected number of steps to return to the same state where it started from. For an MRP, the mean recurrence time of a state can be given as

$$\alpha_i = \frac{1}{\pi_i} \sum_k \pi_k \eta_k, \tag{7.4}$$

where $\eta_i = \sum_{j=1}^{m} p_{ij}\mu_{ij}$ and $\pi_i = (\pi_1, \ldots, \pi_m)$ is the unique stationary distribution of the Markov chain (J_n, $n \geq 0$).

7.4 Results and Discussion

The main objective of this study is to predict the next cyclone based on the historic data as described in Sect. 7.2. We have information on 125 cyclones occurred in Bangladesh coast during 1877–2009, which were categorized as tropical depression (TD), tropical storm (TS), and hurricane (HU). The sojourn times between two successive cyclones are calculated from the data. Based on the types of cyclones

Table 7.2 Estimate of different transition probabilities (the number of transitions is given in parenthesis)

	TD	TS	HU
TD	0.375 (15)	0.450 (18)	0.175 (7)
TS	0.308 (16)	0.404 (21)	0.288 (15)
HU	0.250 (8)	0.406 (13)	0.344 (11)

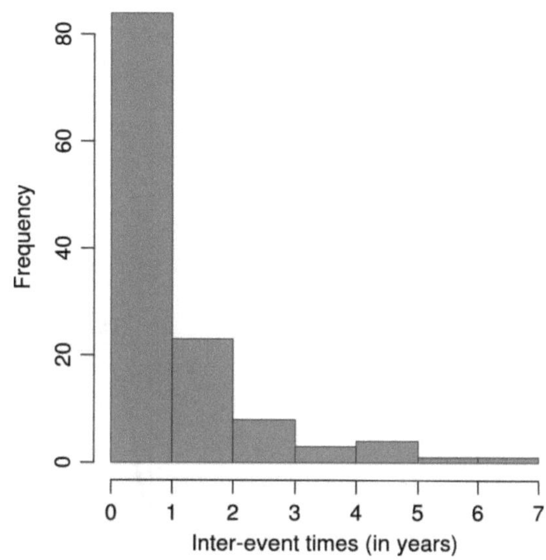

Fig. 7.1 Histogram of inter-event times

and their sojourn times, we develop a Markov renewal model for predicting the probability of future cyclones in Bangladesh. Three conjectures are assumed for fitting the MRP, which are

1. Cyclones occur according to a renewal process.
2. Severity of cyclone is a discrete random variable, and constitutes a homogeneous Markov chain.
3. The longer is the waiting time for transition from a state to another state the higher is the probability that the transition happens.

To construct the Markov chain of the MRP, the three different states tropical depression (TD), tropical storm (TS), and hurricane (HU) are considered and the parameters involving the transition probability matrix are estimated using Eq. (7.2). Estimated transition probabilities corresponding to the number of transitions are given in Table 7.2, which show that the tropical storms are more likely to hit Bangladesh coast irrespective of the immediate last event.

A Markov renewal model also requires specifying the distributions of sojourn times for different transitions and it is extremely important to choose an appropriate distribution of sojourn time because it corresponds to the physical properties of the data taken under consideration. In this study, exponential distribution, the most commonly used distribution for sojourn time, has been chosen for the distribution of sojourn times. Figure 7.1 shows the histogram of the sojourn times of 125 cyclones, which indicates that an exponential distribution of sojourn times is reasonable.

Table 7.3 Estimated mean sojourn times for different transitions

	TD	TS	HU
TD	14.43	18.81	12.16
TS	13.12	8.40	11.66
HU	17.68	11.26	10.61

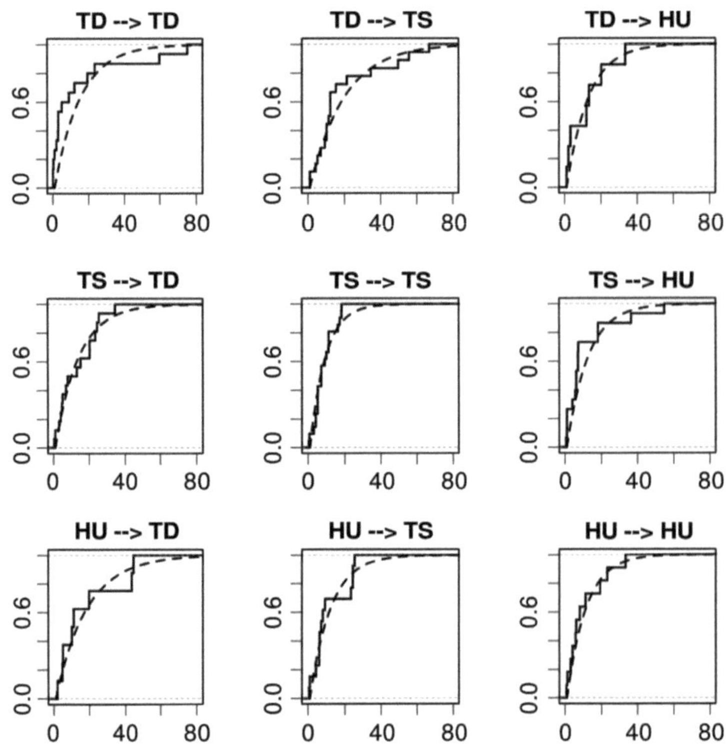

Fig. 7.2 Comparison between empirical (step-function) and estimated (*solid line*) distributions of the sojourn time of cyclone data

Now assuming that sojourn times for different transition follow exponential distribution the sojourn time parameters are estimated by the maximum likelihood method. Table 7.3 shows the estimated mean sojourn times for different transitions which are obtained by maximizing the conditional log-likelihood function in Eq. (7.1) given the estimates of the transition probability matrix in Table 7.2.

Figure 7.2 compares the empirical distribution function (step-function) with the fit (dotted line) under exponential distributions for the sojourn time for different types of transitions. As the fitted line is very close to the empirical line, it can be concluded that the exponential distribution fits the data reasonably well.

Table 7.4 Probability of occurrence of next event of given a state, conditioned to a given state of last event occurred, evaluated for different t^* and different t_0

		TD-TD	TD-TS	TD-HU	TS-TD	TS-TS	TS-HU	HU-TD	HU-TS	HU-HU
0 year	1 year	0.2117	**0.2122**	0.1098	0.1844	**0.3071**	0.1854	0.1232	**0.2663**	0.2328
	2 year	0.3039	0.3244	0.1507	0.2583	0.3807	0.2516	0.1857	0.3580	0.3080
	3 year	0.3441	0.3836	0.1659	0.2879	0.3983	0.2753	0.2174	0.3896	0.3322
	4 year	0.3615	0.4149	0.1716	0.2998	0.4025	0.2838	0.2334	0.4005	0.3400
	5 year	0.3691	0.4315	0.1737	0.3045	0.4035	0.2868	0.2416	0.4043	0.3425
1 year	1 year	0.1977	**0.2405**	0.0878	0.2287	0.2277	0.2050	0.1654	0.2429	0.1989
	2 year	0.2838	**0.3676**	0.1205	0.3203	0.2823	0.2782	0.2494	0.3266	0.2631
	3 year	0.3213	**0.4347**	0.1327	0.3570	0.2954	0.3044	0.2919	0.3554	0.2838
	4 year	0.3376	**0.4702**	0.1372	0.3717	0.2985	0.3138	0.3135	0.3653	0.2905
	5 year	0.3447	**0.4889**	0.1389	0.3776	0.2993	0.3171	0.3245	0.3687	0.2927
2 year	1 year	0.1816	0.2681	0.0690	**0.2706**	0.1612	0.2163	0.2137	0.2130	0.1635
	2 year	0.2606	**0.4097**	0.0947	0.3790	0.1998	0.2936	0.3221	0.2864	0.2162
	3 year	0.2950	**0.4846**	0.1043	0.4224	0.2091	0.3212	0.3771	0.3117	0.2332
	4 year	0.3100	**0.5241**	0.1079	0.4398	0.2113	0.3311	0.4050	0.3204	0.2387
	5 year	0.3165	**0.5450**	0.1092	0.4468	0.2118	0.3346	0.4191	0.3234	0.2405

Table 7.4 shows the estimated probability that the next cyclone will be the type i, given the last cyclone was of type j for $i, j = \{\text{TD, TS, HU}\}$. The predicted probability of experiencing a transition of the type TD to TS, TS to TS, and HU to TS within 1 year is 0.2122, 0.3071, and 0.2663, respectively. That is, immediately after the last event the chance of occurrence of a TS is the highest within the next 12 month period irrespective of the type of the last event. The pattern of forecast probabilities remains the same for all t^* years. So, if no time is elapsed the probability of occurrence of a TS is the highest. With $t_0 = 1$ year the likelihood of occurrence of a TS is 0.2405, 0.3676, 0.4347, 0.4702, 0.4889, respectively, given that the last state was a TD. When $t_0 = 2$ and $t^* = 1$, the probability of occurrence of a TD is 0.2706 which is higher than TS and HU if the last state was a TS. As the chance of occurrence of a hurricane is one of our main interests, we reveal some important patterns from Table 7.4. Firstly the likelihood of occurring a hurricane (HU) is higher than a TS and TD immediately after a hurricane level cyclone, i.e., with $t_0 = 0$. Secondly the chance of an HU is higher if the last cyclone is a tropical storm for $t_0 = 1$ and 2.

In Bangladesh the last cyclone occurred in the year 2009, which was a hurricane level cyclone (HU), it may be interesting to see the prediction probability for the next event. Table 7.5 gives the predicted probability that the next cyclone will be a TD, TS, or HU after t^* years given that about a period of 3 years and a quarter has elapsed ($t_0 = 3.25$). Within next 1 year period, the probability of occurrence of a TD is the highest 0.2771 given that the last event occurred was a hurricane (HU). The corresponding probabilities of occurring a TS or an HU are 0.1703 and 0.1204. The chances of occurrences of TD, TS, and HU increased consistently as t^* increases. Hence, based on the evidence we have the chance of the next cyclone will be a TD is the highest (Fig. 7.3).

Table 7.5 Probability of occurrence of next event of given a state, conditioned to a given state of last event occurred (HU), evaluated for different t^* with $t = 3.25$ years

	Predicted probability		
t^*	HU-TD	HU-TS	HU-HU
1 year	0.2771	0.1703	0.1204
2 year	0.4177	0.2290	0.1593
3 year	0.4890	0.2492	0.1718
4 year	0.5251	0.2562	0.1759
5 year	0.5435	0.2586	0.1772

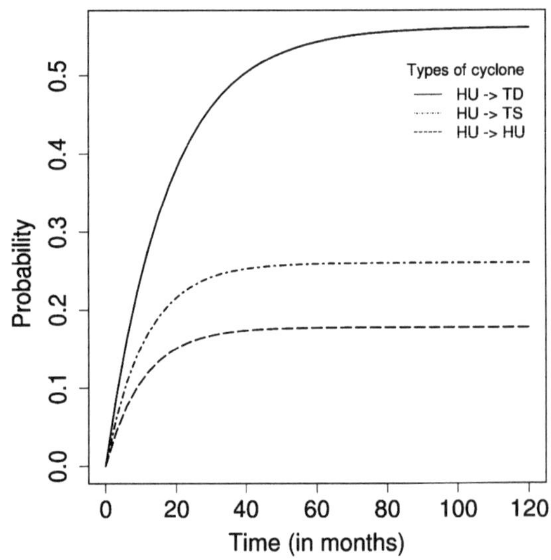

Fig. 7.3 Probability of occurring a TD, TS, and HU given that last event is HU against time in months

Table 7.6 Recurrence periods for each type of cyclone

Types of cyclone	Average recurrence period (in years)
TD	3.44
TS	2.57
HU	4.02

In storm risk analysis, the mean recurrence time plays an important role in terms of the probabilistic hazard assessment. If we assume that the MRP is ergodic, then it is possible to calculate the mean recurrence times of each state. We calculate the mean recurrence times for each type of cyclones (TD/TS/HU) using the fitted Markov renewal model, and the results are presented in Table 7.6. The calculated the mean recurrence times are 3.44, 2.57, and 4.02 years for a tropical depression, a tropical storm, and a hurricane, respectively. These results imply that in the long run tropical storms occur more frequently than tropical depressions and hurricanes with an average period of 2.57 years.

7.5 Conclusion

In this study a Markov renewal model is considered for predicting tropical cyclone occurrence for Bangladesh. A Markov chain for state transition of cyclones and exponential inter-event times are assumed, and parameters of the model are estimated using maximum likelihood method of estimation. A detailed investigation of the occurrence of the three types of cyclones has been performed, and the predicted probabilities based on the model have been presented for varying lengths of time. The model shows that the next event will be a tropical depression with a higher probability than a tropical storm or a hurricane. We also calculated the mean recurrence times for each type of cyclone, and the result showed that in the long run tropical storms will occur more frequently. These conclusions of the study could be helpful in increasing awareness about the tropical cyclones in Bangladesh.

References

Alam M, Hossain MA, Shafee S (2003) Frequency of Bay of Bengal cyclonic storms and depressions crossing different coastal zones. Int J Climatol 23:1119–1125

Ali A (1996) Vulnerability of Bangladesh to climate change and sea level rise through tropical cyclones and storm surges. Water Air Soil Pollut 92(1):171–179

Alvarez E (2005) Estimation in stationary Markov renewal processes, with application to earthquake forecasting in Turkey. Methodol Comput Appl Probab 7(1):119–130

As-Salek JA, Yasuda T (1995) Comparative study of the storm surge models proposed for Bangladesh: last developments and research needs. J Wind Eng Ind Aerodyn 54–55:595–610

Dube S, Sinha P, Roy G (1985) The numerical simulation of storm surges along the Bangladesh coast. Dyn Atmos Oceans 9(2):121–133

Dube S, Sinha P, Roy G (1986) Numerical simulation of storm surges in Bangladesh using a bay-river coupled model. Coast Eng 10(1):85–101

Fiorentino M, Versace P, Rossi F (1984) Two component extreme value distribution for flood frequency analysis. Water Resour Res 20(7):847–856

Garavaglia E, Pavani R (2011) About earthquake forecasting by Markov renewal processes. Methodol Comput Appl Probab 13(1):155–169

Gupta V, Duckstein L (1975) A stochastic analysis of extreme droughts. Water Resour Res 11(2): 221–228

Haque C (1995) Climatic hazards warning process in Bangladesh: experience of, and lessons from, the 1991 April cyclone. Environ Manag 19(5):719–734

Haque C, Blair D (1992) Vulnerability to tropical cyclones: evidence from the April 1991 cyclone in coastal Bangladesh. Disasters 16(3):217–229

Islam T, Peterson R (2009) Climatology of landfalling tropical cyclones in Bangladesh 1877–2003. Nat Hazards 48(1):115–135

Jagger T, Niu X, Elsner J (2002) A space-time model for seasonal hurricane prediction. Int J Climatol 22(4):451–465

Janssen J, Manca R (2007) Semi-Markov risk models for finance, insurance and reliability. Springer, New York

Khalil G (1992) Cyclones and storm surges in Bangladesh: some mitigative measures. Nat Hazards 6(1):11–24

Lardet P, Obled C (1994) Real-time flood forecasting using a stochastic rainfall generator. J Hydrol 162(3–4):391–408
Lu Y, Garrido J (2005) Doubly periodic non-homogeneous Poisson models for hurricane data. Stat Methodol 2(1):17–35
Mooley DA (1980) Severe cyclonic storms in the Bay of Bengal, 1877–1977. Mon Weather Rev 108:1647–1655
Mooley D (1981) Applicability of the Poisson probability model to the severe cyclonic storms striking the coast around the Bay of Bengal. Sankhya Indian J Stat Ser B 43(2):187–197
Murty TS, Flather RA, Henry RF (1986) The storm surge problem in the Bay of Bengal. Prog Oceanogr 16(4):195–233
Ogata Y (1988) Statistical models for earthquake occurrences and residual analysis for point processes. J Am Stat Assoc 83(401):9–27
Ogata Y (1998) Space-time point-process models for earthquake occurrences. Ann Inst Stat Math 50(2):379–402
Raghavendra VK (1973) A statistical analysis of the number of tropical storms and depressions in the Bay of Bengal during 1890–1969. Indian J Meteorol Geophys 24:125–130
Rai Sircar NC (1956) A climatological study of storms and depressions in the Bay of Bengal. Indian J Meteorol Geophys 7:157–160
Rumpf J, Weindl H, Hoppe P, Rauch E, Schmidt V (2007) Stochastic modelling of tropical cyclone tracks. Math Method Oper Res 66(3):475–490
Sinha P, Dube S, Roy G, Jaggi S (1986) Numerical simulation of storm surges in Bangladesh using a multi-level model. Int J Numer Methods Fluids 6(5):305–311

Chapter 8
Performance of Wavelet Transform on Models in Forecasting Climatic Variables

Md. Jahanur Rahman and Md. Al Mehedi Hasan

Abstract An attempt has been made to show whether the recently developed wavelet transformation in forecasting the climatic time series in Bangladesh improves the performance of existing forecasting models, such as ARIMA. These models are applied to forecast the humidity of Rajshahi, Bangladesh. Then the wavelet transformation has been used to decompose the humidity series into a set of better-behaved constitutive series. These decomposed series and inverse wavelet transformation are used as a pre-processing procedure of forecasting humidity series using the same models in two approaches. Finally, the forecasting ability of these two models with and without wavelet transformation is compared using the statistical forecasting accuracy criteria. The results show that the use of wavelet transformation as a pre-processing procedure of forecasting climatic time series improves the performance of forecasting models. The reason is the better behavior of the constitutive series for the filtering effect of the wavelet transform.

Keywords Wavelet transformation • ARIMA models • Forecasting

8.1 Introduction

Time series forecasting is very popular and plays an important role in various fields such as economics, engineering, environment, and bioinformatics. The basic idea behind time series forecasting involves the development of models that estimate the future values of a series based on its past values. There are many forecasting models

---

M.J. Rahman (✉)
Department of Statistics, University of Rajshahi, Rajshahi 6205, Bangladesh
e-mail: jahan.stat@ru.ac.bd

M.A.M. Hasan
Department of Computer Science and Engineering, Rajshahi University of Engineering and Technology, Rajshahi, Bangladesh

T. Islam et al. (eds.), *Computational Intelligence Techniques in Earth and Environmental Sciences*, DOI 10.1007/978-94-017-8642-3_8,

that have been used in the forecasting literature. The models mainly follow two approaches: nonlinear and linear models. Nonlinear models like artificial intelligence (AI)-based methods employing neural networks (NNs) have been proposed by different researchers. Second type of models are linear models like univariate autoregressive (AR), autoregressive moving average (ARMA), autoregressive integrated moving average (ARIMA), multivariate time series models like transfer function and dynamic regression, and generalized autoregressive conditional heteroskedastic (GARCH) model. In order to provide estimates for the future, these models analyze the historical data. Usually time series are not deterministic series. In fact, in many cases the researchers considered the series to be stationary time series. One way to model any time series is to consider it as a deterministic function plus white noise. The white noise in any time series process can be minimized by some procedures which are called the de-noising. Then a better model can be obtained. Consequently, to obtain a good de-noising, there are some mathematical models that can be applied such as Fourier transformation (FT) and wavelet transformation (WT) (Yao et al. 2000; Strang 1993). WT seems to be ideal for time series forecasting since time information is preserved in the transformed variables. Moreover, WT is a very effective technique for local representation of the time series in both time and frequency domains (Yevgeniy et al. 2005). WT is used to split up the time series into one low-frequency subseries (approximation part) and some high-frequency subseries (detailed part) in the wavelet domain. In models mentioned above, after appropriate decomposition, the prediction was made in wavelet domain and then inverse WT was applied to obtain the actual value of the predicted variable.

Wavelet transform has been used in many fields in forecasting models. Among them Wadi et al. (2011) and Arino and Vidakovic (1995) perform wavelet transform in forecasting financial time series based on ARIMA model and neural network-based model, respectively. Rocha et al. (2010) and Henriques and Rocha (2009) have used wavelet transform in NN model to predict acute hypotensive episodes. Gang et al. (2008), Aggarwal et al. (2008), and Antonio et al. (2005) have decomposed electricity price series using wavelet transformation for more efficient forecasting based on ARIMA, artificial neural network, and regression-based techniques. In most cases they decomposed the historical time series data into wavelet domain constitutive subseries using wavelet transform, and then combined with the other time domain variables to perform the set of input variables for the proposed forecasting model (Conejo et al. 2005). Based on statistical analysis the behavior of the wavelet domain constitutive series has been studied. It has been observed that forecasting accuracy can be improved by the use of wavelet transforms in forecasting models. Alrumaih and Al-Fawzan (2002) used Saudi stock index to illustrate that wavelet transformation is better than the other forecasting technique in predicting the de-noising of the financial time series.

Thus, the recently developed wavelet theory has proven to be a useful tool in the time series forecasting methods in different fields. However, the potential of this theory for analyzing and forecasting climatic time series has not been fully exploited yet. The accurate forecasting of climatic variables in Bangladesh is an important issue in disaster management policy-making due to the effects of recently happened climate change. Our objective in this chapter is to check whether the use of the wavelet transformation as a preprocessor in forecasting climatic data improves the predicting behavior of any forecasting model. As forecasting models, we have used the widely used and more popular ARIMA models. Humidity of Rajshahi, Bangladesh, is used as a climatic time series in this chapter. This is the way the comparison is performed with and without the wavelet transform, not across techniques. The fundamental and novel contribution of this chapter is to use the wavelet transformation to decompose the humidity series into a set of better-behaved constitutive series. These decomposed series and inverse wavelet transformation are used as a pre-processing procedure of forecasting humidity series using the same models in two approaches. Finally, the forecasting results based on wavelet transform and ARIMA model (hereafter called Wavelet-ARIMA model) will be compared with the forecasting values based on ARIMA model using some statistical criteria.

This chapter is organized as follows. Section 8.2 gives the brief description of the wavelet transformation. Section 8.3 provides the details of data processing and forecasting framework using wavelet transformation. Forecasting accuracy criteria, which are used to compare the performances of forecasting ability of the models, are defined in Sect. 8.4. Empirical results of a case study based on performance of wavelet transformation in forecasting humidity of Rajshahi are shown in Sect. 8.5. Finally, Sect. 8.6 provides some relevant conclusions.

8.2 Wavelet Transformation

In this section we briefly review the discrete wavelet transform (DWT), which is the wavelet counterpart to the discrete Fourier transform. Then we show the splitting of a time series into cyclical components by using wavelet analysis. As in Fourier analysis, there are continuous and discrete versions of wavelet analysis (Nason and Silverman 1994). Since we will be dealing with discrete data sets, our focus will be on the DWT. Good references on wavelet transformation are in Mallat (1989) and Percival and Walden (2000).

The time series under study is independently decomposed by DWT, which is defined as

$$\Psi_{j,k} = \frac{1}{\sqrt{s_0^j}}\psi\left(\frac{t - k\tau_0 s_0^j}{s_0^j}\right) \tag{8.1}$$

where the parameters j and k are integers that control, respectively, the wavelet dilatation (scale) and translation (time). The value $s_0 > 1$ is a fixed dilation step and the translation factor τ_0 depends on the dilation step. The most common and simplest choice for the parameters s_0 and τ_0 is 2 and 1 (time steps), respectively,

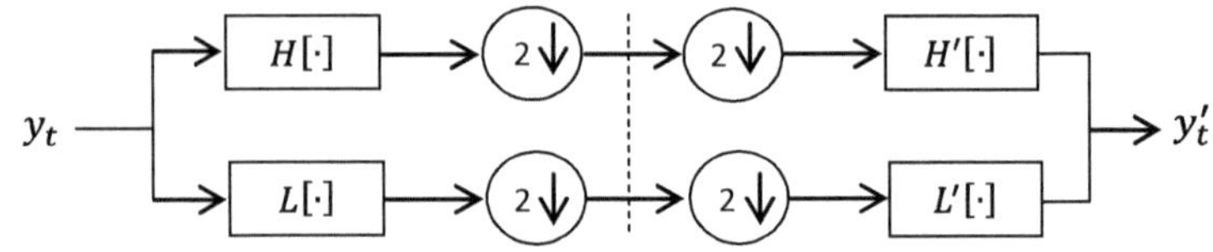

Fig. 8.1 Filter bank for discrete wavelet transformation

known as dyadic grid arrangement. In this case, the coefficients of DWT decomposition are given by

$$W_{j,k} = \frac{1}{2^{j/2}} \sum_{t=0}^{N-1} y_t \psi\left(\frac{t-k}{2^j}\right) \tag{8.2}$$

where the $W_{j,k}$ are the wavelet coefficients corresponding to the scale $S = 2^j$ and the location $\tau = 2^j k$. This dyadic arrangement can be implemented by using a filter bank scheme developed by Mallat (1989), as depicted in Fig. 8.1.

In Fig. 8.1, $H[\cdot]$, $L[\cdot]$ and H' $[\cdot]$, L' $[\cdot]$ are the high-pass and low-pass filters for wavelet decomposition and reconstruction, respectively. In the decomposition phase, the low-pass filter removes the higher frequency components of the series and high-pass filter picks up the remaining parts. Then, the filtered series are down-sampled by two and the results are called approximation and detail coefficients. The major advantage of decimation is that just enough information is kept to allow exact reconstruction of the input data. The reconstruction is just the inverse process of the decomposition and, for perfect reconstruction by filter bank, we should have $y_t = y'_t$. Using this approach, signal can be decomposed by cascade algorithm as shown in the following:

$$\begin{aligned} y_t &= a1_t + d1_t \\ &= a2_t + d2_t + d1_t \\ &= a3_t + d3_t + d2_t + d1_t \\ &\;\;\vdots \\ &= an_t + dn_t + d(n-1)_t + \cdots + d1_t \end{aligned} \tag{8.3}$$

where dn_t and an_t are the detail and approximation coefficients at level n, respectively. These coefficients allow for the identification of changes in the trends at different scales.

A wavelet function of type Daubechies of order 5 and decomposition level 3 is used in this case study (Daubechies 1992). This wavelet offers an appropriate tradeoff between wavelength and smoothness. This results in an appropriate behavior for climate series prediction. In the next section we show how to use the decomposed series in forecasting model.

8.3 Data Processing and Forecasting Framework

In order to illustrate the effectiveness of wavelet transform in forecasting models, the climatic data on humidity of Rajshahi, Bangladesh, is selected. The data is collected from the website of Bangladesh Agricultural Research Council (BARC), Ministry of Agriculture. We consider the monthly humidity series for the time period from January 1964 to December 2008 (1964:1 to 2008:12). The data set is divided into two sub-data sets: (1) a training set to estimate the model parameters and (2) a test set to evaluate these models by calculating error functions. There are 540 observations in the humidity series. The first 528 observations from 1964:1 to 2007:12 are used to build the model, and the last 12 observations from 2008:1 to 2008:12 to check the forecast ability of the models.

We first need to decompose the series under study using wavelet transformation. For this purpose, we have applied the DWT to the humidity series. Many wavelet families exist, where Daubechies family of wavelets, which are compactly supported orthonormal wavelets, is the most popular one and has been used in this work. Thus, a wavelet function of type Daubechies of order 5 and decomposition level 3 is used in this case study. The wavelet transform applied to climatic series y_t, $t = 1, 2, \ldots, T$ results in four series denoted by $d1_t$, $d2_t$, $d3_t$, and $a3_t$ and can be defined by

$$y_t = a3_t + d3_t + d2_t + d1_t \tag{8.4}$$

Series $d1_t$, $d2_t$, and $d3_t$ are denominated detail series, while $a3_t$ is denominated approximation series. This approximation series constitutes the main component of the transformation, while the three detail series provide "small" adjustments. A graph of the original series and its decomposed series is shown in Fig. 8.2a, b. These series present a better behavior (more stable variance and no outliers) than the original humidity series and, therefore, they can be predicted more accurately. The reason for the better behavior of the constitutive series is the filtering effect of the wavelet transform.

We have used this decomposed series in the forecasting models in two approaches; (1) first approach includes $d2_t$, $d3_t$, and $a3_t$ series in the analysis, whereas, (2) in the second approach, all decomposed series are used in forecasting models. Two approaches are outlined as follows.

8.3.1 Approach-1

The steps of modeling the decomposed series by Wavelet-ARIMA and Wavelet-NN techniques are given below.

Step-11: The wavelet transformation of type Daubechies-5 and decomposition level 3 is applied to the humidity series Y_t ($t = 1, 2, \ldots, T$) for the training period 1964:1

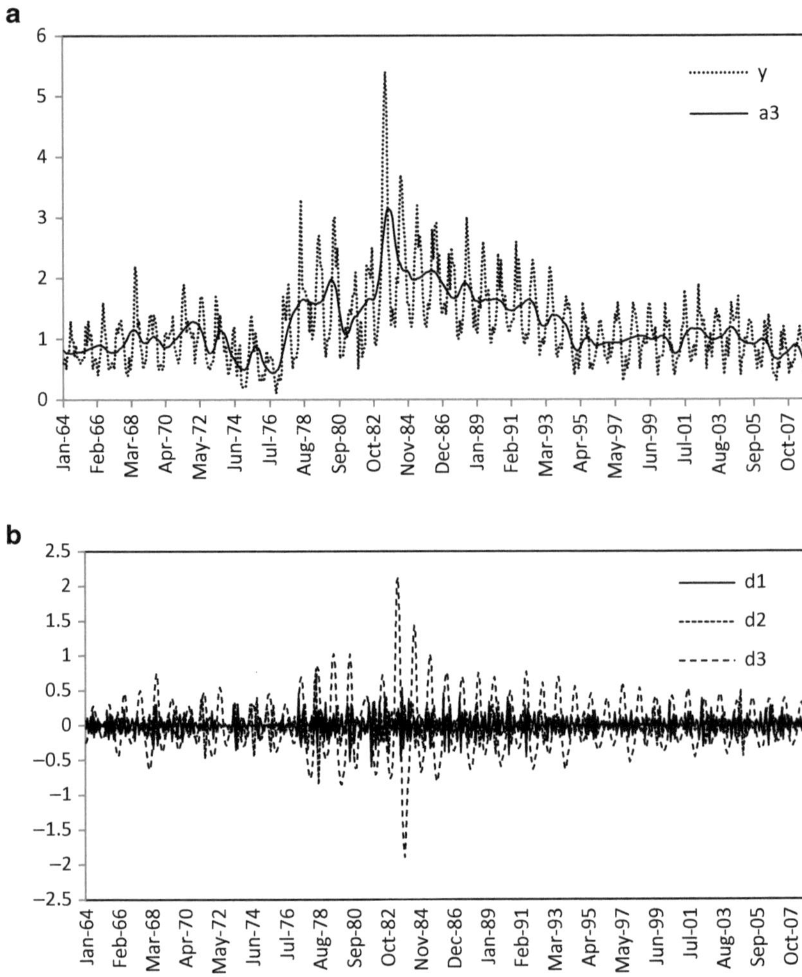

Fig. 8.2 (**a**) Original series and approximation series over 1964:1 to 2008:12. (**b**) Detail series over 1964:1 to 2008:12

to 2007:12 which results in four series denoted by $a3_t$, $d3_t$, $d2_t$, and $d1_t$; $t = 1, 2, \ldots, T$. That is,

$$\mathrm{WT}(y_t; t = 1, 2, \ldots, T) = \{a3_t, d3_t, d2_t, d1_t; t = 1, 2, \ldots, T\} \tag{8.5}$$

Step-12: The decomposed series $d1_t$ contains the highest frequency components among the others and hence is outlier prone. Therefore, series corresponding to $d1_t$ has been discarded and only series $a3_t$, $d3_t$, and $d2_t$ have been used to reconstruct the original series using inverse wavelet transformation as follows:

$$\mathrm{WT}^{-1}\{a3_t, d3_t, d2_t; t = 1, 2, \ldots, T\} = y_t^*; \quad t = 1, 2, \ldots, T \tag{8.6}$$

Step-13: Use the appropriate ARIMA and NN model to the reconstructed series to forecast the future values in the test period 2008:1 to 2008:12;

$$\left\{y_t^*; t = 1, 2, \ldots, T\right\} \xrightarrow[\text{forecast}]{\text{ARIMA/NN}} \left\{y_t^f; t = T + 1, \ldots, T + n\right\} \quad (8.7)$$

We call these forecasting values obtained from Wavelet-ARIMA/Wavelet-NN model using approach-1.

8.3.2 Approach-2

The steps of modeling the decomposed series by Wavelet-ARIMA and Wavelet-NN techniques are given below:

Step-21: The wavelet transformation of type Daubechies-5 and decomposition level 3 is applied to the humidity series Y_t ($t = 1, 2, \ldots, T$) for the full period 1964:1 to 2008:12 which results in four series denoted by $a3_t$, $d3_t$, $d2_t$, and $d1_t$; $t = 1, 2, \ldots, T$.

$$\mathrm{WT}(y_t; t = 1, 2, \ldots, T) = \{a3_t, d3_t, d2_t, d1_t; t = 1, 2, \ldots, T\} \quad (8.8)$$

Step-22: Then, specific ARIMA and NN methods are used to each one of the constitutive series for the training period 1964:1 to 2007:12. The best fitted model is then used to forecast its n future values in the test period which are denoted by $\widehat{d1}_t, \widehat{d2}_t, \widehat{d3}_t$, and $\widehat{a3}_t$; $t = T + 1, T + 2, \ldots, T + n$, respectively. That is,

$$\begin{aligned} &\{a3_t, d3_t, d2_t, d1_t; t = 1, 2, \ldots, T\} \xrightarrow[\text{forecast}]{\text{ARIMA/NN}} \\ &\left\{a3_t^f, d3_t^f, d2_t^f, d1_t^f; t = T + 1, \ldots, T + n\right\} \end{aligned} \quad (8.9)$$

Step-23: Finally, we use the inverse wavelet transform to estimate the forecasting values of the original series using the forecasting values of the constitutive series. The inverse wavelet transform is used in turn to reconstruct the forecasting series for original series, i.e.,

$$\left\{\widehat{d1}_t, \widehat{d2}_t, \widehat{d3}_t \text{ and } \widehat{a3}_t\right\}_{t=T+1}^{T+n} \xrightarrow[\text{Transformation}]{\text{Inverse Wavelet}} \{\hat{y}_t\}_{t=T+1}^{T+n} \quad (8.10)$$

We call these forecasting values obtained from Wavelet-ARIMA/Wavelet-NN model using approach-2. The forecasting performance of Wavelet-ARIMA model is compared with the ARIMA model to forecast the original climatic series using the forecasting accuracy criteria discussed in the next section.

8.4 Comparison of Forecasting Performance

To assess and compare the forecasting performance of the models, three types of forecasting accuracy criteria of the test sets data have been adopted. They are the mean absolute error (MAE), root mean square error (RMSE), and mean absolute percentage error (MAPE), which are defined by

$$\mathrm{MAE} = \frac{1}{n} \sum_{t=T+1}^{T+n} \left| y_{\mathrm{real},t} - y_{\mathrm{forecast},t} \right| \tag{8.11}$$

$$\mathrm{RMSE} = \sqrt{\frac{\sum_{t=T+1}^{T+n} \left(y_{\mathrm{real},t} - y_{\mathrm{forecast},t} \right)^2}{\sum_{t=T+1}^{T+n} \left(y_{\mathrm{real},t} - \bar{y} \right)^2}} \tag{8.12}$$

$$\mathrm{MAPE} = \frac{1}{n} \sum_{t=T+1}^{T+n} \left| \frac{y_{\mathrm{real},t} - y_{\mathrm{forecast},t}}{y_{\mathrm{real},t}} \right| \times 100 \tag{8.13}$$

where $y_{\mathrm{real},t}$ and $y_{\mathrm{forecast},t}$ are the real and forecast data point at time t, respectively, $\bar{y}$ is the mean of $y_{\mathrm{real},t}$, T is the number of observation in the trail series, and n is the number of data points forecasted in the test series. Lower values of the criteria imply the better forecast of the model.

8.5 Empirical Results

The main objective of this chapter is to show the forecasting performance of the time series models using the original data and the decomposed data. The original series is decomposed using wavelet transformation. Time series ARIMA models are used as forecasting models. First, we use these models to the original series over the training period to select an appropriate model. Then the selected models is used to forecast the data points of the test period. Secondly, we use these models to the decomposed series by wavelet transformation and forecast the test data as mentioned in the previous two approaches. Finally, we compare the performance of these forecasting series using the forecasting accuracy criteria discussed in Sect. 8.4.

8.5.1 Forecasting Based on Original Series

Here the original series over the training period is used to select the appropriate ARIMA model. Then, the model is used to forecast the series over the test period. Good references on ARIMA models and standard forecasting techniques are in Box and Jenkins (1976), Pankratz (1991), and Granger and Newbold (1986).

Table 8.1 Original and forecasting values for the test period 2008:1 to 2008:12

	Original series	ARIMA	W-ARIMA (approach-1)	W-ARIMA (approach-2)
Jan-08	0.6	0.5362	0.5679	0.5341
Feb-08	0.6	0.5817	0.6165	0.7561
Mar-08	0.9	0.8037	0.7620	0.8652
Apr-08	0.9	1.3359	1.0557	1.0272
May-08	1.0	1.4512	1.2395	1.1695
Jun-08	1.2	1.4006	1.1891	1.1528
Jul-08	1.0	1.2856	1.1158	1.0662
Aug-08	1.0	1.0912	1.0668	0.8480
Sep-08	0.8	0.9218	0.9019	0.5884
Oct-08	0.6	0.4575	0.6098	0.4135
Nov-08	0.4	0.4183	0.4373	0.2792
Dec-08	0.5	0.5346	0.5341	0.2336

We have used the famous Box and Jenkins (1976) modeling philosophy for choosing an appropriate ARIMA model for the monthly humidity series over the period 1964:1 to 2007:12. The ARIMA(0,1,1)(0,1,1) model shows the more robust coefficients, white-noise error, and the smallest forecasting errors among the competitive models. The out of sample forecasting errors are calculated using the series over the period 2008:1 to 2008:12. The model is

$$\nabla_{12}\nabla_{1}y_t = \underset{(-6.73)}{(1-0.283L)}\ \underset{(-80.35)}{\left(1-0.953L^{12}\right)}\varepsilon_t \tag{8.14}$$

where the operators L^k and ∇_k are defined by $L^k y_t = y_{t-k}$ and $\nabla_k = 1 - L^k$. The parentheses under the model contain the value of t-statistic of each coefficient. Monthly forecasts according to model (8.14) together with their actual values are presented in Table 8.1. The values of out of the sample or test period forecasting accuracy criteria MAE, RMSE, and MAPE are 0.1633, 0.2187, and 18.566, respectively. Figure 8.3 shows a graph of the humidity for the period 2006:1 to 2008:12 and the forecast values of ARIMA model from 2008:1 to 2008:12 along with the forecasting values using wavelet transformation.

We compare these forecasts with the forecasts made after decomposing the data set with the wavelet methodology.

8.5.2 *Forecasting Based on Decomposed Series Using Wavelet Transformation*

For using the Wavelet-ARIMA model to forecast, we first need to decompose the series under study using wavelet transformation. For that purpose, we have applied the DWT to the humidity series. A wavelet function of type Daubechies of order

5 and decomposition level 3 is used in this case study. The wavelet transform applied to climatic series y_t, $t = 1, 2, \ldots, T$ results in four series denoted by $d1_t$, $d2_t$, $d3_t$, and $a3_t$. Series $d1_t$, $d2_t$, and $d3_t$ are denominated detail series, while $a3_t$ is denominated approximation series.

8.5.2.1 Approach-1

In approach-1, as described in Sect. 8.3.1, we have used DWT to decompose the original series into four constitutive series as mentioned above. The decomposed detail series $d1_t$ contains the highest frequency components among the others and hence is outlier prone. Therefore, series corresponding to $d1_t$ has been discarded and only series $a3_t$, $d3_t$, and $d2_t$ have been used to reconstruct the original series using inverse wavelet transformation as follows:

$$y_t^* = \mathrm{WT}^{-1}\{a3_t, d3_t, d2_t; t = 1, 2, \ldots, T\}; \quad t = 1, 2, \ldots, T$$

Then, we have chosen an appropriate ARIMA model for the series y_t^* following the Box–Jenkins modeling philosophy which has the lowest forecasting error according to the three forecasting accuracy criteria mentioned in Sect. 8.4. The selected model is ARIMA(2,1,1)(0,1,1) which is defined as

$$\left(1 - \underset{(11.27)}{0.4256}L + \underset{(-14.64)}{0.5520}L^2\right) \nabla_{12}\nabla_1 y_t^* = \left(1 + \underset{(50.85)}{0.9457}L\right) \left(1 - \underset{(-19.18)}{0.6929}L^{12}\right) \varepsilon_t \quad (8.15)$$

The parentheses under the model contain the value of t-statistic of each coefficient, which shows the estimates of the parameters are highly significant. The forecasting values over the test period 2008:1 to 2008:12 are shown in Table 8.1. The values of test period forecasting accuracy criteria for model (8.15) MAE, RMSE, and MAPE are 0.0655, 0.0799, and 9.1473, respectively.

8.5.2.2 Approach-2

Here the DWT is performed to the original series over the full period 1964:1 to 2008:12. Then, a specific ARIMA model is fitted for each constitutive series over the training period 1964:1 to 2007:12. The ARIMA model for each series is chosen based on the smallest forecasting error over the test period 2008:1 to 2008:12 with significant coefficients and white-noise error as outlined in Box–Jenkins method. In fact, the best ARIMA models for the series $d1_t$, $d2_t$, $d3_t$, and $a3_t$ are ARIMA(2,0,2)(0,0,0), ARIMA(0,0,4)(0,1,0), ARIMA(0,1,2)(0,1,1), and ARIMA(8,2,8)(0,0,1), respectively. The estimates of the ARIMA model for $d1_t$ are

$$\underset{(-12.29)}{\left(1+0.866L\right.}+\underset{(-14.08)}{\left.0.566L^2\right)}d1_t=\underset{(-8.56)}{\left(1-0.671L\right.}-\underset{(-4.12)}{\left.0.326L^2\right)}\varepsilon_t \tag{8.16}$$

The estimates of the ARIMA model for $d2_t$ are

$$\nabla_{12}d2_t=\left(1+\underset{(80.19)}{1.701L}-\underset{(-6.34)}{0.130L^2}-\underset{(-82.87)}{1.736L^3}-\underset{(-41.34)}{0.885L^4}\right)\varepsilon_t \tag{8.17}$$

The estimates of the ARIMA model for $d3_t$ are

$$\nabla_{12}\nabla_1 d3_t=-\underset{(-1.82)}{0.004}+\left(1+\underset{(134.20)}{1.918L}+\underset{(66.25)}{0.937L^2}\right)\left(1-\underset{(-26.89)}{0.753L^{12}}\right)\varepsilon_t \tag{8.18}$$

The estimates of the ARIMA model for $a3_t$ are

$$\left(1+\underset{(23.79)}{0.634L}-\underset{(-13.82)}{0.357L^8}\right)\nabla_1^2 a3_t=\left(1+\underset{(5.98)}{0.095L}-\underset{(-56.48)}{0.930L^8}\right)\left(1-\underset{-14.48}{0.566L^2}\right)\varepsilon_t \tag{8.19}$$

Using the above best models, the forecasting values over the test period 2008:01 to 2008:12 for each transformed series are evaluated. The forecasting series are denoted by $\widehat{d1}_t, \widehat{d2}_t, \widehat{d3}_t$, and $\widehat{a3}_t$ which are shown in Table 8.2 along with their respective forecasting errors.

Finally, the inverse DWT is applied to the series $\widehat{d1}_t, \widehat{d2}_t, \widehat{d3}_t$, and $\widehat{a3}_t$ to get the forecasting values of the humidity series over the test period. The forecasting values are denoted by $\hat{y}_t$, $t = 2008:1, 2008:2, \ldots, 2008:12$. The forecasting series $\hat{y}_t$ using Wavelet-ARIMA model is shown in Table 8.1. The values of test period forecasting accuracy criteria for this Wavelet-ARIMA model MAE, RMSE, and MAPE are 0.1336, 0.1591, and 15.356, respectively.

8.5.3 *Comparison*

Now we have got the results to compare the performances of the forecasting models with and without wavelet transformation. The forecasting values for the test period of the models with the original series are shown in Table 8.1. For convenience, these forecasting values are depicted in Fig. 8.3 with the original humidity series. The forecasting ability of these models is compared by using three forecasting accuracy criteria—MAE, RMSE, and MAPE. Table 8.3 contains the values of these criteria for the models.

Table 8.3 shows that all three measurements of forecasting accuracy criteria are sufficiently smaller when wavelet transformation is used in the models than that of without wavelet transformation. However, the Wavelet-ARIMA model with

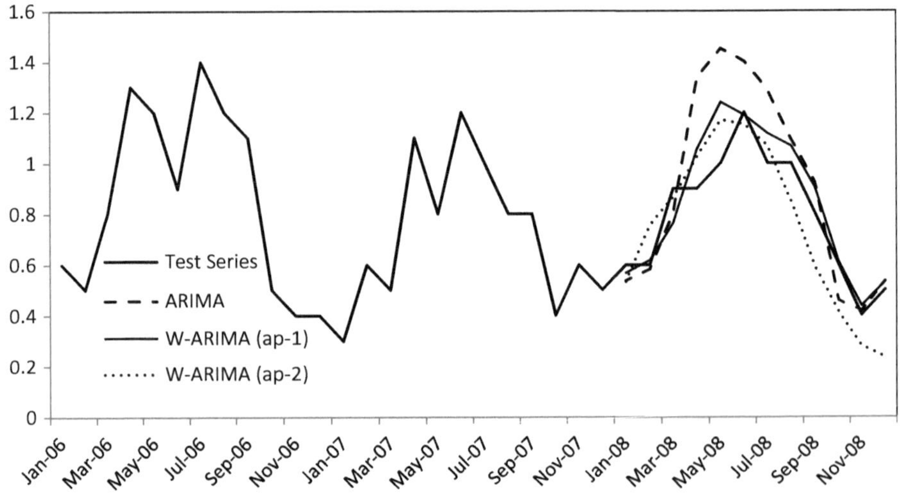

Fig. 8.3 Original series along with the forecasting series of ARIMA and Wavelet-ARIMA (approach-1 and approach-2) models (for visual convenience the figure shows data from 2006:1)

Table 8.2 Forecasting values for the test period (2008:1 to 2008:12) of detail and approximation series

Time	*d*1	*d*2	*d*3	*a*3
Jan-08	−0.0429	−0.0441	−0.2515	0.8726
Feb-08	0.0469	0.0393	−0.2188	0.8886
Mar-08	−0.0163	0.0940	−0.1117	0.8993
Apr-08	−0.0123	0.0279	0.1077	0.9038
May-08	0.0199	−0.0318	0.2825	0.8989
Jun-08	−0.0103	−0.0435	0.3244	0.8822
Jul-08	−0.0023	−0.0390	0.2530	0.8546
Aug-08	0.0078	0.0196	0.0060	0.8144
Sep-08	−0.0054	0.0585	−0.2423	0.7777
Oct-08	0.0002	0.0171	−0.3586	0.7548
Nov-08	0.0028	−0.0222	−0.4438	0.7424
Dec-08	−0.0026	−0.0449	−0.4646	0.7459
MAE	0.0584	0.0319	0.1737	0.1018
RMSE	0.0698	0.0377	0.2505	0.1337
MAPE	116.803	227.753	567.449	17.261

Table 8.3 Forecasting accuracy criteria for the model without and with wavelet transformation

Model	MAE	RMSE	MAPE
ARIMA	0.1633	0.2187	18.566
Wavelet-ARIMA (approach-1)	0.0654	0.0799	9.1472
Wavelet-ARIMA (approach-2)	0.1336	0.1591	15.356

approach-1 shows the smallest forecasting errors than approach-2. From Fig. 8.3, it obviously reveals that the forecasting values from Wavelet-ARIMA (approach-1) are very close to the original series followed by the Wavelet-ARIMA (approach-1) and ARIMA model, respectively. Thus, Wavelet-ARIMA model forecasts humidity series of Rajshahi more accurately than the direct ARIMA model.

8.6 Conclusions

The study has been conducted to show whether the recently developed wavelet transformation in forecasting the climatic time series in Bangladesh improves the performance of existing forecasting models, such as ARIMA. These models are applied to forecast the humidity of Rajshahi, Bangladesh. Then the wavelet transformation has been used to decompose the humidity series into a set of better-behaved constitutive series. These decomposed series and inverse wavelet transformation are used as a pre-processing procedure of forecasting humidity series using the same models in two approaches. Finally, the forecasting ability of these two models with and without wavelet transformation is compared using the statistical forecasting accuracy criteria.

The results show that the use of wavelet transformation as a pre-processing procedure of forecasting climatic time series improves the performance of forecasting models. The reason for the better behavior of the constitutive series is the filtering effect of the wavelet transform. Therefore, the forecasting using the existing models under wavelet transformed series is better than forecasting directly, and also it gives more accurate results.

Thus, the hybrid Wavelet-ARIMA model proposed in this chapter is both novel and effective in forecasting climatic time series, specially using approach-1.

References

Yao S, Song Y, Zhang L, Cheng X (2000) Wavelet transform and neural networks for short-term electrical load. Energy Convers Manag 41:1975–1988

Strang G (1993) Wavelet transforms versus Fourier transforms. Bull New Ser Am Math Soc 28(2):288–305

Yevgeniy B, Lamonova M, Vynokurova O (2005) An adaptive learning algorithm for a wavelet neural network. Expert Syst 22(5):235–240

Wadi SA, Ismail MT, Alkhahazaleh MH, Karim SAA (2011) Selecting wavelet model in forecasting financial time series data based on ARIMA model. Appl Math Sci 5(7):315–326

Arino MA, Vidakovic B (1995) On wavelet scalograms and their applications in economic time series. Discussion paper 95-21, ISDS, Duke University, Durham

Rocha T, Paredes S, Carvalho P, Harris HM (2010) Wavelet based time series forecast with application to acute hypotensive episodes prediction. In: 32nd annual international conference of the IEEE EMBS, Buenos Aires, Argentina

Henriques J, Rocha T (2009) Prediction of acute hypotensive episodes using neural network multi-models. Computers in Cardiology, 36

Gang D, Shi-Sheng Z, Yang L (2008) Time series prediction using wavelet process neural network. J Chin Phys B 17:6

Aggarwal SK, Saini LM, Kumar A (2008) Price forecasting using wavelet transformation and LSE based mixed model in Australian electricity market. Int J Energy Sector Manage 2:521–546

Antonio JC, Plazas MA, Espinola R, Molina AB (2005) Day-ahead electricity price forecasting using the wavelet transform and ARIMA models. IEEE Trans Power Syst 20:1035–1042

Conejo AJ, Plazas A, Espinola R, Molina AB (2005) Day-ahead electricity price forecasting using the wavelet transformation and ARIMA models. IEEE Trans Power Syst 20(2):1035–1042

Alrumaih M, Al-Fawzan A (2002) Time series forecasting using wavelet denoising an application to Saudi stocks index. J King Saudi Univ 14:221–234

Nason GP, Silverman BW (1994) The discrete wavelet transform in S. J Comput Graph Stat 3:163–191

Mallat S (1989) A theory for multi-resolution signal decomposition: the wavelet representation. IEEE Trans Pattern Anal Mach Intell 11:674–693

Percival DB, Walden AT (2000) Wavelet methods for time series analysis. Cambridge University Press, Cambridge

Daubechies I (1992) Ten lectures on wavelets. SIAM, CBMS-NST conference series, vol 61, Philadelphia

Box GEP, Jenkins GM (1976) Time series analysis: forecasting and control, revth edn. Holden Day, San Francisco

Pankratz A (1991) Forecasting with dynamic regression models. Wiley Interscience, New York

Granger CWJ, Newbold P (1986) Forecasting economic time series, 2nd edn. Academic Press, San Diego

Chapter 9
Analysis of Inter-Annual Climate Variability Using Discrete Wavelet Transform

Md. Khademul Islam Molla, A.T.M. Jahangir Alam, Munmun Akter, A.R. Shoyeb Ahmed Siddique, and M. Sayedur Rahman

Abstract This chapter presents a data adaptive filtering technique to extract annual cycles and the analysis of inter-annual climate variability based on different climate signals using discrete wavelet transform (DWT). The annual cycle is considered as higher energy trend in a climate signal and separated by implementing a threshold-driven filtering technique. The fractional Gaussian noise (fGn) is used here as a reference signal to determine adaptive threshold without any prior training constraint. The climate signal and fGn are decomposed into a finite number of subband signals using the DWT. The subband energy of the fGn and its confidence intervals are computed. The upper bound of the confidence interval is set as the threshold level. The energy of individual subband of a climate signal is compared with the threshold. The lowest order subband of which the energy is greater than the threshold level is selected yielding the upper frequency limit of the trend representing annual cycle. All the lower frequency subbands starting from the selected one are used to reconstruct the annual cycle of the corresponding climate signal. The distance between adjacent peaks in the extracted cycles refers to the inter-annual variation of the climate condition. The experimental results illustrate

M.K.I. Molla (✉)
Department of Information and Communication Engineering, The University of Tokyo, Bunkyo, Tokyo, Japan

Department of Computer Science and Engineering, The University of Rajshahi, Rajshahi, Bangladesh
e-mail: khademul.cse@ru.ac.bd

A.T.M.J. Alam
Department of Environmental Sciences, Jahangirnagar University, Savar, Dhaka, Bangladesh

M. Akter • A.R.S.A. Siddique
Department of Computer Science and Engineering, The University of Rajshahi, Rajshahi, Bangladesh

M.S. Rahman
Department of Statistics, The University of Rajshahi, Rajshahi, Bangladesh

T. Islam et al. (eds.), *Computational Intelligence Techniques in Earth and Environmental Sciences*, DOI 10.1007/978-94-017-8642-3_9,

the efficiency of the proposed data adaptive approach to separate the annual cycle and the quantitative analysis of climate variability.

Keywords Climate signal • Discrete wavelet transform • Fractional gaussian noise • Multiband decomposition • Time domain filtering

9.1 Introduction

The term "climate variability" denotes the inherent characteristic of climate which manifests itself in changes of climate with time. The degree of climate variability can be described by the differences between long-term statistics of meteorological elements calculated for different periods (Molla et al. 2006). Climate variability is often used to denote deviations of climate statistics over a given period of time such as a specific month, season, or year from the long-term climate statistics relating to the corresponding calendar period. To mitigate the effects of climate change, risk assessments are being required more frequently by policy makers (Dairaku et al. 2004). Climate is currently changing in ways that mirror the effects of global warming. There is also increasing demand for climate change information, particularly from policy makers for impact assessment studies (Bates et al. 1998). Several linear statistical models have been applied to climate analysis, but the answers are not conclusive due to the high sensitivity of model results to model parameters (Rajagopalan et al. 1997, 1999; Harrison and Larkin 1997), especially when stochastic processes are taken into account. Various approaches have been employed to develop climate change scenarios at different scales (Mpelasoka et al. 2001). The inter-annual climate variability refers to the variation of climate between adjacent years.

The annual cycle of any climate signals represents a good measure of climate variability (Molla et al. 2011). If we consider the seasonal cycle for a year, it can be termed as annual cycle. The variation of such cycle from the mean stream can be used to illustrate the inter-annual variation of climate. It is commonly estimated from observational data or model output by taking the average of all Januaries, all Februaries, and so forth. If the observational record is long enough and conditions are stationary (i.e., there is no significant long-term trend), a meaningful annual cycle will result that can be used to calculate an anomaly time series (Mak 1995). In the analysis of climate variability, the cycle refers to every 12-month (consecutive) length of data.

In this chapter, the annual cycle of several climate signals are extracted using data adaptive time domain filtering approach. The annual cycle is treated as the trend in the climate signal. It is required to develop a method to extract the trend representing the cycle using a data adaptive technique without affecting the other parts of the signal (Mak 1995). It needs a mathematical function used to divide a given function or time series into different scale components, i.e., subbands (Oh et al. 2003). The underlying assumption to separate the trend of annual cycle using multiband approach is that some lower frequency subbands contain relatively higher energies. A filtering technique is required to identify and separate those subband components constructing the annual cycle. The traditional approach—Fourier transform (FT) can be used to decompose the climate time series into several subbands to extract the trend of the signal (Broughton and Bryan 2008). Although the FT has a strong mathematical model, it is not suitable for decomposing non-stationary signals, e.g., climate signals. The discrete wavelet transform (DWT) is the representation of a function by wavelet which is very efficient to decompose the signal in a data adaptive nature (Mallat 2008). Wavelet transforms have advantages over traditional Fourier transforms for representing functions that have discontinuities and sharp peaks, and for accurately deconstructing and reconstructing finite, nonperiodic, and/or non-stationary signals. In this study we attempt to obtain a better understanding of the climate variability through the analysis of annual climate cycles of different climate signals say rainfall, humidity using wavelet transform. A reference signal—fractional Gaussian noise (fGn) and the climate signal is decomposed into several subbands using wavelet. The subbands which represent the trend are determined by comparing their energies with that of the reference signal. A data adaptive thresholding approach is developed to extract the annual climate cycle and its inter-annual variation matrix is derived.

9.2 Multiband Decomposition of Climate Signals

The climate signal is decomposed into multiple subbands (multiband) signals to extract the annual cycles. Such multiband decomposition can be achieved the applying the filter bank technique on the climate signals. In signal processing, a filter bank is an array of bandpass filters that separates the input signal into multiple components, each one carrying a frequency subband of the original signal. The process of decomposition performed by the filter bank is called analysis (the analysis of signal in terms of each subband); the output of analysis is referred to as a subband signal with as many subbands as there are filters in the filter bank (Broughton and Bryan 2008). The reconstruction process is called synthesis, meaning reconstitution of a complete signal resulting from the filtering process. The both analysis and synthesis of the climate signal are required to separate the annual cycles.

Consider an input signal $x(n)$ which might have come from a climate sensor. As discussed briefly, the spectrum of a signal gives the distribution of signal energy as a function of frequency. In order to separate energy from a frequency region of a signal's spectrum, a bandpass filter may be used. An ideal bandpass filter rejects all input signal energy outside of a desired frequency range, while giving as output of all input signal energy within that range. The range of accepted frequencies is often referred to as the band, or passband. The frequency boundaries defining the band, f_{cl} and f_{ch}, are known as the lower and upper cutoff frequencies, respectively. These

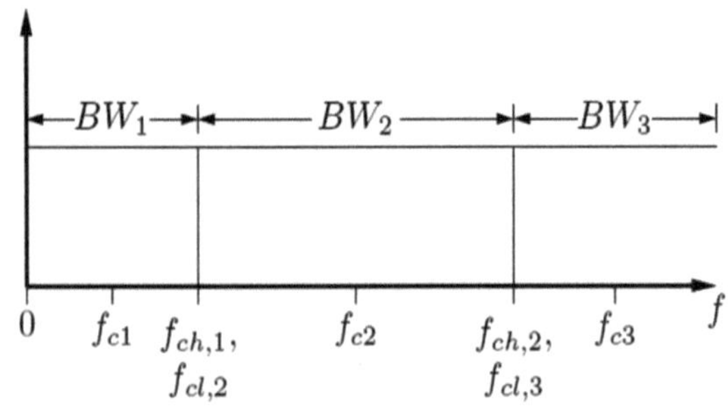

Fig. 9.1 The bands of a three-band filter bank, with adjacent band edges touching but not overlapping. Together, the three bands span the frequency range from $f_{cl,1} = 0$ Hz to $f_{ch,3} = f_{max}$ where f_{max} is the maximum frequency of interest (not shown here)

are also referred to as the band edges. The difference between the upper and lower cutoff frequencies is known as the bandwidth: BW $= f_{ch} - f_{cl}$. The midpoint of the band edges is known as the center frequency f_c of the bandpass filter.

A filter bank is a system that divides the input signal $x(n)$ into a set of analysis signals $x_1(n)$, $x_2(n)$, ... each of which corresponds to a different region in the spectrum of $x(n)$. Typically, the regions in the spectrum given by the analysis signals collectively span within the specified regions. Also, the regions usually do not overlap, but are lined up one after the other, with edges, touching, as shown in Fig. 9.1. The analysis signals $x_1(n)$, $x_2(n)$, ... may be obtained using a collection of bandpass filters with bandwidths BW_1, BW_2, ... and center frequencies f_{c1}, f_{c2}, ..., respectively.

9.2.1 Fourier-Based Filter Bank and Its Limitations

The Fourier transform widely used to implement filter bank is a mathematical operation with many applications in physics and engineering. It expresses a function of time as a function of frequency, known as its frequency spectrum. The function of time and frequency is often called time domain and frequency domain representation, respectively. The inverse Fourier transform expresses a frequency domain function in the time domain. Each value of the function is usually expressed as a complex number (called *complex amplitude*) that can be interpreted as a magnitude and a phase component. The term "Fourier transform" refers to both the transform operation and to the complex-valued function it produces.

In the case of a periodic function, such as a continuous, but not necessarily sinusoidal, the Fourier transform can be simplified to the calculation of a discrete set of complex amplitudes. Also, when a time-domain function is sampled to facilitate storage or computer processing, it is still possible to recreate a version of the original Fourier transform according to the Poisson summation formula, also known as discrete-time Fourier transform (Broughton and Bryan 2008).

Fig. 9.2 M-band analysis filter bank

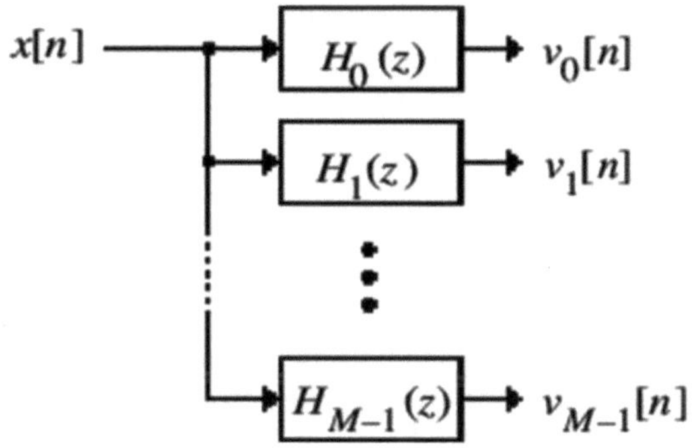

The continuous Fourier transform converts a time-domain signal of infinite duration into a continuous spectrum composed of an infinite number of sinusoids. In astronomical observations we deal with signals that are discretely sampled, usually at constant intervals, and of finite duration or periodic. For such data, only a finite number of sinusoids are needed and the *Discrete Fourier Transform* (DFT) is appropriate. For almost every Fourier transform theorem or property, there is a related theorem or property for the DFT. The DFT of N uniformly sampled data points x_n (where $n = 0, \ldots, N - 1$) and its inverse are defined by:

$$X_k = \sum_{n=0}^{N-1} x_n e^{-2\pi jnk/N} \quad \text{and} \quad x_n = \frac{1}{N}\sum_{k=0}^{N-1} X_k e^{2\pi jnk/N} \tag{9.1}$$

Once again, sign and normalization conventions may vary, but the definition is most common. The result of the DFT of an N-point input time series is an N-point frequency spectrum, with Fourier frequencies k ranging from $-(N/2 - 1)$, through the 0-frequency or the so-called DC component, and up to the highest Fourier frequency $N/2$. Each bin number represents the integer number of sinusoidal periods present in the time series. The amplitudes and phases represent the amplitudes A_k and phases $\boldsymbol{\phi}_k$ of those sinusoids. In summary, each bin can be described by $X_k = A_k e^{j\varphi_k}$.

The digital filter bank is set of bandpass filters with either a common input or a summed output. An M-band analysis filter bank is shown below (Fig. 9.2):

The subfilters $H_k(z)$ in the analysis filter bank are known as *analysis filters*. The analysis filter bank is used to decompose the input signal $x[n]$ into a set of *subband signals* with each subband signal $v_k[n]$ occupying a portion of the original frequency band. The mentioned M filters can be used as the analysis filters in the analysis filter bank or as the synthesis filters in the synthesis filter bank. Since the magnitude responses of all M filters are uniformly shifted version of that of the prototype filter, the filter bank obtained is called a uniform filter bank. The signal is synthesized by summing up all the subbands. The use of Fourier transform-based filter bank introduces a noticeable reconstruction error during synthesis.

It is well known from Fourier theory that a signal can be expressed as the sum of a, possibly infinite, series of sines and cosines. This sum is also referred to as a Fourier expansion. The principal disadvantage of a Fourier transform is that it has only frequency resolution and no time resolution. This means that although we might be able to determine all the frequencies present in a signal, we do not know when they are present. To overcome this problem in the past decades, several solutions have been developed which are more or less able to represent a signal in the time and frequency domain at the same time. The Fourier transform-based filter bank is not suitable for non-stationary climate signals.

The idea behind these time-frequency joint representations is to cut the signal of interest into several parts and then analyze the parts separately (Mallat 2008). It is clear that analyzing a signal this way will give more information about the when and where of different frequency components, but it leads to a fundamental problem as well: how to cut the signal? Suppose that we want to know exactly all the frequency components present at a certain moment in time. We cut out only this very short time window using a Dirac pulse, transform it to the frequency domain and something is very wrong. The problem here is that cutting the signal corresponds to a convolution between the signal and the cutting window. The convolution in the time domain is identical to multiplication in the frequency domain. Since the Fourier transform of a Dirac pulse contains all possible frequencies the frequency components of the signal will be smeared out all over the frequency axis. In fact this situation is the opposite of the standard Fourier transform since we now have time resolution but no frequency resolution whatsoever.

The underlying principle of the phenomena just described is Heisenberg's uncertainty principle, which, in signal processing terms, states that it is impossible to know the exact frequency and the exact time of occurrence of this frequency in a signal. In other words, a signal can simply not be represented as a point in the time-frequency space. The uncertainty principle shows that it is very important how one cuts the signal.

9.2.2 Wavelet-Based Filter Bank

The *wavelet transform* is the solution to overcome the shortcomings of the Fourier transform. In wavelet analysis the use of a fully scalable modulated window solves the signal-cutting problem. The window is shifted along the signal and for every position the spectrum is calculated (Strang and Nquyen 1996). Then this process is repeated many times with a slightly shorter (or longer) window for every new cycle. In the end the result will be a collection of time-frequency representations of the signal, all with different resolutions. Because of this collection of representations, we can speak of a multiresolution analysis. In the case of wavelets we normally do not speak about time-frequency representations but about time-scale representations, scale being in a way the opposite of frequency, because the term frequency is reserved for the Fourier transform.

A time-scale representation of a digital signal is obtained by using digital filtering techniques. Recall that the continuous wavelet transform (CWT) is a correlation between a wavelet at different scales and the signal with the scale (or the frequency) being used as a measure of similarity. The CWT was computed by changing the scale of the analysis window, shifting the window in time, multiplying by the signal, and integrating over all times. In the discrete case, filters of different cutoff frequencies are used to analyze the signal at different scales (Mallat 2008). The signal is passed through a series of high pass filters to analyze the high frequencies, and it is passed through a series of low pass filters to analyze the low frequencies. The resolution of the signal, which is a measure of the amount of detail information in the signal, is changed by the filtering operations, and the scale is changed by upsampling and downsampling (subsampling) operations. Subsampling a signal corresponds to reducing the sampling rate, or removing some of the samples of the signal. For example, subsampling by two refers to dropping every other sample of the signal. Subsampling by a factor p reduces the number of samples in the signal p times. The upsampling of any signal corresponds to increasing the sampling rate of a signal by adding new samples to the signal. For example, upsampling by two refers to adding a new sample, usually a zero or an interpolated value, between every two samples of the signal. Upsampling a signal by a factor of p increases the number of samples in the signal by a factor of p.

The climate signal is considered as fully nonlinear and non-stationary and hence wavelet transform is more suitable technique to be analyzed. The DWT is also computed by changing the scale of the analysis window, shifting the window in time, multiplying by the signal, and integrating over all times. The filters of different cutoff frequencies are used to analyze the signal at different scales (Strang and Nquyen 1996). The coefficients are usually sampled on a dyadic grid, i.e., $s_0 = 2$ and $t_0 = 1$, yielding $s = 2^j$ and $t = k*2^j$.

The DWT analyzes the signal at different frequency bands with different resolutions by decomposing the signal into a coarse approximation and detail information. The DWT employs two sets of functions, called scaling functions and wavelet functions, which are associated with low pass and high pass filters, respectively. The decomposition of the signal into different frequency bands is simply obtained by successive high pass and low pass filtering of the time domain signal. The original signal $x(n)$ is first passed through a half band high pass filter $g(n)$ and a low pass filter $h(n)$. After the filtering, half of the samples can be eliminated according to the Nyquist's rule, since the signal now has a highest frequency of $\pi/2$ radians instead of π. The signal can therefore be subsampled by 2, simply by discarding every other sample. This constitutes one level of decomposition and can mathematically be expressed as follows:

$$\begin{aligned} y_{\mathrm{h}}(k) &= \sum_n x(n) \cdot g(2k - n) \\ y_{\mathrm{l}}(k) &= \sum_n x(n) \cdot h(2k - n) \end{aligned} \tag{9.2}$$

where $y_h(k)$ and $y_l(k)$ are the outputs of the high pass and low pass filters, respectively, after subsampling by 2. This decomposition halves the time resolution since only half the number of samples now characterizes the entire signal. However, this operation doubles the frequency resolution, since the frequency band of the signal now spans only half the previous frequency band, effectively reducing the uncertainty in the frequency by half.

The above procedure, which is also known as the subband decomposition, can be repeated for further decomposition. At every level, the filtering and subsampling will result in half the number of samples (and hence half the time resolution) and half the frequency band spanned (and hence double the frequency resolution) (Strang and Nquyen 1996). The number of detail subbands is equal to the decomposition levels and one approximation, hence total of $(k + 1)$ subbands for k decomposition levels. The frequencies that are most prominent in the original signal will appear as high amplitudes in that region of the DWT signal that includes those particular frequencies. The difference of this transform from the Fourier transform is that the time localization of these frequencies will not be lost. However, the time localization will have a resolution that depends on which level they appear. A toy signal and its different subbands obtained by applying the DWT are shown in Fig. 9.3.

The perfect reconstruction is important issue in the proposed method of climate variability analysis. The reconstruction in this case Eq. (9.2) is very easy since half band filters form orthonormal bases. The above procedure is followed in reverse order for the reconstruction. The signals at every level are upsampled by two, passed through the synthesis filters $g(n)$, and $h(n)$ (high pass and low pass, respectively), and then added. The interesting point here is that the analysis and synthesis filters are identical to each other, except for a time reversal. Therefore, the reconstruction formula becomes (for each layer):

$$x(n) = \sum_k [y_h(k) \cdot g(-n + 2k)] + [y_l(k) \cdot h(-n + 2k)] \tag{9.3}$$

However, if the filters are not ideal half band, then perfect reconstruction becomes difficult to be achieved. Although it is not possible to realize ideal filters, under certain conditions it is possible to find filters that provide perfect reconstruction.

The most famous wavelet bases are the ones developed by Ingrid Daubechies, and they are known as Daubechies' wavelets. The reconstructed signal by wavelet synthesis is illustrated in Fig. 9.4. The energy of the error signal (sample-wise difference) is negligible (in the order of 10^{-11}) and hence it is stated that the perfect reconstruction is possible from the DWT-based subband decomposition. From the above mentioned explanation, it is clear that the wavelet-based subband decomposition is to be used as data adaptive filter bank technique with perfect reconstruction. The climate signal is a nonharmonic time series and hence this type of data adaptive method is suitable for decomposition and reconstruction.

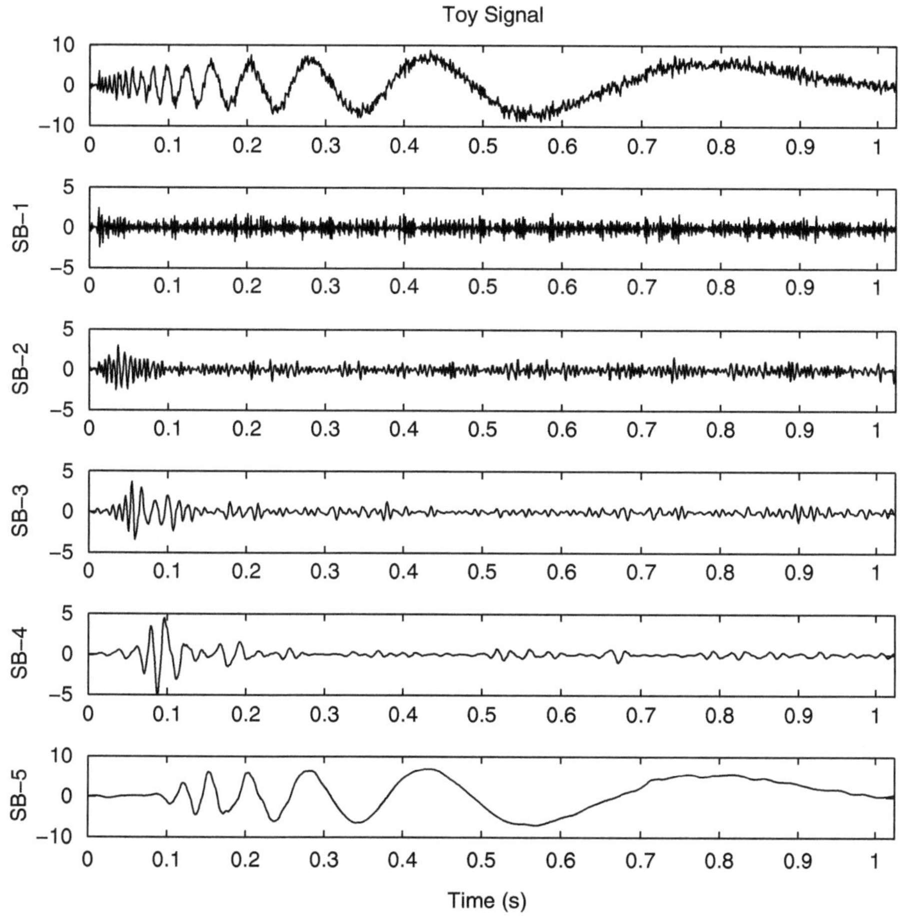

Fig. 9.3 A toy signal and different subbands obtained by the DWT

9.3 Annual Cycle Extraction

The annual cycle is considered as the low frequency and relatively higher energy trend appeared in the climate signals. The trends of the recorded climate signals are detected using the energy distribution of the signal over the individual subbands. The cycle is retrieved here by partial reconstruction of the subbands in the wavelet domain. Usually the modes contain a mixture of frequencies and these mixed modes are much more difficult to interpret. Since we are not interested in intra-annual variations, a minimal amount of pre-smoothing is performed so that single or multiple modes collectively contain the annual cycle. The subbands will generally be ordered from high to low frequency. It is important to determine the significance of the subbands. We should not expect all bands to be significant to separate the annual cycle.

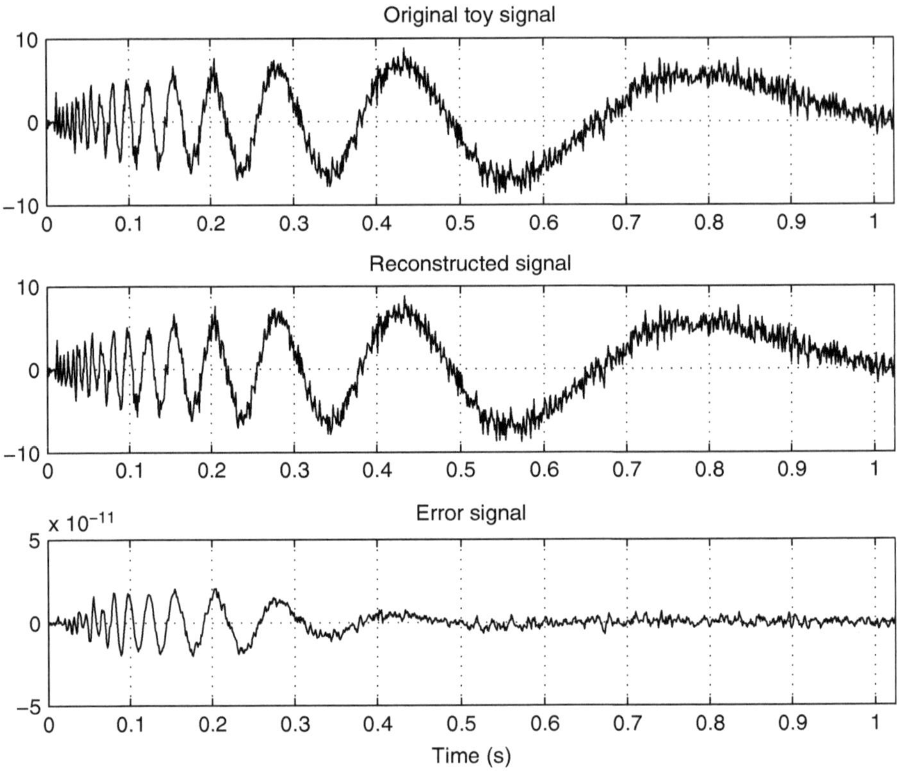

Fig. 9.4 The original toy signal (*top*), the reconstructed signal using wavelet synthesis (*middle*), and the error between those two (*bottom*)

The aim of the first step of the analysis is to find low frequency signal from the original climate signal. The original signal is decomposed into subband signals using the DWT. The analyzing climate signal $x(n)$ consists a slowly varying trend $y(n)$ superimposed to a fluctuating process $\zeta(n)$, the trend is expected to be captured by subband signals of large indices. A process of de-trending $x(n)$, which corresponds to estimating $\zeta(n)$, may therefore relate to compute the partial, fine-to-coarse, reconstruction

$$\zeta(n) = \sum_{k=1}^{K} d_k(n) \tag{9.4}$$

where K is the largest subband index prior the remaining subbands representing signal trend contamination. For the subbands $d_k(n)$; $k = 1, 2, \ldots, K$, a rule of thumb, so the choice of K is based on observation of the evolution of the $\zeta(n)$ energy as a function of a test order k. The optimized $k = K$ is chosen when the energy index departs significantly from the energy of the reference signals (Flandrin et al. 2004). The starting index of the subband to separate the trend, i.e., the low frequency components of the climate signal is determined by comparing the subband energy with that of the reference signal. The fGn is used here as the reference signal. There is a subband of climate signal exceeding the upper limit of 95 % confidence interval (CI) of the corresponding subbands' energies of the fGn. That subband is selected as the lower bound of the low frequency component of the climate signal. All the higher order (lower frequency) subbands starting from the selected one are summed to construct the low frequency trend $y(n) = x(n) - \zeta(n)$ representing the annual cycle.

The fGn is a generalization of ordinary white noise. It is a versatile model of homogeneously spreading broadband noise without any dominant frequency band, is an intrinsically discrete-time process, and may be described as the increment process of fractional Brownian motion (fBm) since the fBm is the only self-similar Gaussian process with stationary increments (Flandrin et al. 2004). Consequently the statistical properties of the fGn are entirely determined by its second-order structure, which depends solely upon one single scalar parameter, Hurst exponent. The energies of the subbands of the fGn are decreased linearly with increasing its order. Such property of the fGn is very much applicable to determine the trend of the analyzing signal by comparing the energies of its subbands. The normalized fGn is decomposed only once using the DWT and then each of the climate signals is decomposed into multiple subbands (using the DWT). The scaling factor of each signal is reused to obtain the original scale in the wavelet domain.

The steps of the proposed algorithm to separate annual cycle are as follows:

1. Apply the DWT-based subband decomposition on the fGn. Compute Log energy of its individual subband and then its upper and lower bound with 95 % confidence interval (CI).
2. Apply the similar decomposition on the climate signal.
   (a) Compute the Log energy of its subbands.
   (b) Find the highest frequency subband which has the energy higher than the upper bound of 95 % confidence interval derived in step 1, say it qth subband. Thus selected qth subband is the starting index to constructing trend, i.e., annual cycle.
   (c) The annual cycle is separated by summing up all the higher order (lower frequency) subbands starting from qth one of the climate signals.

Thus obtained low frequency trend with higher energy represents the annual cycle of the respective climate signal. It illustrates the non-stationary nature of the climate variable. The inter-annual climate variability is defined by the distance between two consecutive peaks of the extracted annual cycle.

9.4 Results and Discussion

The performance of the proposed annual cycle extraction method as well as the analysis of climate variability are evaluated by real climate data collected at different regions in Bangladesh. Only the data collected in Dhaka region from

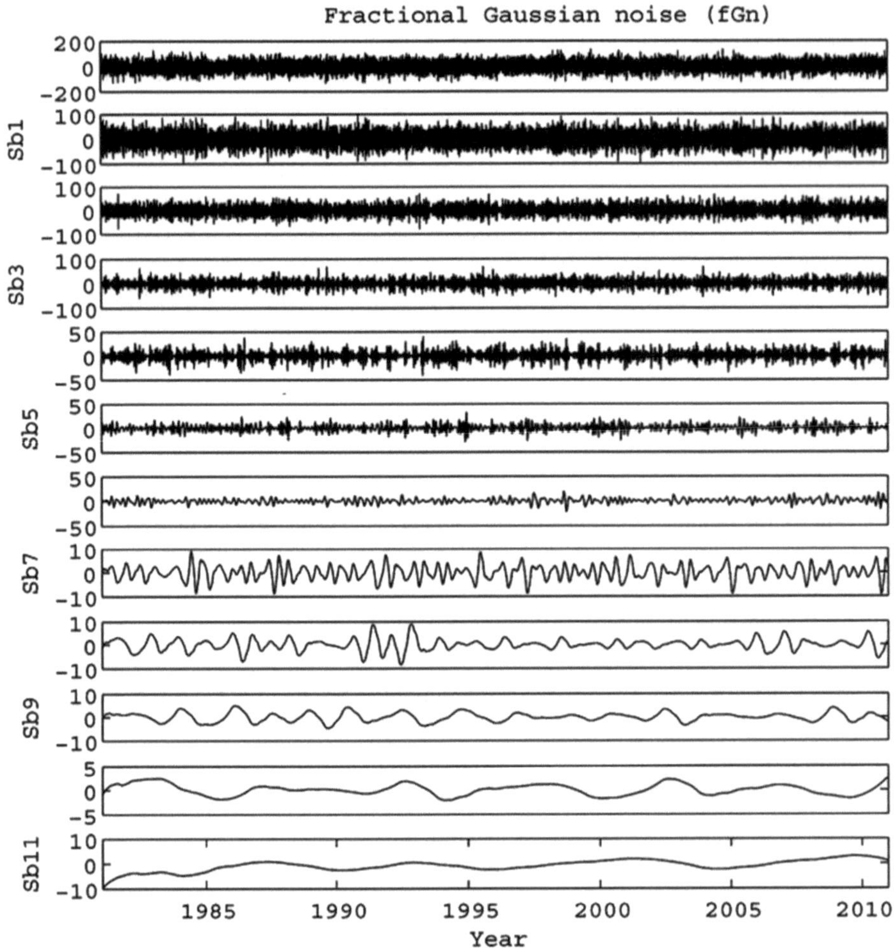

Fig. 9.5 The fGn and its different subbands obtained by the DWT

January 1981 to December 2011 are used in this analysis. Two climate variables—daily average rainfall (in mm) and average humidity (in %) are taken into consideration. All the weather parameters are measured with sufficiently efficient instruments. The fGn and both of the climate signals (daily rainfall and humidity) are decomposed into 11 subband (ten levels decomposition) signals using the DWT with db4 wavelet as illustrated in Figs. 9.5, 9.6, and 9.7, respectively.

According to the proposed algorithm, the energy of each subband of the fGn is computed then the 95 % confidence interval (CI) of the energy curve (of fGn) is determined. The energy of each subband of the climate signal (daily rainfall) is compared with the upper bound of the CI which is used here as data adaptive threshold. The subband with energy exceeding the upper limit of the CI is selected; say it qth band which is considered as the starting index of detecting the trend,

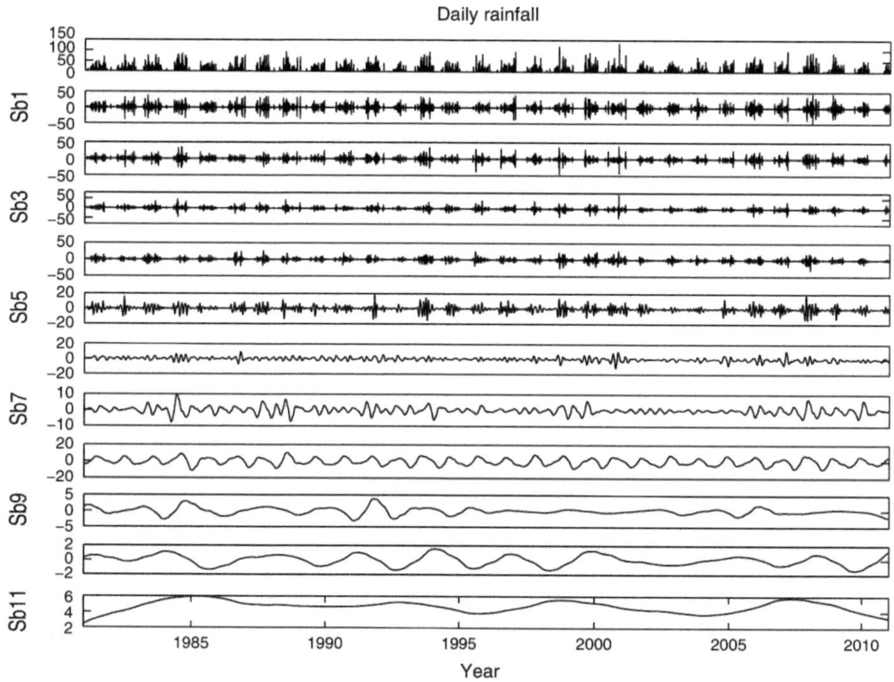

Fig. 9.6 The daily average rainfall (mm) data and its different subbands obtained by the DWT

i.e., the annual cycle. The energy-based thresholding to find the qth subband is shown in Fig. 9.8 for the daily rainfall. The eighth subband of the daily rainfall signal is selected as the starting index of trend detection. Similarly, the starting index determination of daily humidity signal is illustrated in Fig. 9.9 where the eighth one is selected as the starting subband of seasonal cycle.

The annual cycle is separated by summing up the higher order subband signals starting from qth one up to the last subband of climate signals. The daily rainfall, the separated seasonal cycle, and the residue signals (suppressing the annual cycle from the original signal) are illustrated in Fig. 9.10. The separation of annual cycle of daily humidity is shown in Fig. 9.11.

The proposed annual cycle extraction method is fully data adaptive. No training is required to determine the threshold. Thus obtained cycle carries necessary properties for further processing of the climate variables. The annual cycle is considered here as the principal parameter to explain the climate variability.

It is considered that when there is a climate change, there is a good chance of non-stationarity in the annual cycle of the climate signals. The inter-annual distances (defined as the distances between two adjacent peaks of annual cycle) of the two climate signals (daily average rainfall and humidity) are shown in Fig. 9.12.

It is expected that the inter-annual distance is close to 360 days whereas such distances for the mentioned two climate signals are varied up to 2 months

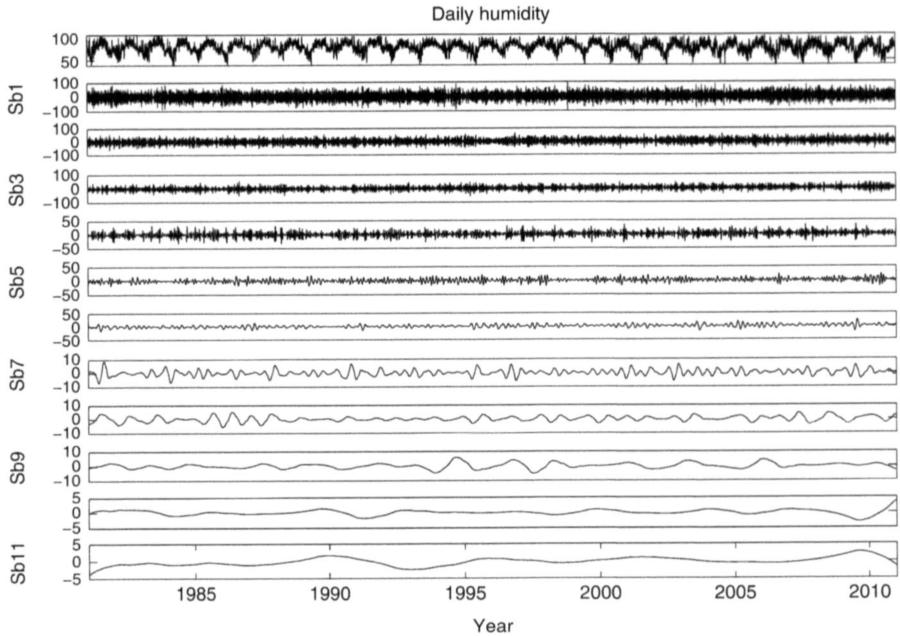

Fig. 9.7 The daily average humidity (%) data and its different subbands obtained by the DWT

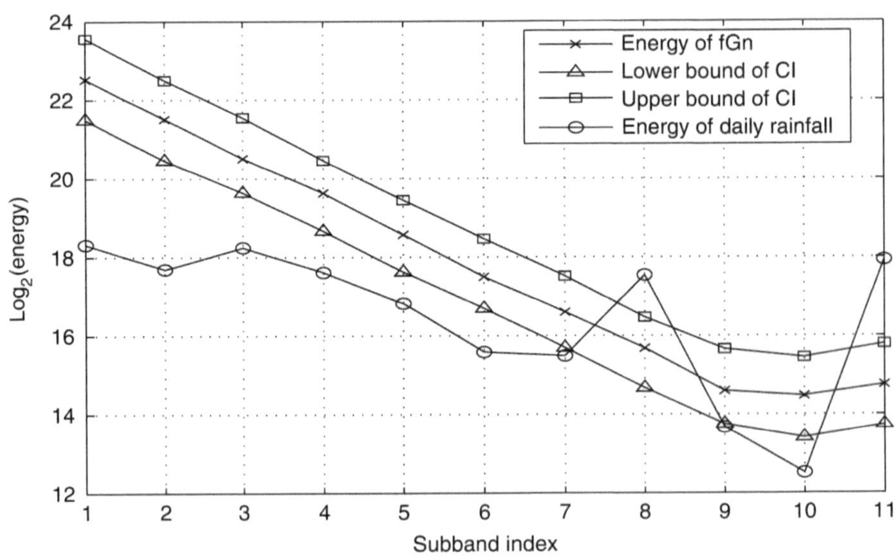

Fig. 9.8 Selection of starting subband to extract annual cycle of daily rainfall. The eighth subband is selected as the starting index

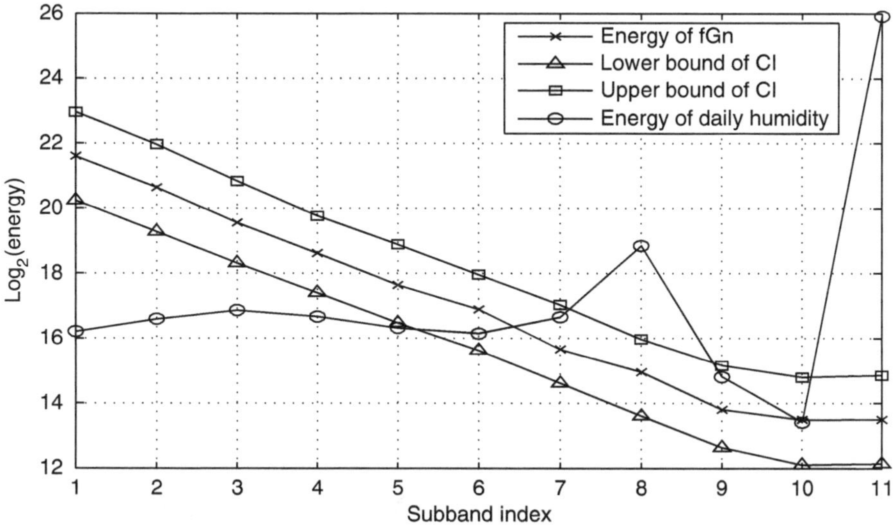

Fig. 9.9 Selection of starting subband to extract annual cycle of daily humidity. The eighth subband is selected as starting index

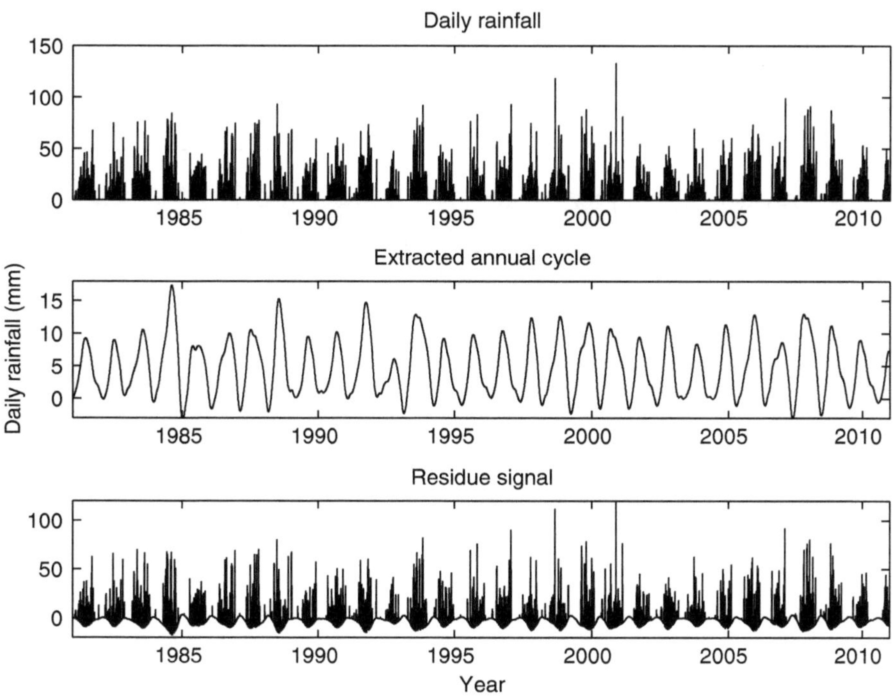

Fig. 9.10 Annual cycle separation results; daily rainfall data (*top*), extracted annual cycle (*middle*), and residue after suppressing the cycle (*bottom*)

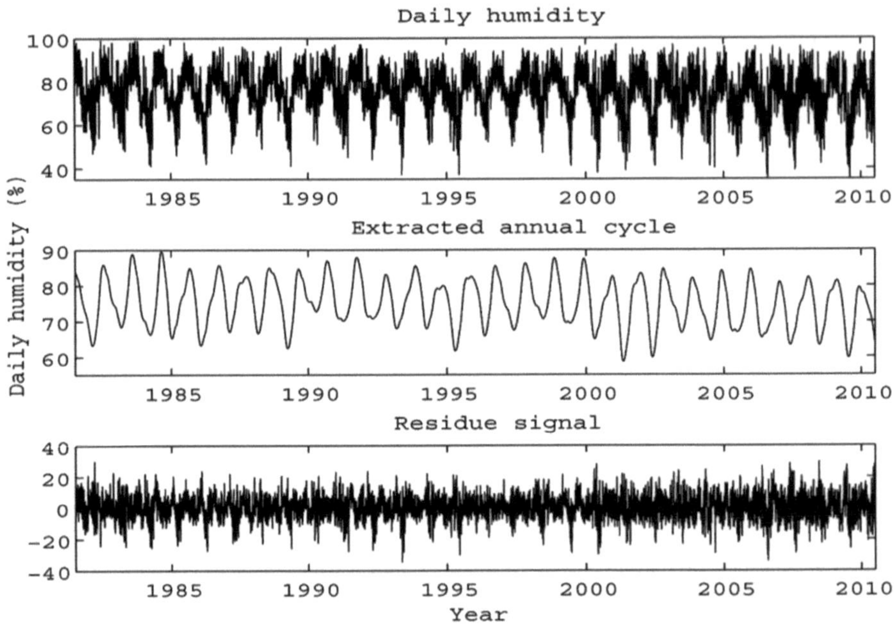

Fig. 9.11 Annual cycle separation results; daily humidity data (*top*), extracted annual cycle (*middle*), and residue after suppressing the cycle (*bottom*)

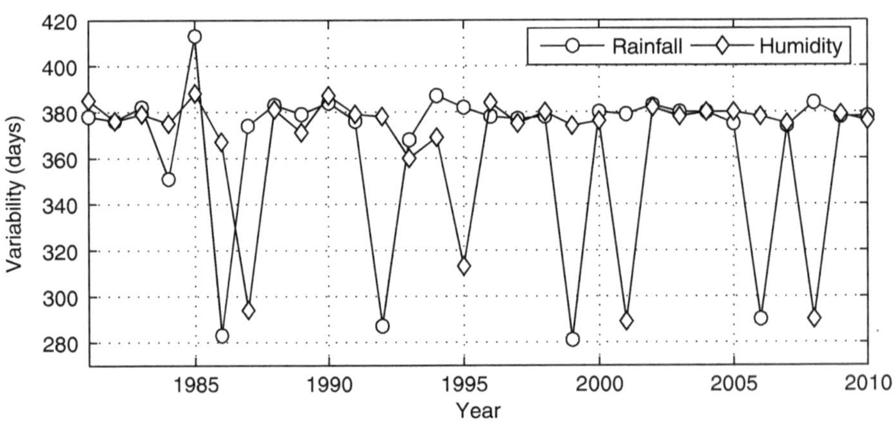

Fig. 9.12 The inter-annual distances of two climate signals

(as illustrated in Fig. 9.12). Hence there is a noticeable amount of climate change/variation with those climate signals. Climate signals always represent non-stationary data. It is not possible to assume such data as the sum of harmonics and hence Fourier-based transformation is not suitable for the investigation of climate data. The DWT is highly data adaptive and efficient for analyzing nonlinear and non-stationary climate signals.

9.5 Conclusions

The wavelet-based data adaptive time domain filtering technique is implemented to extract the annual cycle of different climate signals. The main superiority of this method is that the use of the DWT method yields subband signals based on local properties of the signal. It eliminates the need of harmonic assumption to represent nonlinear and non-stationary signals. The DWT is a well-known approach to many researchers in climate research. This study plays a vital role for analyzing the properties of nonlinear and non-stationary daily rainfall and humidity time series data. The annual cycle is separated from the climate signals without any prior information. It adopts a fully data adaptive approach to express the climate variation.

References

Bates BC, Charles SP, Hughes JP (1998) Stochastic downscaling of numerical climate model simulations. Environ Model Softw 13(3–4):325–333

Broughton SA, Bryan KM (2008) Discrete Fourier analysis and wavelets: applications to signal and image processing, 1st edn. John Wiley & Sons, Inc., Hoboken, New Jersey

Dairaku K, Emori S, Nozawa T, Yamazaki N, Hara M, Kawase H (2004) Hydrological change under the global warming in Asia with a regional climate model nested in a general circulation model. In: Proceedings of the third international workshop on monsoons (IWM-III), Hangzhou, China

Flandrin P, Rilling G, Goncalves P (2004) Empirical mode decomposition as a filter bank. IEEE Signal Process Lett 11(2):112–114

Harrison DE, Larkin NK (1997) Darwin sea level pressure, 1876–1996: evidence for climate change? Geophys Res Lett 24(14):1779–1782

Mak M (1995) Orthogonal wavelet analysis: inter-annual variability in the sea surface temperature. Bull Am Meteorol Soc 76:2179–2186

Mallat S (2008) A wavelet tour of signal processing, 3rd edn. Academic Press, Orlando

Molla MKI, Rahman MS, Sumi A, Banik P (2006) Empirical model decomposition analysis of climate changes with special reference to rainfall data. Discrete Dyn Nat Soc 2006:1–17

Molla MKI, Ghosh PR, Hirose K (2011) Bivariate EMD-based data adaptive approach to the analysis of climate variability. Discrete Dyn Nat Soc 2011:1–21

Mpelasoka FS, Mullan AB, Heerdegen RG (2001) New Zealand climate change information derived by multivariate statistical and artificial neural networks approaches. Int J Climatol 21(11):1415–1433

Oh HS, Ammann CM, Naveau P, Nychka D, Otto-Bliesner BL (2003) Multi-resolution time series analysis applied to solar irradiance and climate reconstructions. J Atmos Solar Terr Phys 65:191–201

Rajagopalan B, Lall U, Cane MA (1997) Anomalous ENSO occurrences: an alternate view. J Climate 10(9):2351–2357

Rajagopalan B, Lall U, Cane MA (1999) Comment on reply to the comments of Trenberth and Hurrell. Bull Am Meteorol Soc 80(12):2724–2726

Strang G, Nquyen T (1996) Wavelets and filter banks. Willesley-Cambridge University Press, Willesley

Part IV
Hybrid Intelligence Techniques in Earth and Environmental Sciences

Chapter 10
Modeling of Suspended Sediment Concentration Carried in Natural Streams Using Fuzzy Genetic Approach

Özgür Kişi and Halil İbrahim Fedakar

Abstract This chapter proposes fuzzy genetic approach so as to predict suspended sediment concentration (SSC) carried in natural rivers for a given stream cross section. Fuzzy genetic models are improved by combining two methods, fuzzy logic and genetic algorithms. The accuracy of fuzzy genetic models was compared with those of the adaptive network-based fuzzy inference system, multilayer perceptrons, and sediment rating curve models. The daily streamflow and suspended sediment data belonging to two stations, Muddy Creek near Vaughn (Station No: 06088300) and Muddy Creek at Vaughn (Station No: 06088500), operated by the US Geological Survey were used as case studies. The root mean square errors and determination coefficient statistics were used for evaluating the accuracy of the models. The comparison results revealed that the fuzzy genetic approach performed better than the other models in the estimation of the SSC.

Keywords Suspended sediment concentration • Fuzzy genetic approach • Adaptive network-based fuzzy inference system • Multilayer perceptrons • Sediment rating curve

10.1 Introduction

Correct estimation of suspended sediment concentration (SSC) carried in natural streams is very important with regard to channel navigability, pollution, reservoir filling, hydroelectric equipment longevity, fish habitat, and science interests.

Ö. Kişi (✉)
Faculty of Engineering and Architecture, Civil Engineering Department,
Canik Başarı University, Samsun, Turkey
e-mail: okisi@basari.edu.tr

H.İ. Fedakar
Faculty of Engineering and Natural Sciences, Civil Engineering Department,
Abdullah Gül University, Kayseri, Turkey

T. Islam et al. (eds.), *Computational Intelligence Techniques in Earth and Environmental Sciences*, DOI 10.1007/978-94-017-8642-3_10,

All surface water reservoirs are designed to a volume known as "the dead storage" to accommodate sediment income that will accumulate over a specified period called the economic life. The underestimation of sediment load gives rise to insufficient reservoir capacities while the overestimation will bring about over-capacity reservoirs. For this reason, to determine sediment load accurately is vital in order to accomplish an appropriate reservoir design and operation. In environmental engineering, the estimation of the SSC has an additional significance, especially if the particles also transport pollutants (Guldal and Muftuoglu 2001).

Fuzzy logic has been successfully used for the estimation of suspended sediment during recent years. Tayfur et al. (2003)employed a fuzzy logic algorithm for runoff-induced sediment transport from bale soil surfaces. Kisi (2004a) developed fuzzy models in order to anticipate daily suspended sediments. He compared the fuzzy estimates with those of the sediment rating curves (SRCs) and found that the fuzzy models performed better than the SRCs. Kisi et al. (2006) showed that fuzzy rule-based models outdo the SRC models in forecasting of the SSC. Lohani et al. (2007) used fuzzy logic for deriving stage–discharge–sediment concentration relationships.

Also, genetic algorithm (GA) has been utilized in the estimation of suspended sediment. Altunkaynak (2009) anticipated relationship between discharge and sediment load using the GA. Reza Pour et al. (2011) used the GA to identify the relation between streamflow discharge and sediment loads.

Adaptive network-based fuzzy inference system (ANFIS) has also been successfully applied in prediction of the SSC in recent years. Kisi (2005) used an ANFIS model for daily suspended sediment estimation. Kisi et al. (2008) investigated the accuracy of an ANFIS in modeling of daily suspended sediment of rivers in Turkey. Kisi et al. (2009)anticipated monthly suspended sediments using the ANFIS models. Rajaee et al. (2009) utilized an ANFIS model for the SSC simulation. Cobaner et al. (2009) used the ANFIS in forecasting of daily SSC. Firat and Gungor (2010) investigated the accuracy of the ANFIS approach in modeling of monthly suspended sediment.

In the hydrological predicting context, recent experiments have reported that multilayer perceptrons (MLP) may present a promising alternative for modeling suspended sediment. Jain (2001) used a single MLP approach to establish the sediment–discharge relationship and found that the MLP model could outperform the SRC. Cigizoglu (2004) investigated the accuracy of a single MLP in the prediction and anticipation of daily suspended sediment data. Kisi (2004b) used different artificial neural networks (ANNs) methods for estimating daily SSC, and he denoted that the MLP models outperformed the generalized regression neural networks and radial basis function networks. Kisi (2005) developed the MLP models in order to model suspended sediment and compared the MLP results with those of the SRC and multi-linear regression models. Kisi (2008) successfully used a data-driven algorithm for constructing the MLP-based sediment prediction models. Tayfur and Guldal (2006) anticipated total suspended sediment from precipitation. Kisi et al. (2008) used different ANN techniques in suspended sediment estimation and compared their results with the SRCs.

Cobaner et al. (2009) modeled daily SSC using hydrometeorological data. To the knowledge of the authors, there isn't any published work indicating the input–output mapping capability of fuzzy genetic technique in modeling of suspended sediments.

This chapter is concerned with the implementation of fuzzy genetic approach for estimating the SSC using the daily streamflow and suspended sediment time series data belonging to two stations on Muddy Creek, Montana, USA. The accuracy of fuzzy genetic is compared with those of the ANFIS, MLP, and SRC.

10.2 Methods

10.2.1 Fuzzy Logic

The concepts of the fuzzy logic were first introduced by Zadeh (1965).These concepts and operational algorithms are given in many textbooks (Kosko 1993; Ross 1995). Fuzzy logic allows for something to be partly this and partly that, rather than having being either all this or all that. A membership number between 0 and 1 can describe the degree of "belongingness" to a set or category numerically. Fuzzy membership functions (MFs) can take many forms (Kisi 2010).

Fuzzy categories can be used to set up rules of the following form for control purposes: "If the value of variable x_1 is 'high' and variable x_2 is 'medium,' then the result, y is 'low.'" It is believed that these rules more closely resemble the way we think than do more explicit mathematical rules. There are two main ways in which fuzzy logic programming can be used: (1) to relate a set of outputs to a set of inputs in a "model-free" way the "fuzzy inference method", and (2) to try to model the behaviour of a human expert (Russel and Campbell 1996).

Sets of input data along with the corresponding outputs are provided to the fuzzy system in the fuzzy inference method. And the fuzzy system "learns" how to transform a set of inputs to the corresponding set of outputs through a fuzzy associative map (FAM), sometimes called fuzzy associative memory (Kosko 1993). While the ANNs can perform the same function such as regression, these tend to be "black box" approaches. A fuzzy logic system is more flexible and more transparent. It is possible to open the lid, see how it works, and adjust it if necessary by using the black box analogy (Russel and Campbell 1996).

In this chapter, fuzzy logic is employed for the prediction of the SSC. Its name means fuzzy logic does not provide a rigorous approach for developing or combining fuzzy rules that can be achieved through many ways. The approach utilized in the present chapter is explained as below.

First the input and output variables are divided into a number of subsets with *Gaussian combination membership functions*. Generally, there are c^n fuzzy rules where c and n denote the numbers of subsets and input variables, respectively. As the number of subsets increase so does the possible accuracy but the rule base gets larger, which is more difficult to construct (Sen 1998). In the case of one input, x, with n subsets, the rule base takes the form of an output y_j ($j = 1, 2, \ldots, n^2$). If there is one input variable as x with "low," "medium," and "high" fuzzy subsets, then consequently there will be three rules as follows:

Rule 1: IF x has low THEN y_1
Rule 2: IF x has medium THEN y_2
Rule 3: IF x has high THEN y_3

Membership degree, w_j, for x is computed to be assigned to the corresponding output y_j for each rule triggered. Hence the weighted average of the outputs from three rules results a single weighted output, y, as follows:

$$y = \frac{w_1 \cdot y_1 + w_2 \cdot y_2 + w_3 \cdot y_3}{w_1 + w_2 + w_3} = \frac{\sum_{j=1}^{3} w_j \cdot y_j}{\sum_{j=1}^{3} w_j} \tag{10.1}$$

Hence, the values of the output y can be computed from Eq. (10.1) for any combination of input variable fuzzy subsets after setting up the rule base. To use sample data and derive the necessary rule base by the fuzzy inference procedure is a very common method in deciding about the fuzzy rule base (Kisi 2010).

A fuzzy rule base can be achieved step-by-step from sets of input and output data as follows:

1. Divide the range of each input variable into a number of subintervals with a membership function.
2. Compute the membership value (w_j) for x in each of the fuzzy subsets for each data point n (one value for x and y).
3. Store the output y_j beside the complete set of rule weights w_j.
4. Repeat for all the other data points.
5. Calculate the weighted average with an expression similar to Eq. (10.1) (Kiszka et al. 1985a, b).

One of the main problems in designing any fuzzy system is construction of the fuzzy subsets because all changes in the subsets will affect the performance of the fuzzy model directly. For this reason, optimum determination of the membership functions has an important role for the successful optimum modeling. Here, the genetic algorithm is employed for the adjustment of membership functions. Next, the GA, fuzzy genetic, ANFIS, MLP, and SRC are briefly explained (Kisi 2010).

10.2.2 Genetic Algorithm

Based on the mechanics of natural genetic and natural selection, the genetic algorithms (GAs) are search algorithms. In the genetic search procedure, the basic elements of natural genetics such as reproduction, crossover, and mutation are employed. The GA comprises the following steps (Burn and Yulianti 2001):

1. Choosing an initial population of strings.
2. Evaluating the fitness of each string.
3. Choosing strings from the current population to mate.
4. Performing crossover (mating) for the selected strings.
5. Performing mutation for selected string elements.
6. Repeating steps 2–5 for the required number of generations.

Genetic algorithm is a powerful method in terms of searching the optimum solution to complex problems such as the selection of the membership functions where it is hard or almost impossible to test for optimality (Ahmed and Sarma 2005; Kisi 2010).

The main importance of a GA formulation is to denote design or decision variable of the problem in a string (or chromosome). Each decision variable of the string can be coded in real or binary number. The strings comprised binary bits in early GAs (Wang 1991). Real value coding was proved to be more effective in many problems than binary coding (Ahmed and Sarma 2005). There is no discretization of decision variable space in real value coding. To find the fitter strings, a GA searches from a population of strings not from a single string. Strings go through genetic operators to produce the fitter chromosomes. Selection, crossover, and mutation are the genetic employed in the reproductive process.

Various selection schemes have been compared by Goldberg and Deb (1990) and tournament selection scheme was investigated. A group of the population of strings is chosen at random and allowed to compete in the tournament where the fittest string is selected to enter mating pool in the tournament selection scheme. The process is finished when the size of mating pool is equal to the size of the population of the strings. Strings in the mating pool are mated at random after selection and then each pair of strings goes through crossover with some specified probability usually in the range of 0.5–1.0. Next step is the mutation after crossover. To foreclose premature convergence to local optima, mutation includes random alteration of bit value for an allele. Mutation occurs according to the specified probability. For search of new strings, the crossover operator is mostly responsible even though a mutation operator is also used for this purpose sparingly. The power of the GAs results from a simple assumption (Oliveira and Loucks 1997): the best solutions are more promising to be found in the regions of the solutions space consisting of high proportions of good solutions.

The fundamental differences between the GAs and traditional optimization methods are (Goldberg 1989):

1. The parameter sets are coded in the GAs not the parameters.
2. Local optimum is searched from a population of points in the GAs not of a single point.
3. The objective function information is used in the GAs not derivatives or other auxiliary knowledge.
4. Probabilistic transition rule is used in the GAs not deterministic rules.

The algorithm in the GA, searches for the best possible solutions of a problem from available set of solutions. The problem is changed into binary form along with the solutions. Then the solutions are allowed to crossover and mate with a given criterion to generate the best possible solution to the problem. The basic of GA have already been explained by many authors like Wang (1991) and Ahmed and Sarma (2005). Thus, the present chapter is not focused on the details of the basic procedures of the GA.

10.2.3 Fuzzy Genetic Approach

Fuzzy system is composed of fuzzy IF-THEN rules and degree of these rules obtained from input membership functions (Şen 2004). The rule base of a Sugeno fuzzy system with n inputs, one output, and m rules is expressed as

$$R_i = \text{IF } x_1 \text{ is } f_{i,1} \text{ AND } x_j \text{ is } f_{i,j} \text{ AND} \ldots x_n \text{ is } f_{i,n} \text{ THEN } y_i \text{ is } g_i$$

where x_j is jth input, $f_{i,j}$ is the membership function of the ith rule, y_i is the output, and g_i is the constant or linear output of the ith rule. Each rule can be considered as a sub-model of the fuzzy system and combination of the fuzzy rules makes the model nonlinear (Ozger 2009).

In the present study, fuzzy rule base was automatically set alike to the ANFIS system. The parameters of the input and output membership functions were optimized using genetic algorithm. The GA mechanism is illustrated in Fig. 10.1. In the GA, parameters are optimized by minimizing the objective function which is an error between model results and observed values. In this study, mean square error (MSE) was used as an objective function. The MSE is shown as

$$\text{MSE} = \frac{1}{M}\sum_{i=1}^{M}\left(y_{\text{observed}} - y_{\text{model}}\right)^2 \tag{10.2}$$

where M denotes the number of data set. Detailed information about Sugeno fuzzy system is given in the following section.

10.2.4 Adaptive Network-Based Fuzzy Inference System

The ANFIS, first introduced by Jang (1993), is a universal approximator and, as such, is capable of approximating any real continuous function on a compact set to any degree of accuracy (Unal et al. 2010). Therefore, in parameter prediction, where the given data are such that the system associates measurable system variables with an internal system parameter, a functional mapping may be constructed

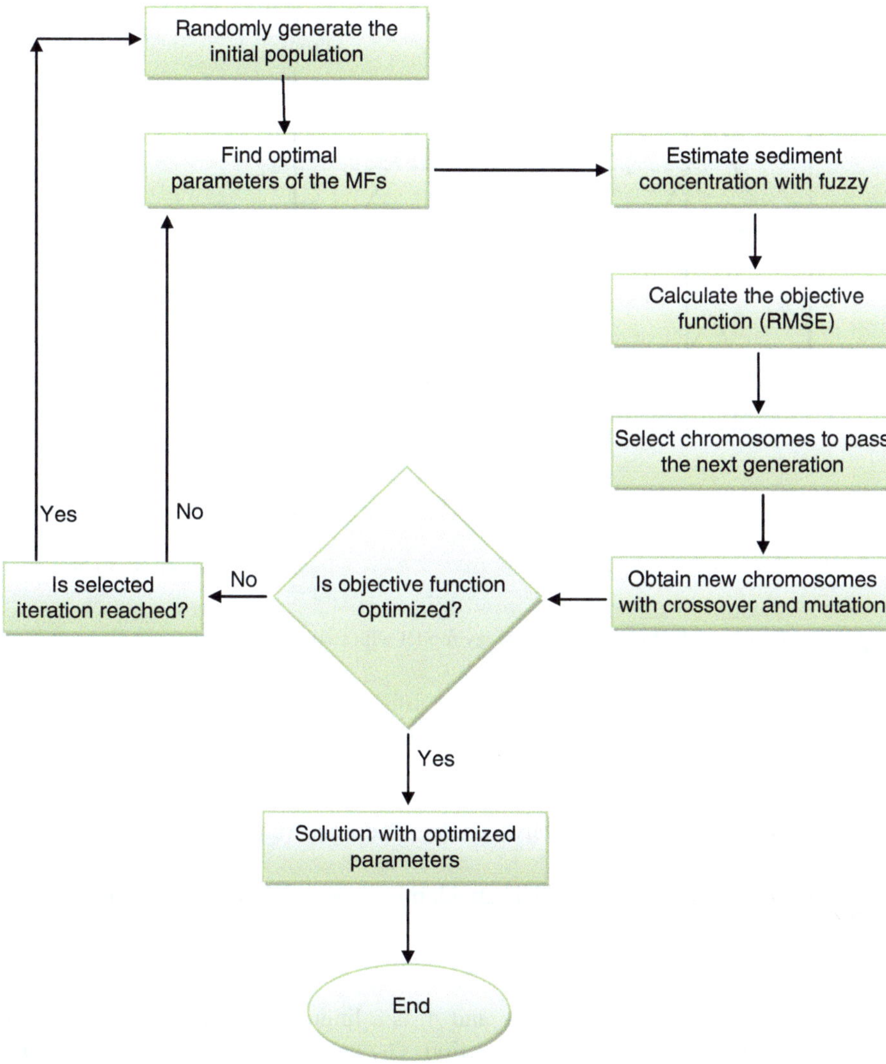

Fig. 10.1 The flowchart of the fuzzy genetic model

by the ANFIS that approximates the process of estimation of the internal system parameter (Kiszka et al. 1985b).

As a simple example, a fuzzy inference system with two inputs x and y and one output f is assumed. The first-order Sugeno fuzzy model, a typical rule set with two fuzzy IF-THEN rules can be expressed as follows:

$$\text{Rule } 1: \text{IF } x \text{ is } A_1 \text{ AND } y \text{ is } B_1 \text{ THEN } f_1 = p_1 x + q_1 y + r_1 \tag{10.3}$$

$$\text{Rule } 2: \text{IF } x \text{ is } A_2 \text{ AND } y \text{ is } B_2 \text{ THEN } f_2 = p_2 x + q_2 y + r_2 \tag{10.4}$$

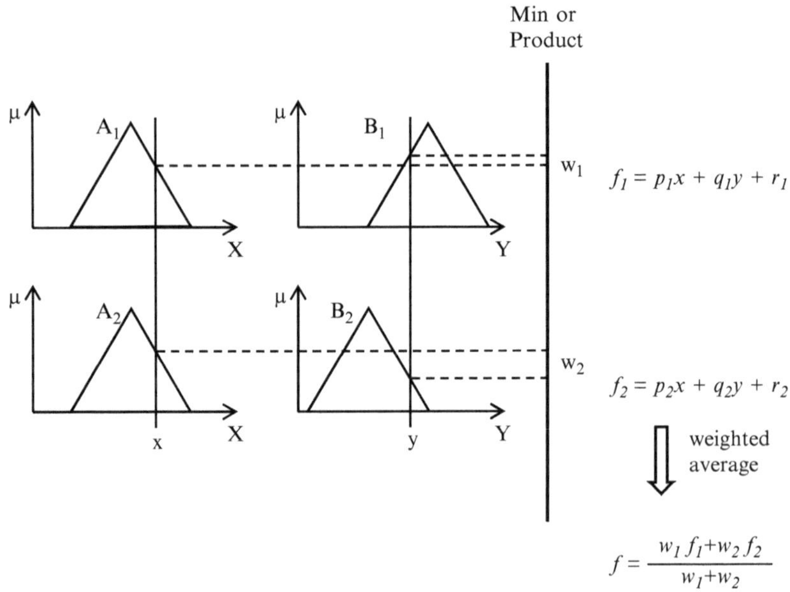

Fig. 10.2 Two-input first-order Sugeno fuzzy model with two rules

Here, the output f is the weighted average of the individual rules outputs and is itself a crisp value. The resulting Sugeno fuzzy reasoning system for two inputs is shown in Fig. 10.2.

The ANFIS is composed of five layers. The node functions in each layer are described next.

Every square node i in layer 1 is an adaptive node with a node function

$$O_{l,i} = \varphi A_i(x), \quad \text{for } i = 1, 2 \tag{10.5}$$

where x is the input to the ith node and A_i is a linguistic label (such as "low" or "high") associated with this node function. $O_{l,i}$ is the membership function (MF) of a fuzzy set A and it specifies the degree to which the given input x satisfies the quantifier A_i. $\varphi A_i(x)$ is usually chosen to be Gaussian function with maximum equal to 1 and minimum equal to 0

$$\varphi A_i(x) = \exp\left(-\left(\frac{x - a_i}{b_i}\right)^2\right) \tag{10.6}$$

where $\{a_i, b_i\}$ is the parameter set. As the values of these parameters change, the Gaussian function varies accordingly, thus exhibiting various forms of the MFs on linguistic label A_i (Jang 1993). Parameters of this layer are known as premise parameters.

Every circle node in layer 2 is labeled Π which multiplies the incoming signals and sends the product out. For instance,

$$w_i = \varphi A_i(x)\varphi B_i(y)\varphi C_i(z), \quad i = 1,2 \tag{10.7}$$

The output of each node represents the firing strength of a rule.

Every circle node in layer 3 is labeled N. Here the ratio of the ith rule's firing strength to the sum of all rules' firing strengths is calculated by ith node as

$$\overline{w}_i = \frac{w_i}{w_1 + w_2}, \quad i = 1,2 \tag{10.8}$$

Every square node in layer 4 has a node function

$$O_{4,i} = \overline{w}_i f_i = \overline{w}_i(p_i x + q_i y + r_i) \tag{10.9}$$

where $\overline{w}_i$ is the output of layer 3, and $\{p_i, q_i, r_i\}$ is the parameter set. Each parameter of this layer is referred to as consequent parameter.

The single circle node in layer 5 labeled Σ computes the final output as the summation of all incoming signals

$$O_{5,i} = \sum_{i=1} \overline{w}_i f_i = \frac{\sum_i w_i f_i}{\sum_i w_i} \tag{10.10}$$

Thus, an ANFIS network has been constructed which is functionally equivalent to the first-order Sugeno FIS. The ANFIS method uses linear or constant functions for the output. More information on the ANFIS can be found in Jang (1993).

10.2.5 Neural Networks

The ANNs are based on the present understanding of biological nervous system, although much of the biological detail is neglected. The ANNs are massively parallel systems that comprise many processing elements connected by links of variable weights. Of the many ANN paradigms, the MLP is by far the most popular (Lippman 1987). The network comprises layers of parallel processing elements, called neurons, with each layer being fully connected to the proceeding layer by interconnection strengths, or weights, W. Figure 10.3 illustrates a three-layer neural network comprising layers i, j, and k, with the interconnection weights W_{ij} and W_{jk} between layers of neurons. Initial assigned weight values are progressively rectified during a training process which compares anticipated outputs to known outputs, and back propagates any errors (from right to left in Fig. 10.3) to determine the appropriate weight adjustments necessary to minimize the errors.

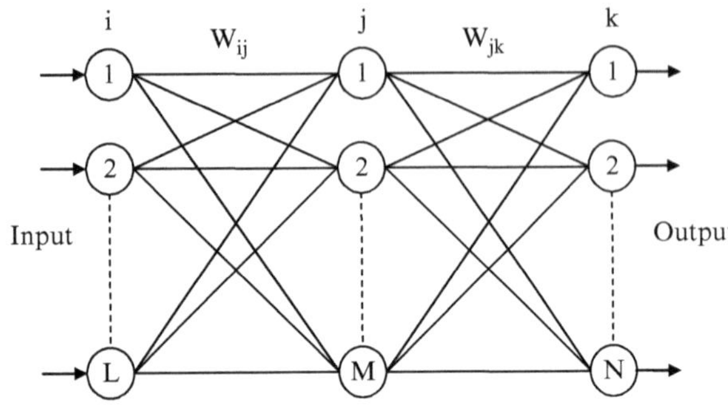

Fig. 10.3 A three-layer MLP structure

The Levenberg–Marquardt training algorithm was employed here in order to adjust the weights. The numbers of hidden layer neurons were found by using simple trial-error method (Altun et al. 2008).

As can be seen from Fig. 10.3, the neurons in the hidden layer receive summed inputs (SI) from input layer's neurons as

$$\mathrm{SI} = \sum_{i=1}^{L} W_{ij} O_{pi} + \varnothing_j \tag{10.11}$$

where L is number of the input set, $\varnothing_j$ is bias, W_{ij} are weights between the layers i and j, O_{pi} is the output of the layer i for sample p. The SI values are passed through a nonlinear activation function. Logistic activation function which is commonly used is given as

$$f(\mathrm{SI}) = \frac{1}{1 + \mathrm{e}^{-\mathrm{SI}}} \tag{10.12}$$

In training stage, total error of the network based on the squared difference between the predicted and observed outputs for pattern p is computed as

$$\mathrm{ER}_p = \sum_{k=1}^{N} \left(O_{pk} - P_{pk}\right)^2 \tag{10.13}$$

where O_{pk} and P_{pk} refer to observed and predicted outputs of the layer k for pattern p. The goal of the training period is to minimize the global error E_g as given in Eq. (10.14)

$$E_\mathrm{g} = \sum_{p=1}^{P} E_p \tag{10.14}$$

In the present study, the Levenberg–Marquardt training algorithm was used for the ANN models. In this algorithm, each weight is updated according to the following equation:

$$W_{ij}^{\text{yeni}} = W_{ij}^{\text{eski}} - \left[J^{\text{T}}J + \mu I\right]J^{\text{T}}H_{\text{p}} \tag{10.15}$$

where J is Jacobian matrix including the derivatives of the errors with respect to weights, J^{T} is transpose of the Jacobian matrix, I is unit matrix, and μ is a constant parameter which affects the convergence speed.

10.2.6 Sediment Rating Curve

A rating curve comprises an equation or graph, relating sediment discharge or concentration to streamflow discharge, which can be used to forecast daily SSC from streamflow record. The SRCs generally represent a functional relationship of the form:

$$S = aQ^b \tag{10.16}$$

where S is suspended sediment concentration and Q is streamflow discharge (Sandy 1990). a and b values for particular stream are determined from data by establishing a linear regression between (log S) and (log Q).

10.3 Applications and Results

The flow–sediment time series data from Muddy Creek near Vaughn (Station No: 06088300, Latitude 47°37′30″, Longitude 111°38′05″) and Muddy Creek at Vaughn (Station No: 06088500, Latitude 47°33′42″, Longitude 111°32′33″) operated by the US Geological Survey (USGS) were used in this chapter. The stations (Station No: 06088300 and 06088500) have drainage area 730.38 and 1,012.69 km^2 and gauge data 1,051.56 and 1,015.17 m above sea level, respectively. The daily time series of river flow and sediment concentration for these stations were downloaded from the web server of the USGS (http://co.water.usgs.gov/sediment). Data collected from September 1, 1978 to August 31, 1980 were used for calibration, and the data from September 1, 1980 to August 31, 1981 and September 1, 1981 to August 31, 1982 were chosen for testing and validation for both stations, respectively.

The statistical parameters of discharge and sediment concentration data for Muddy Creek stations are given in Table 10.1. In this table, the x_{mean}, S_{x}, C_{v}, C_{sx}, x_{max}, and x_{min} demonstrate the mean, standard deviation, variation and skewness coefficients, maximum and minimum, respectively. It is clear from

Table 10.1 The statistical parameters of data set for the stations

Station no	Data set	Data type	X_{mean}	S_x	C_v	C_{sx}	X_{max}	X_{min}	X_{max}/X_{mean}
06088300	Training	Discharge (m^3/s)	3.143	3.074	0.978	2.738	31.1	0.425	9.90
		Sediment (mg/L)	212.8	516.3	2.426	10.36	9,110	14	42.8
06088500	Training	Discharge (m^3/s)	4.110	4.034	0.981	2.128	34	0.453	8.27
		Sediment (mg/L)	527.0	882.6	1.675	5.828	13,000	16	24.7
06088300	Testing	Discharge (m^3/s)	3.414	3.603	1.055	4.632	43.6	0.311	12.8
		Sediment (mg/L)	259.8	781.9	3.010	11.31	12,000	19	46.2
06088500	Testing	Discharge (m^3/s)	4.418	4.539	1.027	3.501	49.3	0.425	11.2
		Sediment (mg/L)	669.1	1,482	2.215	9.206	21,100	31	31.5
06088300	Validation	Discharge (m^3/s)	3.142	3.086	0.982	1.710	19.2	0.538	6.11
		Sediment (mg/L)	245.9	504.4	2.051	6.045	5,540	13	22.5
06088500	Validation	Discharge (m^3/s)	4.078	4.092	1.003	1.509	23.8	0.566	5.84
		Sediment (mg/L)	603.9	984.7	1.631	3.171	6,820	31	11.3
06088300	Whole data	Discharge (m^3/s)	3.213	3.216	1.001	3.197	43.6	0.311	13.6
		Sediment (mg/L)	232.7	590.9	2.539	10.91	12,000	13	51.6
06088500	Whole data	Discharge (m^3/s)	4.183	4.178	0.999	2.429	49.3	0.425	11.8
		Sediment (mg/L)	581.9	1,087	1.868	8.008	21,100	16	36.3

the table that the sediment and flow data show a significantly high skewed distribution. The ratio between the standard deviation and the mean, C_v, is also high. The maximum–mean ratios ($X_{max}/X_{mean} = 51.6$ and 36.3) for sediment series are also quite high. All these statistics show the complexity of the discharge–sediment phenomenon as also stated by Kisi (2005). In the calibration sediment data, X_{min} and X_{max} values of both stations fall in the ranges 14–9,110 and 16–13,000 mg/L. However, the testing flow data set extremes are $X_{min} = 19$ mg/L, $X_{max} = 12{,}000$ mg/L and $X_{min} = 31$ mg/L, $X_{max} = 21{,}100$ mg/L, respectively. The value of X_{max} for the calibration sediment data is lower than that for the corresponding testing set. This may cause extrapolation difficulties in prediction of high sediment values.

In the present chapter, different fuzzy genetic (FG) models were tried in terms of the number of membership functions, type of input and output membership functions, population sizes, and generations. The FG models were compared with the ANFIS, MLP, and SRC models. Different program codes, including fuzzy logic, genetic algorithm, and neural network toolboxes, were written in MATLAB language for the simulations of the FG, ANFIS, MLP, and SRC techniques. The SRC formulas for the Muddy Creek Stations (06088300 and 06088500, respectively) were obtained as below:

$$S = 53.61Q^{0.829} \tag{10.17}$$

$$S = 70.65Q^{1.161} \tag{10.18}$$

Table 10.2 The performances of the ANFIS, FG, MLP, and SRC models in the testing and validation phases, 06088300

		Testing		Validation	
Model inputs	Model	RMSE (mg/L)	R^2	RMSE (mg/L)	R^2
Q_t	ANFIS (3, Gbellmf)	322.9	0.834	315.2	0.616
	FG (3, Gaussmf)	367.9	0.815	317.0	0.620
	MLP	368.8	0.779	381.0	0.561
	SRC	719.8	0.437	462.4	0.365
S_{t-1}	ANFIS (3, Pimf)	467.8	0.668	466.9	0.398
	FG (2, Pimf)	347.6	0.830	357.3	0.497
	MLP	415.3	0.743	362.2	0.499
Q_t and S_{t-1}	ANFIS (2, Gbellmf)	415.5	0.724	352.0	0.515
	FG (3, Gbellmf)	275.1	0.877	301.5	0.647
	MLP	279.8	0.874	235.1	0.786
S_{t-1} and S_{t-2}	ANFIS (3, Trimf)	377.9	0.773	402.0	0.470
	FG (3, Gaussmf)	365.0	0.828	352.3	0.518
	MLP	387.6	0.813	399.6	0.458
Q_t, S_{t-1}, and S_{t-2}	ANFIS (2, Trimf)	333.1	0.821	396.9	0.470
	FG (3, Pimf)	284.6	0.894	300.0	0.695
	MLP	306.9	0.851	352.1	0.526
Q_t, Q_{t-1}, Q_{t-2}, and S_{t-1}	ANFIS (2, Gaussmf)	249.9	0.936	230.6	0.819
	FG (3, Gauss2mf)	238.9	0.920	205.2	0.840
	MLP	242.3	0.926	196.4	0.856
Q_t, Q_{t-1}, Q_{t-2}, S_{t-1}, and S_{t-2}	ANFIS (2, Gbellmf)	424.8	0.772	561.1	0.524
	FG (2, Gbellmf)	242.8	0.919	239.3	0.775
	MLP	227.2	0.917	229.7	0.793

Root mean square errors (RMSE) and determination coefficient (R^2) statistics were used as the comparing criteria in this study. The RMSE and R^2 statistics are shown as below:

$$\mathrm{RMSE} = \sqrt{\frac{1}{n}\sum_{j=1}^{n}\left[(\mathrm{SSC})_{\mathrm{observed},j} - (\mathrm{SSC})_{\mathrm{predicted},j}\right]^2} \tag{10.19}$$

$$R^2 = \frac{\sum_{j=1}^{n}\left\{(\mathrm{SSC})_{\mathrm{observed},j} - \overline{(\mathrm{SSC})}_{\mathrm{observed}}\right\}^2 - \sum_{j=1}^{n}\left\{(\mathrm{SSC})_{\mathrm{observed},j} - (\mathrm{SSC})_{\mathrm{predicted},j}\right\}^2}{\sum_{j=1}^{n}\left\{(\mathrm{SSC})_{\mathrm{observed},j} - \overline{(\mathrm{SSC})}_{\mathrm{observed}}\right\}^2} \tag{10.20}$$

in which n is the number of data, the SSC is the suspended sediment concentration, and $\overline{\mathrm{SSC}}$ is mean of the SSC.

The RMSE and R^2 statistics of the FG, ANFIS, MLP, and SRC models for the Muddy Creek stations are represented in Tables 10.2 and 10.3. In these tables,

Table 10.3 The performances of the ANFIS, FG, MLP, and SRC models in the testing and validation phases, 06088500

		Testing		Validation	
Model inputs	Model	RMSE (mg/L)	R^2	RMSE (mg/L)	R^2
Q_t	ANFIS (2, Trimf)	531.1	0.637	675.0	0.551
	FG (3, Gbellmf)	556.6	0.870	676.9	0.543
	MLP	719.0	0.873	691.9	0.516
	SRC	1,074.5	0.767	774.4	0.484
S_{t-1}	ANFIS (2, Trimf)	946.2	0.599	472.5	0.772
	FG (3, Gauss2mf)	970.6	0.586	476.6	0.766
	MLP	994.2	0.574	471.4	0.771
Q_t and S_{t-1}	ANFIS (2, Gbellmf)	716.2	0.840	424.7	0.823
	FG (3, Gauss2mf)	367.7	0.941	424.6	0.822
	MLP	409.2	0.929	406.0	0.843
S_{t-1} and S_{t-2}	ANFIS (3, Pimf)	979.3	0.614	475.0	0.767
	FG (2, Gauss2mf)	963.3	0.593	458.3	0.799
	MLP	964.7	0.625	435.6	0.806
Q_t, S_{t-1}, and S_{t-2}	ANFIS (2, Trimf)	891.9	0.651	463.0	0.785
	FG (3, Gbellmf)	403.6	0.938	430.5	0.817
	MLP	462.8	0.903	475.5	0.774
Q_t, Q_{t-1}, Q_{t-2}, and S_{t-1}	ANFIS (2, Gaussmf)	698.4	0.790	380.3	0.873
	FG (2, Gauss2mf)	307.6	0.960	291.3	0.912
	MLP	391.6	0.944	310.0	0.901
Q_t, Q_{t-1}, Q_{t-2}, S_{t-1}, and S_{t-2}	ANFIS (2, Gaussmf)	594.5	0.846	298.4	0.909
	FG (2, Gbellmf)	279.3	0.965	284.1	0.917
	MLP	404.1	0.935	284.9	0.916

the ANFIS (3, Gbellmf) indicates an ANFIS model having three Gbellmf membership functions (MFs) for the input Q_t and FG (3, Gaussmf) refers an FG model comprising three Gaussmf MFs for the input Q_t. Table 10.2 reveals that the MLP model having five input parameters performs better than the FG, ANFIS, and SRC models from the RMSE viewpoint. The ANFIS (2, Gaussmf) model has the highest R^2 value for testing phase. Note that the R^2 term provides information for linear dependence between observations and corresponding estimates. Therefore, it is not always expected that R^2 is in agreement with performance criteria such as the RMSE. For example, in the case of two time series such as ($X_i = 1, 2, 3, \ldots, 10$; $Y_i = 20, 40, 60, \ldots, 200$) the R^2 between these two series is equal to 1 whereas the RMSE value is quite high. An R^2 value equal to 1 does not guarantee that a model captures the behaviour of the investigated time series (Kisi et al. 2008). In the validation phase, however, the FG (3, Gauss2mf) model having two Gaussian MFs for the input combination Q_t, Q_{t-1}, Q_{t-2}, and S_{t-1} outperforms the optimal ANFIS (2, Gaussmf), MLP, and SRC models.

For the second station, the FG (2, Gbellmf) model having two gumbell MFs for the input combination Q_t, Q_{t-1}, Q_{t-2}, S_{t-1}, and S_{t-2} performs better than the other models in terms of the RMSE and R^2 in the testing and validation phases.

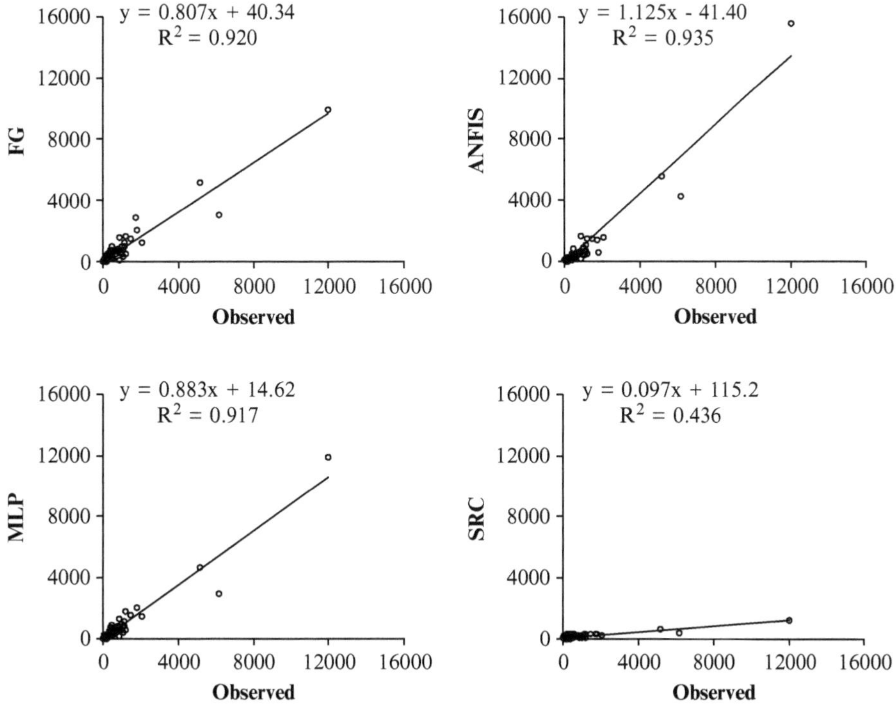

Fig. 10.4 Plotting of prediction performances for the testing phase using the FG, ANFIS, MLP, and SRC-Muddy Creek Station, 06088300

The comparison of the observed and predicted SSCs in testing phase is shown in Fig. 10.4 for the first station (Station no: 06088300) in the form of scatterplot. It can be obviously seen from Fig. 10.4 that the estimates of the ANFIS are closer to the exact fit line with higher R^2 value than the FG, MLP, and SRC in the testing phase. Figure 10.5 illustrates the suspended sediment estimates of the models in logarithm scaled. Low- and high sediment estimates can be clearly seen from these double logarithm graphs. It is clear from Fig. 10.5 that the ANFIS model is not as successful as the FG model in low sediment estimation for the test phase.

The observed and estimated SSC values of the first station (Station no: 06088300) in the validation phase are illustrated in Fig. 10.6. It can be clearly seen from Fig. 10.6 that the predictions of the FG model are closer to the exact fit line than the ANFIS, MLP, and SRC models. These confirm the statistics given in Table 10.2.

The comparison of the observed and predicted SSCs in testing phase is shown in Fig. 10.7 for the second station (Station no: 06088500) in the form of scatterplot. It can be obviously seen from Fig. 10.7 that the estimates of the FG are closer to the exact fit line with higher R^2 value than the ANFIS, MLP, and SRC in the testing phase. Figures 10.8 and 10.9 illustrate the plotting of prediction performances of the

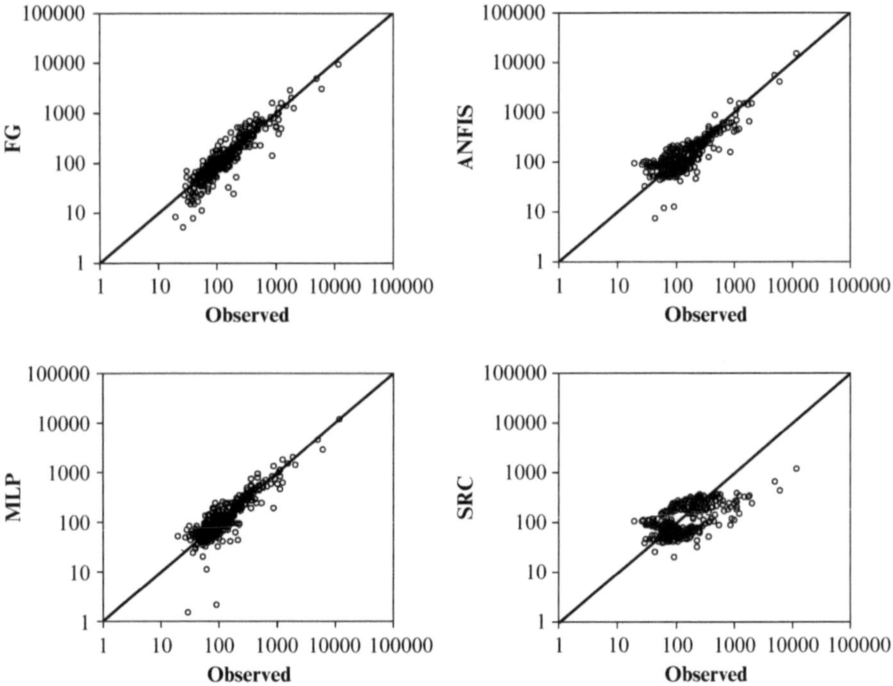

Fig. 10.5 Plotting of prediction performances for the testing phase using the FG, ANFIS, MLP, and SRC (logarithm scaled)-Muddy Creek Station, 06088300

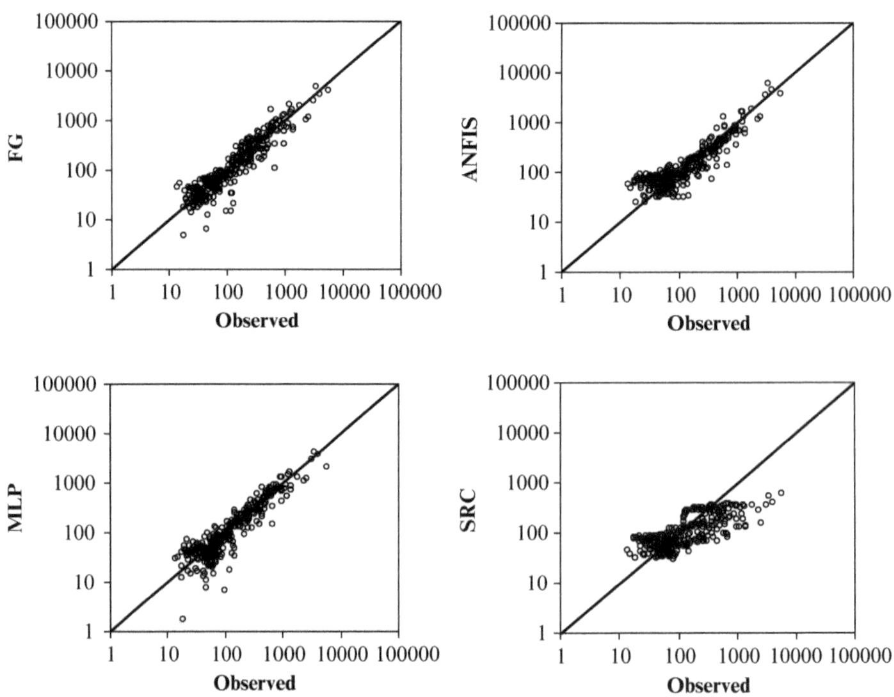

Fig. 10.6 Plotting of prediction performances for the validation phase using the FG, ANFIS, MLP, and SRC (logarithm scaled)-Muddy Creek Station, 06088300

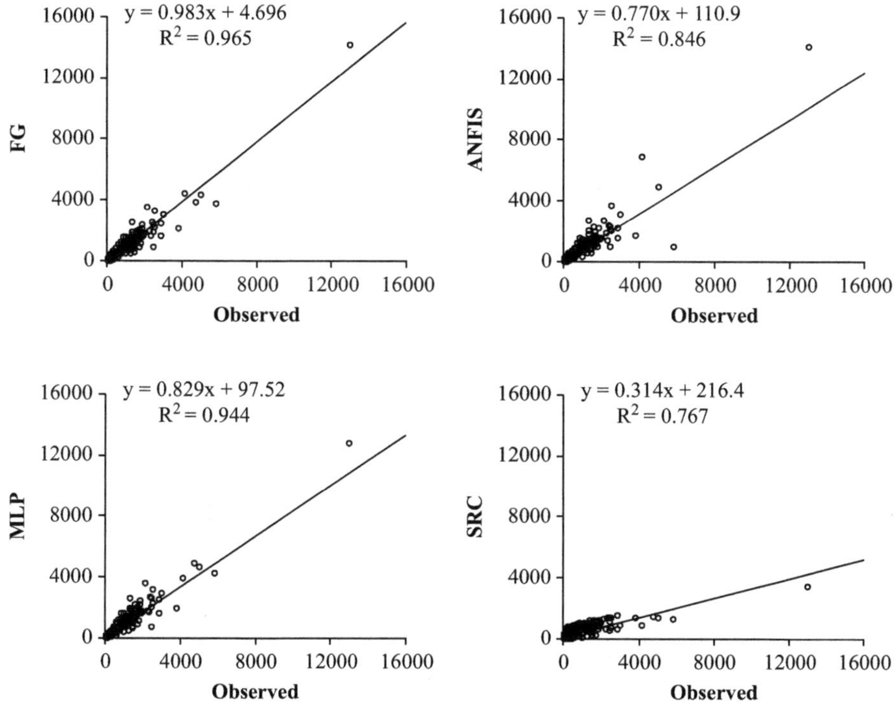

Fig. 10.7 Plotting of prediction performances for the testing phase using the FG, ANFIS, MLP, and SRC-Muddy Creek Station, 06088500

FG, ANFIS, MLP, and SRC for the second station (Station no: 06088500) in the testing and validation phases. It can be seen from Fig. 10.8, the FG model estimates are closer to the exact line than those of the ANFIS and MLP models especially for the low sediment values in the testing phase. The underestimations of the high values are obviously seen for the SRC model. Validation estimates of the models shown in Fig. 10.9 indicate that both the FG and MLP models give estimates closer to the exact line than those of the ANFIS and SRC models especially for the low sediment values in testing phase for the second station (Station no: 06088500). From Table 10.3, however, it can be said that the FG model performs slightly better than the MLP and both the ANFIS and SRC models.

The peak sediment estimates of the FG, ANFIS, MLP, and SRC models are given in Tables 10.4 and 10.5 for the validation phase of each station, respectively. In Table 10.4, in general, the MLP and FG models seem to have almost equal accuracy in estimating sediment peaks for the first station. The ANFIS shows better accuracy than the FG and MLP for the first two peaks both in the validation phase, but, for the other peaks, the FG shows better accuracy than the ANFIS. From Table 10.5, it is clear that the FG model has closer estimates than the ANFIS and MLP models. Here, the ANFIS model is ranked as the second best. The ANFIS

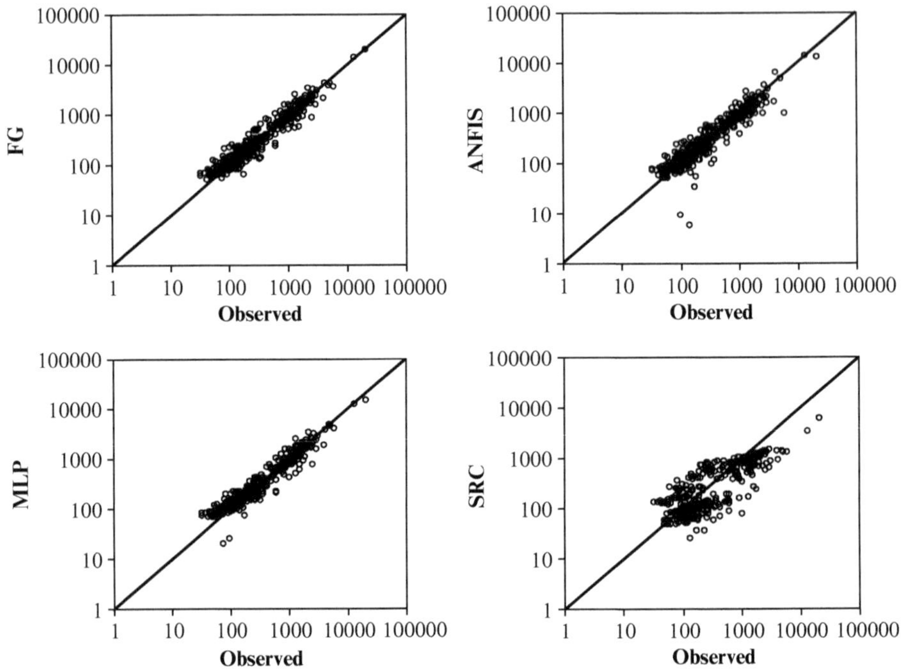

Fig. 10.8 Plotting of prediction performances for the testing phase using the FG, ANFIS, MLP, and SRC (logarithm scaled)-Muddy Creek Station, 06088500

model has a better accuracy than the FG and MLP models for the second and third peaks. However, in general, the FG shows better accuracy than the ANFIS and MLP in estimating sediment peaks for the second station. It is clear from the tables that the SRC model significantly underestimates the observed sediment peaks for both the stations.

The total sediment load estimates of the FG, ANFIS, MLP, and SRC models in the testing and validation phases are given in Table 10.6. It is clear from this table that the total sediment estimates of the FG model are closer to the corresponding observed values than the other models in both the stations.

The performances of the optimal the FG, ANFIS, MLP, and SRC models in the testing and validation phases for each station are given in Table 10.7. For the first station, the FG and MLP seem to give better predictions than the ANFIS and SRC in the test phase. The MLP has a lower RMSE value than the FG in the testing phase but, in the validation phase, the FG model has lower RMSE and higher R^2 values than the MLP and ANFIS models. Therefore, it seems from this table that, in general, the FG gives slightly better predictions than the MLP. For the second station, the FG seems to give better predictions than the ANFIS, MLP, and SRC in both the testing and validation phases.

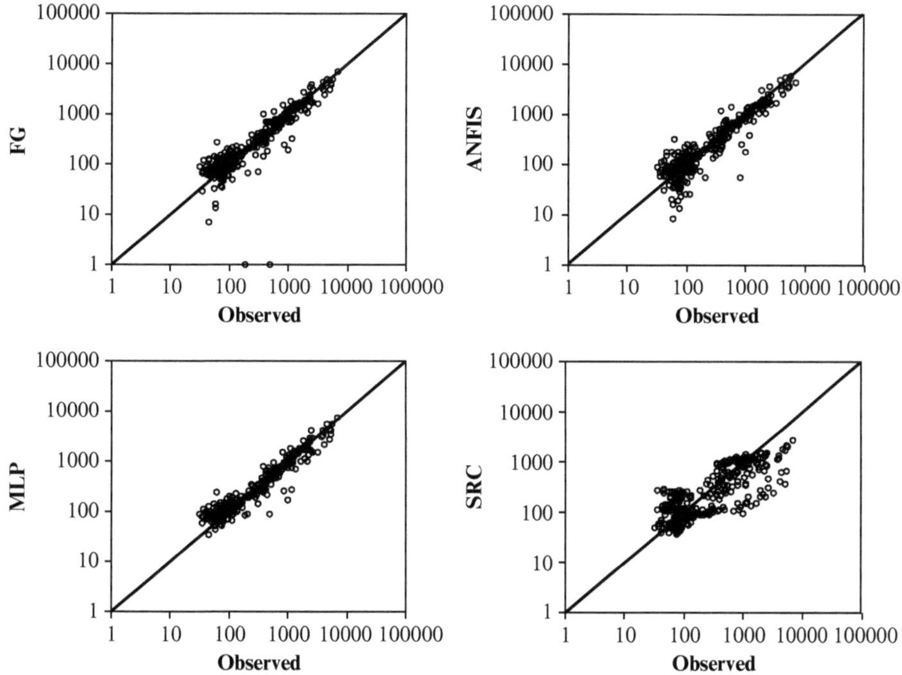

Fig. 10.9 Plotting of prediction performances for the validation phase using the FG, ANFIS, MLP, and SRC (logarithm scaled)-Muddy Creek Station, 06088500

Table 10.4 The comparison of the FG, ANFIS, MLP, and SRC peak-estimates for the validation phase, 06088300

Peaks >2,000 mg/L	FG	ANFIS	MLP	SRC	Relative error FG	Relative error ANFIS	Relative error MLP	Relative error SRC
2,480	1,255	1,360	1,317.5	168.8	−49.3	−45.1	−46.9	−93.1
2,210	1,107	1,252	1,169.5	293.9	−49.8	−43.3	−47.1	−86.6
3,090	2,639	3,740	3,104.3	364.6	−14.5	21.0	0.5	−88.2
3,930	3,571	4,659	4,077.3	429.3	−9.12	18.5	3.7	−89.0
5,540	4,169	3,901	2,187.0	621.0	−24.7	−29.5	−60.5	−88.7
3,280	5,117	6,241	4,544.5	558.6	56.0	90.2	38.6	−82.9
		Total (absolute) =			203.42	247.6	197.3	528.5

Table 10.5 The comparison of the FG, ANFIS, MLP, and SRC peak-estimates for the validation phase, 06088500

Peaks >2,000 mg/L	FG	ANFIS	MLP	SRC	Relative Error FG	Relative Error ANFIS	Relative Error MLP	Relative Error SRC
2,600	1,707.4	1,658.6	1,572.2	306.6	−34.3	−36.2	−39.5	−88.2
3,130	1,666.4	1,765.6	1,503.1	245.1	−46.8	−43.6	−52.0	−92.2
4,940	3,096.9	3,496.3	2,779.8	382.1	−37.3	−29.2	−43.7	−92.3
5,390	3,936.8	3,757.1	3,552.7	2,235.7	−27.0	−30.3	−34.1	−58.5
2,330	2,262.0	2,208.8	2,412.2	1,400.4	−2.9	−5.2	3.5	−39.9
2,290	2,282.8	1,815.5	2,397.2	1,388.0	−0.3	−20.7	4.7	−39.4
		Total (absolute) =			148.6	165.2	177.5	410.5

Table 10.6 The observed and predicted total SSC values in the testing and validation phases for both the stations

		Total SSC (mg/L)		Relative error (%)	
Methods	Station no.	Testing phase	Validation phase	Testing phase	Validation phase
Observed	06088300	94,816	89,741	–	–
FG	06088300	91,334	91,085	−3.7	1.5
ANFIS	06088300	91,581	87,642	−3.4	−2.3
MLP	06088300	89,085	84,685	−6.0	−5.6
SRC	06088300	51,321	47,839	−46	−47
Observed	06088500	244,219	220,415	–	–
FG	06088500	241,890	216,634	−1.0	−1.7
ANFIS	06088500	228,598	211,338	−6.4	−4.1
MLP	06088500	238,178	216,172	−2.5	−1.9
SRC	06088500	155,745	142,349	−36	−35

Table 10.7 The performances of the FG, ANFIS, MLP, and SRC models in the testing and validation phases for each station

		Testing phase		Validation phase	
Methods	Station no.	RMSE (mg/L)	R^2	RMSE (mg/L)	R^2
FG	06088300	238.9	0.920	205.2	0.840
ANFIS	06088300	249.9	0.936	230.6	0.819
MLP	06088300	227.2	0.917	229.7	0.793
SRC	06088300	719.8	0.437	462.4	0.365
FG	06088500	279.3	0.965	284.1	0.917
ANFIS	06088500	594.5	0.846	298.4	0.909
MLP	06088500	391.6	0.944	310.0	0.901
SRC	06088500	1,074.5	0.767	774.4	0.484

10.4 Conclusion

The accuracy of the FG approach in prediction of the SSC was investigated in this chapter. The FG models were tested by applying different input combinations of the SSC and flow data from Muddy Creek Stations, Montana. The performance of the FG was compared with those of the ANFIS, MLP, and SRC models. The results obtained using the FG model to predict sediment concentration were found to be better than those reached using the ANFIS and SRC models and taking into consideration both stations, the FG model also gave better predictions than the MLP model. Therefore, they confirmed the ability of this approach to provide a useful tool in solving a specific problem in hydrology, such as forecasting suspended sediment. The SRC technique gave significantly poor estimates relative to the FG, ANFIS, and MLP models. The results suggest that the FG may provide a superior alternative to the ANFIS, MLP, and SRC approaches for developing input–output simulations and estimation models in situations that do not require modeling of the internal structure of the watershed.

Acknowledgements The data used in this study were downloaded from the web server of the USGS. The authors wish to thank the staff of the USGS who are associated with data observation, processing, and management of USGS web sites.

References

Ahmed JA, Sarma AK (2005) Genetic algorithm for optimal operating policy of a multipurpose reservoir. J Water Resour Manag 19(2):145–161

Altun F, Kisi O, Aydin K (2008) Predicting the compressive strength of steel fiber added lightweight concrete using neural network. Comput Mater Sci 42(2):259–265

Altunkaynak A (2009) Sediment load prediction by genetic algorithms. Adv Eng Softw 40(9): 928–934

Burn DH, Yulianti JS (2001) Waste-load allocation using genetic algorithm. J Water Resour Plann Manag 127(2):121–129

Cigizoglu HK (2004) Estimation and forecasting of daily suspended sediment data by multi layer perceptrons. Adv Water Resour 27:185–195

Cobaner M, Unal B, Kisi O (2009) Suspended sediment concentration estimation by an adaptive neuro-fuzzy and neural network approaches using hydro-meteorological data. J Hydrol 367(1–2):52–61

Firat M, Gungor M (2010) Monthly total sediment forecasting using adaptive neuro fuzzy inference system. Stoch Environ Res Risk Assess 24(2):259–270

Gokmen T and Guldal V (2006) Artificial neural networks for estimating daily total suspended sediment in natural streams. Nord Hydrol 37(1):69–79

Goldberg DE (1989) Genetic algorithms in search, optimization and machine learning. Addison-Wesley Longman, Boston

Goldberg DE, Deb K (1991) A comparative analysis of selection schemes used in genetic algorithms. In Rawlings, G., editor, Foundation of genetic algorithms, pages 69–93, Morgan Kaufman, San Mateo.

Guldal V, Muftuoglu RF (2001) 2D unit sediment graph theory. J Hydrol Eng 6(2):132–140

Jain SK (2001) Development of integrated sediment rating curves using ANNs. J Hydraulic Eng ASCE 127(1): 30–37

Jang JSR (1993) ANFIS: adaptive network based fuzzy inference system. IEEE Trans Syst Manage Cybern 23(3):665–685

Kisi O (2004a) Daily suspended sediment modelling using a fuzzy differential evolution approach. Hydrol Sci J 49(1):183–197

Kisi O (2004b) Multi-layer perceptrons with Levenberg-Marquardt training algorithm for suspended sediment concentration prediction and estimation. Hydrol Sci J 49(6):1025–1040

Kisi O (2005) Suspended sediment estimation using neuro-fuzzy and neural network approaches. Hydrol Sci J 50(4):683–696

Kisi O (2008) Constructing neural network sediment estimation models using a data-driven algorithm. Math Comput Simul 79(1):94–103

Kisi O (2010) Fuzzy genetic approach for modeling reference evapotranspiration. J Irrigat Drain 136(3):175–183

Kisi O, Karahan ME, Sen Z (2006) River suspended sediment modeling using fuzzy logic approach. Hydrol Process 20(20):4351–4362

Kisi O, Yuksel I, Dogan E (2008) Modelling daily suspended sediment of rivers in Turkey using several data driven techniques. Hydrol Sci J 53(6):1270–1285

Kisi O, Haktanir T, Ardiclioglu M, Ozturk O, Yalcin E, Uludag S (2009) Adaptive neuro-fuzzy computing technique for suspended sediment estimation. Adv Eng Softw 40(6):438–444

Kiszka JB, Kochanskia ME, Sliwinska DS (1985a) The influence of some fuzzy implication operators on the accuracy of fuzzy model. Part I Fuzzy Sets and Systems 15:111–128
Kiszka JB, Kochanskia ME, Sliwinska DS (1985b) The influence of some fuzzy implication operators on the accuracy of fuzzy model. Part II Fuzzy Sets Systems 15:223–240
Kosko B (1993) Fuzzy thinking: the new science of fuzzy logic. Hyperion, New York
Lippman R (1987) An introduction to computing with neural nets, IEEE ASSP Mag 4:4–22
Lohani AK, Goel NK, Bhatia KKS (2007) Deriving stage-discharge-sediment concentration relationships using fuzzy logic. Hydrol Sci J 52(4):793–807
Oliveira R, Loucks DP (1997) Operating rules for multireservoir systems. Water Resour Res 33(4):839–852
Ozger M (2009) Comparison of fuzzy inference systems for streamflow prediction. Hydrol Sci J 54(2):261–273
Rajaee T, Mirbagheri SA, Zounemat-Kermani M, Nourani V (2009) Daily suspended sediment concentration simulation using ANN and neuro-fuzzy models. Sci Total Environ 407(17): 4916–4927
Reza Pour OM, Shui LT, Dehghani AA (2011) Genetic algorithm model for the relation between flow discharge and suspended sediment load (Gorgan river in Iran). Electron J Geotech Eng 16:539–553
Ross TJ (1995) Fuzzy logic with engineering applications. McGraw-Hill, New York
Russel SO, Campbell PF (1996) Reservoir operating rules with fuzzy programming. J Water Resour Plann Manag 122:165–170
Sandy R (1990) Statistics for business and economics. McGraw-Hill, New York
Sen Z (1998) Fuzzy algorithm for estimation of solar irrigation from sunshine duration. Sol Energ 63(1):39–49
Şen Z (2004) Fuzzy logic and system models in water sciences. Turkish Water Foundation, Istanbul, Turkey
Tayfur G, Ozdemir S, Singh VP (2003) Fuzzy logic algorithm for runoff-induced sediment transport from bare soil surfaces. Adv Water Resour 26(12):1249–1256
Unal B, Mamak M, Seckin G, Cobaner M (2010) Comparison of an ANN approach with 1-D and 2-D methods for estimating discharge capacity of straight compound channels. Adv Eng Softw 41(2):120–129
Wang QJ (1991) The genetic algorithm and its application to calibrating conceptual rainfall-runoff models. Water Resour Res 27(9):2467–2471
Zadeh LA (1965) Fuzzy sets. Inform Control 8(3):338–353

Chapter 11
Prediction of Local Scour Depth Downstream of Bed Sills Using Soft Computing Models

A. Zahiri, H. Md. Azamathulla, and Kh. Ghorbani

Abstract Bed sill local scour is an important issue in environmental and water resources engineering in order to prevent degradation of river bed and save the stability of grade-control structures. This chapter presents genetic algorithms (GA), gene expression programming, and M5 decision tree model as an alternative approaches to predict scour depth downstream of bed sills. Published data were compiled from the literature for the scour depth downstream of sills. The proposed GA approach gives satisfactory results ($R^2 = 0.96$ and RMSE $= 0.442$) compared to existing predictors.

Keywords Grade-control structures • Local scour • Genetic algorithms • M5 tree decision model • Gene expression programming

11.1 Introduction

Bed sills are a common solution to stabilize degrading bed rivers and channels. They are aimed at preventing excessive channel-bed degradation in alluvial channels. Although their presence limits the general erosion process in the upstream, but the erosive action of the weir overflow and turbulence generated from plunging jet may cause significant local scour at downstream. By this local scour, the structure itself (and many times other structures in vicinity of it, like bridge piers or abutments, or bank revetments) might be undermined (Bormann and Julien 1991; Gaudio and Marion 2003).

A. Zahiri (✉) • K. Ghorbani
Water Engineering Department, Gorgan University of Agricultural Sciences and Natural Resources, Shahid Beheshti Blvd., PO Box 386, Gorgan, Golestan Province, Iran
e-mail: zahiri_reza@yahoo.com

H. M. Azamathulla
REDAC, Universiti Sains Malaysia, Georgetown, Pulau Penang, Malaysia
e-mail: mdazmath@gmail.com

T. Islam et al. (eds.), *Computational Intelligence Techniques in Earth and Environmental Sciences*, DOI 10.1007/978-94-017-8642-3_11,

For practical purposes, designers and civil engineers are often interested in a short-term local scouring and its extent. For instance, they are often required to predict the maximum scour depth at bed sills in the proximity of bridges when a flood occurs. Therefore, most researchers have focused on local scouring at isolated or series bed sill structures. Summaries of research for the bed sills can be found in Lenzi et al. (2002). Bormann and Julien (1991) reviewed experimental studies of scour downstream of hydraulic structures. They also investigated the scour downstream of grade-control structures with large-scale experiments. Gaudio et al. (2000) presented a theoretical framework for calculation of maximum scour depth downstream of bed sills through identification of non-dimensional parameters by Buckingham's π-theorem. They proposed an empirical equation for the estimation of maximum scour depth at bed sills taking into account of morphological effects in low-gradient rivers through clear water laboratory tests. Later, their result was generalized by Lenzi et al. (2002) to cover steep channels. Under similar slopes and hydrological settings, an equation developed through laboratory results to predict the maximum scour holes at grade-control structures in alluvial mountain rivers. Lenzi and Comiti (2003) analyzed local scouring downstream of 29 drop structures. Lenzi et al (2003) investigated the main characteristics of local scouring downstream of bed sills in the form of a staircase-like system in high-gradient streams with nonuniform alluvium. They found that the jet regime plays an important role both for the depth and the length of the scour, and consequently affects the scour shape. They proposed two equations for the estimation of the maximum scour depth and length in low and high gradient streams. Marion et al. (2004) conducted a series of tests to determine the effect of bed sill spacing and sediment grading on the potential erosion by jets flowing over the sills. Tregnaghi (2008) conducted some experimental runs in the case of clear water and live bed scouring at bed sills placed in steep gravel bed streams. He concluded that the percentage reduction in maximum scour depth in the case of sediment feeding compared with the clear water tests is considerable. Chinnarasri and Kositgittiwong (2008) conducted some experimental tests in steep slopes and bed sills with different sill spacing. They proposed a simple equation based on nonlinear regression model for relative maximum scour depth at the equilibrium condition.

Although useful in many circumstances, these empirical formulae have one key shortcoming. Specifically, due to wide ranges of hydraulic and sediment characteristics of flow and also bed slopes in rivers, application of any empirical equation doesn't reflect the complex actual conditions of river and structure itself and also the boundary conditions at the downstream of structures. Owing to rapid increase in successful applications of neural computing, machine learning and evolutionary algorithms in many fields of hydraulic engineering, and also owing to high complexity of scouring phenomena at bed sills, there is a need to explore the applicability of these new methods in prediction of maximum scour depth at bed sills. In this regard, Guven and Gunal (2008) using genetic programming (GP) provides alternative formulation for prediction of local scour downstream of grade-control structures. Azamathulla (2012) presents an equation using gene expression programming (GEP) for prediction of scour depth downstream of sills. In this study, using the 226 experimental data set of maximum scour depth at bed sills from literatures in different canal bed slopes and at clear water scouring, applicability of new methods of the GA, GEP, and M5 tree model have been examined in prediction of relative maximum scour depth at bed sills. The results have been compared with the empirical equations obtained by previous researchers.

11.2 Material and Methods

11.2.1 Physical Definition of Scouring

According to Gaudio et al. (2000), the main variable of interest, equilibrium maximum scour depth (y_s), in the case of uniform sediment beds is mainly dependent on flow and sediment characteristics as follows:

$$y_s = f\left(g, \upsilon, \rho_w, \rho_s, q, q_s, h, D, S_0, S_{eq}, L\right) \tag{11.1}$$

where g = acceleration of gravity, υ = kinematic viscosity of water, ρ_w, and ρ_s = density of water and sediment, respectively, q and q_s = water and sediment discharge per unit width supplied by upstream, respectively, h = water depth of uniform flow condition, D = characteristic grain size, S_0 and S_{eq} = initial and equilibrium bed slopes, respectively, and L = horizontal spacing between sills. This is a general definition of the maximum scour depth for hydraulic, geometric, and sediment properties (see Fig. 11.1).

The application of Buckingham's π-theorem leads to identification of the following dimensionless group (Chinnarasri and Kositgittiwong 2008):

$$\frac{y_s}{H_s} = f_2\left(\frac{a}{H_s}, \frac{a}{\Delta D_{50}}, \frac{L}{H_s}, \frac{D_{50}}{H_s}, S_0\right) \tag{11.2}$$

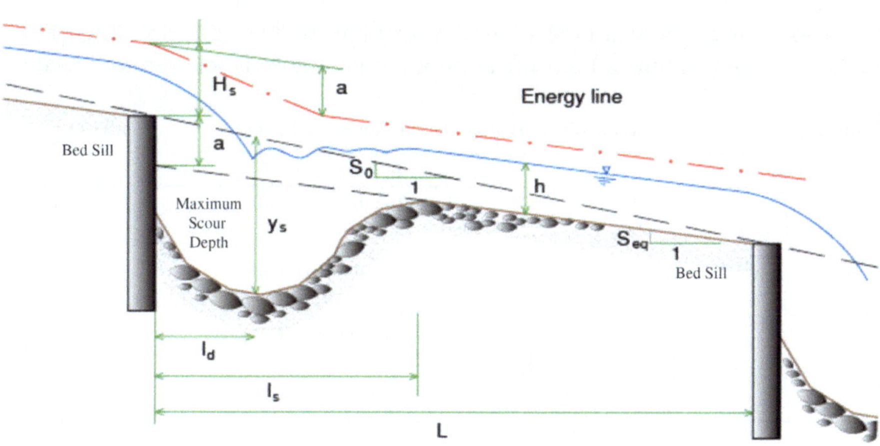

Fig. 11.1 Schematic of scour depth and length downstream of a bed sill (Tregnaghi 2008)

where $\Delta = (\rho_s - \rho_w)/\rho_w$ is the relative submerged density of sediment, $H_s = 1.5\sqrt[3]{q^2/g}$ is critical-specific energy on the sills and a = morphological jump, which was first introduced by Gaudio et al. (2000). This important factor defines a geometrical correspondence between the initial and equilibrium bed slopes and the spacing between sills:

$$a = \left(S_0 - S_{eq}\right)L \tag{11.3}$$

11.2.2 Scouring Prediction at Bed Sills

11.2.2.1 Empirical Equations

According to non-dimensional parameters obtained for maximum scour depth at bed sills, some empirical equations based on regression analysis of experimental data have been developed. These equations are presented in Table 11.1.

11.2.2.2 Genetic Algorithm

Genetic algorithm (GA) technique is capable of solving complex problems that the traditional algorithms have been unable to conquer. This algorithm begins by creating an initial random set of potential solutions for a particular problem. Then, the fittest "parents" are selected and "children" are generated by means of sexual reproduction (crossover) or asexual alteration (mutation). In crossover, two parents swap random pieces of information with each other while in mutation, a piece of information is replaced by another randomly generated piece. Finally, the resulting solutions (children) are evaluated for their fitness (effectiveness) and selected for reproduction. This process is repeated over-successive generations until a stopping criterion is met (Sharifi 2009).

Once the initial population is generated, each chromosome is evaluated and its "goodness" (fitness) is measured using some measure of fitness function. Then, based on the value of this fitness function, a set of chromosomes is selected for breeding.

Table 11.1 Empirical equations for maximum scour depth prediction

Empirical equation	Investigator	Eq. Number
$\frac{y_s}{H_s} = 1.45\left(\frac{a}{H_s}\right)^{0.86} + 0.06\left(\frac{a}{\Delta d_{50}}\right)^{1.49} + 0.44$	Lenzi et al. (2004)	(11.4)
$\frac{y_s}{H_s} = 1.6\left(\frac{a}{H_s}\right)^{0.61} + 1.89\left(\frac{a}{\Delta d_{50}}\right)^{0.21} - 2.03$	Chinnarasri and Kositgittiwong (2008)	(11.5)
$\frac{y_s}{H_s} = 3\left(\frac{a}{H_s}\right)^{0.6} SI^{-0.19}\left(1 - e^{-0.25\frac{L}{H_s}}\right)$	Tregnaghi (2008)	(11.6)

In order to simulate a new generation, genetic operators such as crossover and mutation are applied to the selected parents. The offsprings are evaluated and the members of the next generation population are selected from the set of parents and offsprings. This cycle continues until the termination criterion is met (Sharifi 2009).

In this study, the absolute percentage error of output parameter prediction was selected as the performance measure. The selected objective function is as follows:

$$f = \text{Min} \sum_{i=1}^{i=N} \left(\left(\frac{y_s}{H_s} \right)_{\text{exp}} - \left(\frac{y_s}{H_s} \right)_{\text{cal}} \right)^2 \tag{11.7}$$

where N is the number of sample tests and the subscripts exp and cal refer to experimental value and the predictions obtained using each model selected in this study, respectively.

11.2.2.3 Gene Expression Programming

GEP, which is an extension of the GP (Koza 1992), is a search technique that evolves computer programs of different sizes and shapes encoded in linear chromosomes of fixed lengths. The chromosomes are composed of multiple genes, each gene encoding a smaller subprogram. Furthermore, the structural and functional organization of the linear chromosomes allows the unconstrained operation of important genetic operators such as mutation, transposition, and recombination (Azamathulla 2012).

11.2.2.4 M5 Tree Model

Dividing a complex modeling problem into a number of subproblems and combining their solutions is the main idea in building model trees (MT). In this idea, the parameter space is split into areas (subspaces) and a linear regression model is built in each of them, which is an "expert" for that subspace. The algorithm makes it possible to split the multidimensional parameter space into subspaces and to generate the models automatically for each subspace according to an overall quality criterion. First, the initial tree is built and then the initial tree is pruned (reduced) to overcome the over-fitting problem (that is a problem when a model is very accurate on the training data set and fails on the test set). Finally, the smoothing process is employed to compensate for the sharp discontinuities between adjacent linear models at the leaves of the pruned tree (this operation is not needed in building the decision tree). In smoothing, the adjacent linear equations are updated in such a way that the predicted outputs for the neighboring input vectors corresponding to the different equations are becoming close in value (Witten and Frank 2005).

Table 11.2 Range of geometric and hydraulic parameters for scouring at bed sills

Symbol	variable definition	Variable range	Mean value
L (m)	Sills spacing	0.4–2.5	1.07
S_0	Initial bed slope	0.0059–0.268	0.1099
Q (L/s)	Flow discharge	0.68–30.6	16.5
D_{50} (mm)	Sediment mean diameter	0.6–9.0	6.17
y_s (cm)	Maximum scour depth	2.4–29.8	14.45

Table 11.3 Range of input and output parameters used in this study

Dimensionless parameter	Maximum value	Minimum value	Mean value
a/H_s	9.703	0.096	2.12
$a/\Delta D_{50}$	164.62	0.494	23.906
L/H_s	55.74	0.1531	17.736
D_{50}/H_s	0.4615	0.0136	0.106
S_0	0.268	0.0059	0.1099
y_s/H_s	10.617	0.261	2.12

11.2.2.5 Data Set

Two hundred and twenty six experimental data of maximum scour depth at bed sills in clear water conditions have been collected and used in this chapter. These data are from Lenzi et al. (2002), Gaudio and Marion (2003), Marion et al (2004), Tregnaghi (2008), and Chinnarasri and Kositgittiwong (2008). Range of variations, as well as the mean values of experimental data, is shown in Table 11.2.

11.2.2.6 Selection of Input and Output Parameters

Based on dimensional analysis, the parameters of a/H_s, $a/\Delta D_{50}$, L/H_s, D_{50}/H_s, and S_0 have been selected as input variables and relative maximum scour depth, y_s/H_s, has been selected as output variable. Table 11.3 reports the ranges of input and output parameters, which are used in this study.

11.2.3 Experimental Setup

In this study, GA_SOLVER tool in Microsoft Excel was adopted for the GA modeling. For the GP modeling, the Gene_XPro_Tools software was used. Finally, for the model tree experiment, a model tree was built using the M5 algorithm implemented in WEKA software (Witten and Frank 2005).

11.3 Results

11.3.1 GA Model

Applying GA_Solver for developing a new equation, following relationship has obtained for training data set:

$$\frac{y_s}{H_s} = 0.00076\left(\frac{a}{H_s}\right)^{16.999}\left(\frac{a}{\Delta D_{50}}\right)^{-16.464}\left(\frac{L}{H_s}\right)^{-0.0613}\left(\frac{D_{50}}{H_s}\right)^{-16.595} S_0^{0.0802} \quad (11.8)$$

Results of using the above equation for training and testing data have been presented in Fig. 11.2. As can be seen, in all over the data ranges, this equation shows good agreement between experimental and predicted values of maximum scour depth at bed sills. Obtained R^2 values of 0.96 and 0.94 for training and testing data, respectively, indicate this agreement.

11.3.2 GEP Model

In this study, according to training data, another equation has been developed using GeneXProTools program as following:

$$\frac{y_s}{H_s} = \ln\left(\frac{a}{\Delta D_{50}} + \frac{a}{H_s} + 9.8561\frac{D_{50}}{H_s}\right) + \frac{D_{50}}{H_s}\left(\frac{a}{\Delta D_{50}} - \frac{S_0}{D_{50}/H_s}\right) + \frac{D_{50}/H_s}{\log(a/H_s) - (a/(\Delta D_{50}))^{1/3}} \quad (11.9)$$

This equation with $R^2 = 0.97$ has considerable accuracy. By using this equation for testing data, R^2 is obtained as 0.97. These results have been shown in Fig. 11.3. The detailed information of the GEP model is indicated in Table 11.4.

11.3.3 M5 Tree Equations

For training data, seven linear models have been derived based on mainly variations of bed slope. This bed slope dividing is very important from hydraulic and morphologic point of view which defines two different conditions for bed sill scouring. These linear models, as well as the classification criteria, are as follows:

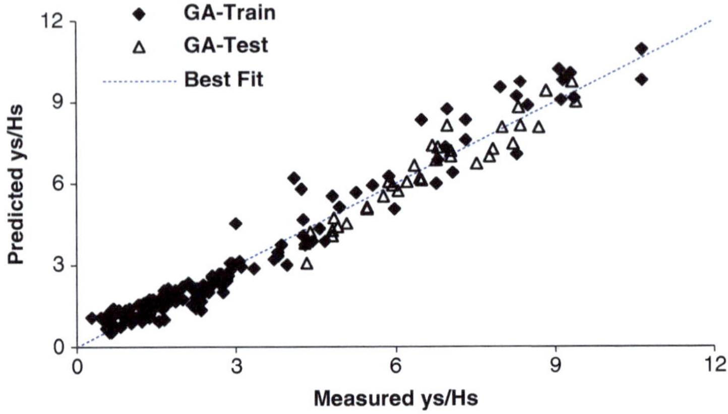

Fig. 11.2 Proposed GA model for relative maximum scour depth prediction

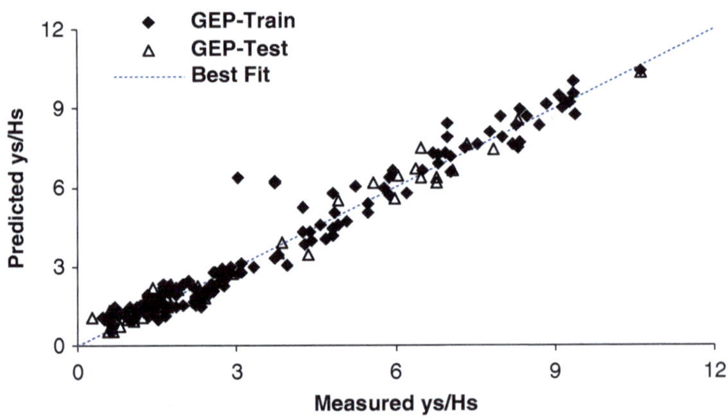

Fig. 11.3 GEP model results of relative maximum scour depth for training and testing data in this study

Table 11.4 Parameters of the optimized GEP model

Parameter	Description of parameter	Parameter amount
P1	Chromosomes	30
P2	Genes	3
P3	Mutation rate	0.044
P4	Inversion rate	0.1
P5	Function set	×, power
P6	One-point recombination rate	0.3
P7	Two-point recombination rate	0.3
P8	Gene recombination rate	0.1
P9	Gene transposition rate	0.1
P10	Linking function	Subtraction

$S_0 <= 0.162$:
| $a/H_s <= 0.579$:
| | $a/H_s <= 0.385$: LM1(37/14.364%)
| | $a/H_s > 0.385$: LM2(28/16.214%)
| $a/H_s > 0.579$:
| | $a/(\Delta d_{50}) <= 4.266$: LM3(17/14.957%)
| | $a/(\Delta d_{50}) > 4.266$: LM4(28/8.399%)
$S_0 > 0.162$:
| $a/H_s <= 3.744$: LM5(25/35.412%)
| $a/H_s > 3.744$:
| | $a/H_s <= 6.218$: LM6(20/53.031%)
| | $a/H_s > 6.218$: LM7(20/62.068%)

LM num 1

$$\frac{y_s}{H_s} = 0.812\left(\frac{a}{H_s}\right) + 0.0285\left(\frac{a}{\Delta d_{50}}\right) + 0.5264S_0 + 0.8136 \tag{11.10}$$

LM num 2

$$\frac{y_s}{H_s} = 0.9265\left(\frac{a}{H_s}\right) + 0.0285\left(\frac{a}{\Delta d_{50}}\right) + 0.5264S_0 + 1.0991 \tag{11.11}$$

LM num 3

$$\frac{y_s}{H_s} = 0.3236\left(\frac{a}{H_s}\right) + 0.1019\left(\frac{a}{\Delta d_{50}}\right) + 0.5264S_0 + 1.4 \tag{11.12}$$

LM num 4

$$\frac{y_s}{H_s} = 0.3263\left(\frac{a}{H_s}\right) + 0.0854\left(\frac{a}{\Delta d_{50}}\right) + 0.5264S_0 + 1.7508 \tag{11.13}$$

LM num 5

$$\frac{y_s}{H_s} = 0.351\left(\frac{a}{H_s}\right) + 0.0029\left(\frac{a}{\Delta d_{50}}\right) + 0.8224S_0 + 3.2061 \tag{11.14}$$

LM num 6

$$\frac{y_s}{H_s} = 0.4555\left(\frac{a}{H_s}\right) + 0.0029\left(\frac{a}{\Delta d_{50}}\right) + 0.8224S_0 + 3.7096 \tag{11.15}$$

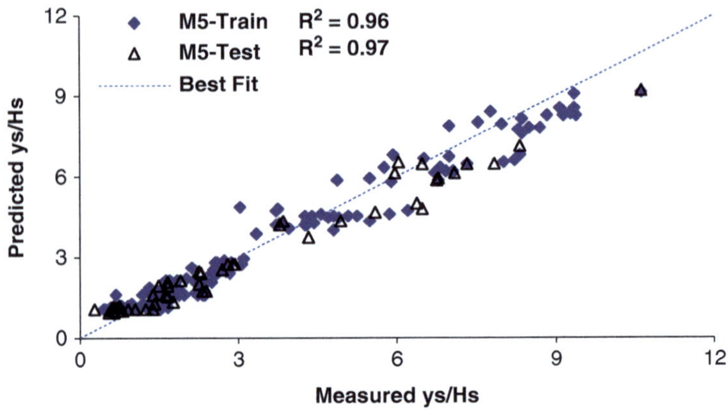

Fig. 11.4 Proposed M5 model relative maximum scour depth obtained for training and testing data

LM num 7

$$\frac{y_s}{H_s} = 0.4544\left(\frac{a}{H_s}\right) + 0.0029\left(\frac{a}{\Delta d_{50}}\right) + 0.8224S_0 + 4.3617 \tag{11.16}$$

Calculation results for training and testing data have been presented in Fig. 11.4. According to R^2 values, good accuracy has been obtained in comparison to experimental maximum relative depth scours.

In Fig. 11.5, all results including selected models, as well as empirical equations, are shown for data set of this study. As can be seen, the overall trend of Eq. 11.4 (Lenzi et al. 2004) is overestimation of maximum scour depth with very large errors especially at high relative scour depths. Also, it is interesting to note that Eq. 11.5 (Chinnarasri and Kositgittiwong 2008) has good agreement with measured data. Owing to simplicity of this equation and more importantly, its dependency to only two dimensionless parameters, it can be proposed as an option for engineers to predict maximum scour depth at bed sills with sufficient accuracy ($R^2 = 0.91$). Equation 11.6 (Tregnaghi 2008) underestimates the scour depth with high errors, especially at low relative scour depths. All selected models in this study (GA, GEP, and M5) have high accuracy through all ranges of experimental data.

11.4 Performance Analysis of Results

To validate the results for the training and testing sets, several common statistical measures are used, such as R^2 (coefficient of determination), RMSE (root mean square error), and AE (the average error) (Azamathulla 2012).

The results of statistical analysis are presented in Table 11.5. Based on this table, it is indicated that among different models considered in this study, Eq. 11.4

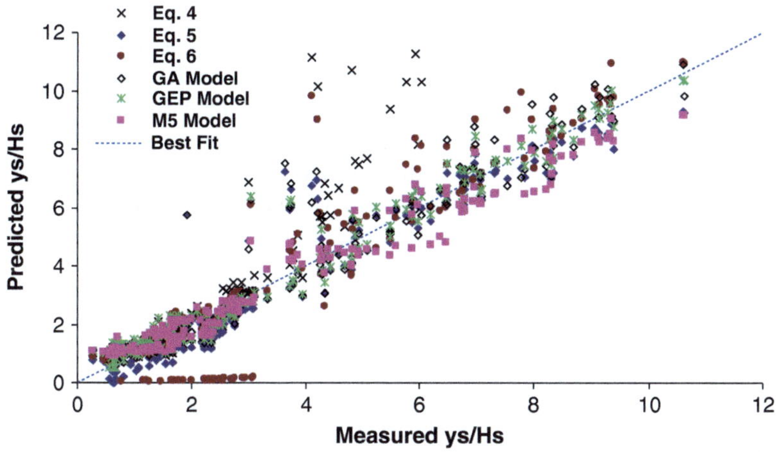

Fig. 11.5 Comparison of all empirical equations and selected models results for prediction of relative maximum scour depth

Table 11.5 Evaluation of empirical equations and selected models for scour depth prediction

	Training			Testing			All data		
Method	R^2	RMSE	%AE	R^2	RMSE	%AE	R^2	RMSE	%AE
Empirical Eqs.									
Eq. 11.4	–	–	–	–	–	–	0.61	26.06	−161
Eq. 11.5	–	–	–	–	–	–	0.91	0.774	5.41
Eq. 11.6	–	–	–	–	–	–	0.82	1.291	4.7
GA model	0.96	0.442	−0.93	0.94	0.938	−1.24	0.95	0.805	−0.98
GEP model	0.97	0.451	−8.46	0.95	0.555	−11.3	0.96	0.535	−8.94
M5 model	0.97	0.537	−4.84	0.96	0.652	−6.68	0.96	0.561	−5.17

(Lenzi et al. 2004) has the worst accuracy and therefore, is not recommended for application. On the other hand, the GA, GEP, M5, and even simple empirical equation of Chinnarasri and Kositgittiwong (11.5) have the best accuracies. By consideration of all statistical parameters, it seems that the GA model can be proposed as an option for prediction of maximum scour depth at bed sills. In addition, the Chinnarasri and Kositgittiwong (2008) equation, with requiring for only two input parameters and also having good accuracy, may be considered as a suitable approach.

11.5 Conclusions

Soft computing tools such as the GA, GEP, and M5 tree approaches were used to derive new expressions for the prediction of scour downstream of bed sills. The proposed GA equation is found to be useful to estimate scour depth for mountain rivers for various bed slopes. Performance of the GA expression is carried out by comparing its predictions with the published data ($R^2 = 0.96$ and RMSE $= 0.442$). The comparison shows that the new expression has the least RMSE and the highest coefficient of determination. The expression is found to be particularly suitable for bed slopes where predictions are very close to the measured scour depth. These models can be further extended for the estimation of scour geometry based on additional prototype data of parameters such as type of rock bed classified as per rock quality designation (RQD) and rock mass rating (RMR).

References

Azamathulla HM (2012) Gene expression programming for prediction of scour depth downstream of sills. J Hydrol 460–461:156–159

Bormann N, Julien PY (1991) Scour downstream of grade-control structures. J Hydraul Eng ASCE 117(5):579–594

Chinnarasri C, Kositgittiwong D (2008) Laboratory study of maximum scour depth downstream of sills. ICE Water Manage 161(5):267–275

Gaudio R, Marion A (2003) Time evolution of scouring downstream of bed sills. J Hydraul Res IAHR 41(3):271–284

Gaudio R, Marion A, Bovolin V (2000) Morphological effects of bed sills in degrading rivers. J Hydraul Res IAHR 38(2):89–96

Guven A, Gunal M (2008) Genetic programming approach for prediction of local scour downstream of hydraulic structures. J Irrig Drain Eng ASCE 134(2):241–249

Koza JR (1992) Genetic programming: on the programming of computers by means of natural selection. MIT Press, Cambridge, MA

Lenzi MA, Comiti F (2003) Local scouring and morphological adjustments in steep channels with check-dams sequences. Geomorphology 55:97–109

Lenzi MA, Marion A, Comiti F, Gaudio R (2002) Local scouring in low and high gradient streams at bed sills. J Hydraul Res IAHR 40(6):731–739

Lenzi MA, Marion A, Comiti F (2003) Interference processes on scouring at bed sills. Earth Surf Process Landf 28(1):99–110

Lenzi MA, Comiti F, Marion A (2004) Local scouring at bed sills in a mountain river: Plima river, Italian alps. J Hydraul Eng ASCE 130(3):267–269

Marion A, Lenzi MA, Comiti F (2004) Effect of sill spacing and sediment size grading on scouring at grade-control structures. Earth Surf Process Landf 29(8):983–993

Sharifi S (2009) Application of evolutionary computation to open channel flow modeling. PhD Thesis in Civil Engineering, University of Birmingham, p 330

Tregnaghi M (2008) Local scouring at bed sills under steady and unsteady conditions. PhD Thesis, University of Padova, p 161

Witten IH, Frank E (2005) Data mining: practical machine learning tools and techniques with Java implementations. Morgan Kaufmann, San Francisco

Chapter 12
Evaluation of Wavelet-Based De-noising Approach in Hydrological Models Linked to Artificial Neural Networks

Vahid Nourani, Aida Hosseini Baghanam, Aida Yahyavi Rahimi, and Farzad Hassan Nejad

Abstract The inherent complexities in hydrologic phenomena have been turned into a barrier to get accurate prediction by conventional linear methods. Therefore, there is an increasing interest toward data-driven black box models. In recent decades artificial neural network (ANN) as a branch of artificial intelligence method has proved its efficiency in providing accurate results to model hydrologic processes, which subsequently leads to provide important information for the urban and environmental planning, land use, flood, and water resources management. The efficiency of any data-driven model (e.g., ANN) largely depends on quantity and quality of available data; furthermore, the occult noises in data may impact the performance of the model. Although ANN can capture the underlying complexity and nonlinear relationship between input and output parameters, there might be a need to preprocess data. In this way, noise reduction of data using an appropriate de-noising scheme may lead to a better performance in the application of the data-driven ANN model. Thereupon, in this chapter, the ANN-based hydrological models (i.e., stream-flow and sediment) were developed by focusing on wavelet-based global soft thresholding method to de-noise hydrological time series on the daily scale. The appropriate selection of decomposition level and mother wavelet type is effective in thresholding results, so that sensitivity analysis was performed over levels and several Daubechies group mother wavelets (*Haar*, *db2*, *db3*, *db4*, and *db5*) to choose the proper variables. In this way, de-noised time series were imposed into an ANN model to forecast flow discharge and sediment values. The comparison of obtained results for both single ANN-based and de-noised-based (i.e., preprocessed) approaches revealed that the outcomes have been improved for the later model. Furthermore, the consequences indicated that the wavelet de-noising was significantly dependent on the chosen mother wavelet whereas forecasting results varied obviously with the alteration of mother wavelets.

V. Nourani (✉) • A.H. Baghanam • A.Y. Rahimi • F.H. Nejad
Faculty of Civil Engineering, Department of Water Resources Engineering, University of Tabriz, Tabriz, Iran
e-mail: vnourani@yahoo.com; hosseinibaghanam@gmail.com

T. Islam et al. (eds.), *Computational Intelligence Techniques in Earth and Environmental Sciences*, DOI 10.1007/978-94-017-8642-3_12,

Eventually, it was resulted that after a specific threshold, no eminent progress in results was obtained unlike the reduction occurred. Overall, the wavelet-based de-noising approach, as a preprocessing method, can be a promising idea to improve the ANN-based hydrological models.

Keywords Artificial neural network • De-noising • Stream-flow • Sediment • Wavelet

12.1 Introduction

Accurate models of the stream-flow–sediment process that are embedded with high complexity, non-stationarity, and nonlinearity in both spatial and temporal scales can provide important information for watershed and water resources management. The stream-flow–sediment yield is a complex hydrological phenomenon to comprehend due to tremendous spatial variability of watershed characteristics and initial soil moisture, land use, slope, etc., making the physical modeling quite complex. The correct assessment of the suspended sediment load (SSL) being transported by a river is a considerable issue for hydrologists as it affects the design, management, and operation of water resources projects. Furthermore, precise estimation of SSL is essential for the study of river improvement and utilization and pollution because sediment can carry pollutants such as radioactive materials, pesticides, and nutrients. The increasing number of extreme and unexpected flood situations on one hand and low flow importance in the fields of water supply management, industrial use of freshwater, optimization of reservoir operations, navigation, and other water-related issues on the other hand make accurate modeling of runoff–sediment more critical than before. Therefore, numerous hydrological models have been developed to simulate these complex and essential processes.

As stated by Jain and Ormsbee (2002), probably the most frequently used forecasting models for hydrological phenomenon are based on linear regression. Most of the time series modeling procedures fall within the framework of multivariate autoregressive moving average (ARMA) models (Salas et al. 1980). However, they are basically the statistical method with linear assuming that leads to consider data as stationary time series, so that they have a limited ability to capture non-stationarities and nonlinearities involved in the hydrologic data. The conventional time series forecasting techniques are deficient in representing the nonlinear dynamics, which are inherent in the stream-flow and SSL processes. The physically based conventional models require extensive series of parameters related to land uses, soil characteristics, soil horizon, watershed treatment, conservation practices, soil moisture variation, topographic data, surface roughness, etc. These parameters fluctuate radically over a space and time, and are very difficult to be monitored. Encountering the deficiency of statistical- and physical-based models, data-driven models are becoming as increasingly popular as a complementary technology for modeling complex natural phenomena (Dawson et al. 2005; Solomatine and Price 2004). The data-driven models are extremely reliant on the available data, and are commonly referred to as grey or black box models. Artificial neural network (ANN) as one of the black box modeling tools has been recently found to show good performance in modeling hydrological and environmental processes. The power of ANN arises from its capability for constructing complicated indicators (nonlinear models) for multivariate time series. Currently, ANN as a self-learning and self-adaptive function approximator has exposed great ability in modeling and forecasting nonlinear hydrologic time series. ANN offers an effective approach for handling large amounts of dynamic, nonlinear, and noisy data, especially when the underlying physical relationships are not fully understood (Zhang et al. 1998). Several ANN configurations have been used for hydrological forecasting with good performance (e.g., Hsu et al. 1995; Tokar and Johnson 1999; Dawson and Wilby 2001; Nourani et al. 2009a, b, 2012b; Maier et al. 2010). In this regard, Abrahart et al. (2012) investigated studies about ANN applications on rainfall–runoff and stream-flow modeling and prepared a review paper on the issue.

Considering ANN success in modeling of hydrological processes, several papers were also presented about SSL modeling. In this light, Jain (2001) used the ANN approach to develop an integrated stage–discharge–sediment concentration relation and showed the superiority of ANN over the conventional methods. Tayfur (2002) applied ANN for sheet sediment transport estimation and tested the performance of the ANNs against physically based models. The comparison results revealed that ANNs performed as well as the physically based models for simulating non-steady-state sediment loads from different slopes. Agarwal et al. (2006) presented an ANN model for simulation of runoff and sediment yield and showed that daily time-scale modeling performed well in both calibration and verification steps. Alp and Cigizoglu (2007) predicted river sediment yield by using generalized regression neural network (GRNN) and feed forward back-propagation (FFBP) neural networks. They stated that both types of neural networks were able to predict daily sediment load, but they found that GRNN was faster and could produce accurate results within shorter time than FFBP model. Kisi (2009) designed neural network model for estimation of suspended sediment concentration of two stations in the USA. The result of the study indicated that the statistical preprocessing of the data could significantly reduce the effort and computational time required in developing the ANN model. Rai and Mathur (2008) proposed an FFBP ANN for computing event-based temporal variation of sediment yield. Based on root mean-squared error (RMSE), correlation coefficient, and Nash efficiency criteria, ANN-based model performed better than the linear transfer function model for the computation of runoff hydrographs and sedimentographs. Nourani (2009) used water discharge at current and previous time steps as input neurons for forecasting sediment load at current time step by ANN. In order to approve the efficiency and ability of the proposed method, the obtained results were compared with the results of two other classic methods (i.e., linear regression and rating curve methods). Melesse et al. (2011) predicted SSL of river systems by using neural network with back-propagation (BP) training algorithms and compared the model performance with multiple linear regressions, multiple nonlinear regression, and autoregressive integrated moving average. The performance of ANN was evaluated using different combinations of input data sets and results showed that ANN predictions for most simulations were superior compared to the aforementioned models. Rajaee et al. (2010) used conjunction of wavelet and neuro-fuzzy (WNF) for suspended sediment prediction. The results illustrated that the predicted values by the proposed WNF model were in good agreement with the observed SSL values and gave better results than other models. Nourani et al. (2012a) developed two ANN models for semi-distributed modeling of the SSL. The first model investigated an integrated ANN model trained by the data of multiple stations inside the watershed, and the second model developed a geomorphology-based ANN model, in which space depended geomorphologic parameters of the sub-basins, extracted by GIS (Geographic Information System) tools and accompanied by time depended meteorological data in order to impose to ANN. The obtained results demonstrated that although the predicted SSL time series by both models were in satisfactory agreement with the observed data, the geomorphological ANN model produced better performance because of employing spatially variable factors as the model inputs.

In spite of suitable flexibility of ANN in modeling hydrologic time series such as stream-flow and sediment, sometimes, there is a shortage in appropriate forecasting results while data consist of noises. This happens because the efficiency of data-driven models is highly dependent on the available data in context of quantity and quality. Recent studies have shown that the noise limits the performance of many techniques used for identification and prediction of deterministic systems. The severity of the noise influence on data analysis techniques largely depends on the level and the nature of noise. In general, there are two types of noises; measurement noise and dynamical noise. Measurement noise refers to the corruption of observations by errors, which are independent of the dynamics. Dynamical noise, in contrast to measurement noise, is a feedback process wherein the system is perturbed by a small random amount at each time step (Schouten et al. 1994). Dynamical noise induces much greater problems in data processing than does measurement noise. Even the approach investigated in this work is to consider dynamical noise reduction as a fundamentally required process to enhance the estimation process of the captured time series. Noise reduction is considered as a continuous mapping process of the noisy input data to a noise-free output data. Decades ago, a number of noise reduction methods were developed by Wiener (1964) and Kalman (1960). Wiener filter and Kalman filter were two of de-noising techniques which were suitable for linear systems. Wiener and Kalman filtering methods were only capable of dealing with linear natural systems, and the results depended on the establishment of state space functions to a great extent (Sang, 2013). Focusing on hydrological processes that are nonlinear, the linear de-noising filters cannot behave effectively. Considering the mentioned shortage, the extended Kalman filter was improved, which could be used for nonlinear dynamics, but had limited application in processes with strong systematic nonlinearity (Reichel et al. 2002). Therefore, the ensemble Kalman filter (EnKF) and a Monte Carlo-based Kalman filter were introduced later. The EnKF has recently gained popularity in hydrology (Martyn et al. 2008) but, in principle, the EnKF is suitable only for Gaussian error statistics and just propagates the first two moments of error statistics and has limited effectiveness for highly nonlinear uncertainty evolutions. Sivakumar et al. (1999) proposed a systematic approach to additive measurement noise reduction in chaotic hydrological data, by coupling a noise level determination method and a noise reduction method and addressed some of the potential problems in applying such methods to chaotic hydrological data, and discussed the effectiveness of estimating the noise level prior to noise reduction. Elshorbagy et al. (2002) investigated commonly used algorithms for noise reduction in order to estimate chaotic hydrologic time series. It was found that the commonly used algorithm for noise reduction in hydrologic data might also remove a significant part of the original signal and recommended that noise reduction algorithm should be applied with caution. Porporato and Ridolfi (1997) investigated the effects of noise on the identification and prediction of the river flow time series by using a simple noise reduction method of Schreiber and Grassberger. The efficiency of their utilized noise reduction method was generally influenced by the precise selection of optimal values of parameters and the number of iterations involved in the method.

The threshold-based wavelet de-noising method, which illuminates the localized characteristics of non-stationary time series both in temporal and frequency domains, is a potential filter (Jansen 2006) in comparison to the so-called methods at literature. With specific regard to de-noising methods based on wavelets, Cannas et al. (2006) and Nourani et al. (2009a) explored the multi-scaling property of wavelets for maximization of the ANN forecasting accuracy in the context of flow forecasting. Guo et al. (2011) used the wavelet de-noising method to reduce or eliminate the noise in runoff time series and improve the performance of support vector machine in runoff prediction model. Nejad and Nourani (2012) applied global soft threshold-based wavelet de-noising method to de-noise daily time series of river stream discharge and subsequently de-noised time series was imposed into the ANN model to forecast flow discharge value one day ahead. Results showed that networks trained with de-noised data, which performed a preprocessing approach, revealed appropriate results in comparison to networks trained via noisy raw time series without any preprocessing. More recently, Nourani et al. (2013) used the wavelet transform to extract dynamic and multi-scale features of the non-stationary runoff time series and removed the observed data noise. The application of wavelet-based global soft thresholding method as a de-noising method for the ANN-based stream-flow and SSL forecasting model is a quite novel methodology presented in the current chapter. On this point, through the present research the influence of dynamical noise at stream-flow and SSL time series on the ANN model performance was investigated; thereupon, global soft threshold-based wavelet de-noising method was applied to de-noise daily time series of stream-flow and suspended sediment, observed at outlet of the Potomac River in the USA. The de-noised time series enhanced in accuracy could then be imposed to ANN with the aim of forecasting processes. Three basic steps were followed to achieve the purpose of study; (1) selection of appropriate mother wavelet, (2) determination of optimum decomposition level, and (3) threshold value determination and de-noising via the soft threshold wavelet-based global de-noising method.

The remaining parts of the chapter were organized as follows. First as the continuum of introduction section, the study area and utilized data were presented. Secondly through the materials and methods section a brief description of wavelet transform and ANN along with evaluation criteria was presented. Then, the proposed methodology for the introduced watershed was comprehensively explained, and the obtained results were discussed in detail. Concluding remarks was the final section of research.

12.2 Study Area and Data

Daily stream-flow and SSL data at outlet of the Potomac River, which is located in Frederick County of Maryland State, USA (latitude 39°16′24.9″, longitude 77°32′35.2″), were used in this study (Fig. 12.1). Approximately 40 % of the Potomac River basin is forested, 33 % is farmland and pasture, and an estimated 27 % is urban area. The average flow of river observed at the outlet is 306 m^3/s. The upstream drainage area is 25,996 km^2. The historical daily stream-flow and SSL data for 20 years (from 1960 to 1980, 7,333 days) which were used in this research are available at the United States Geological Survey (USGS) website, http://waterdata.usgs.gov/usa/nwis/uv?01638500).

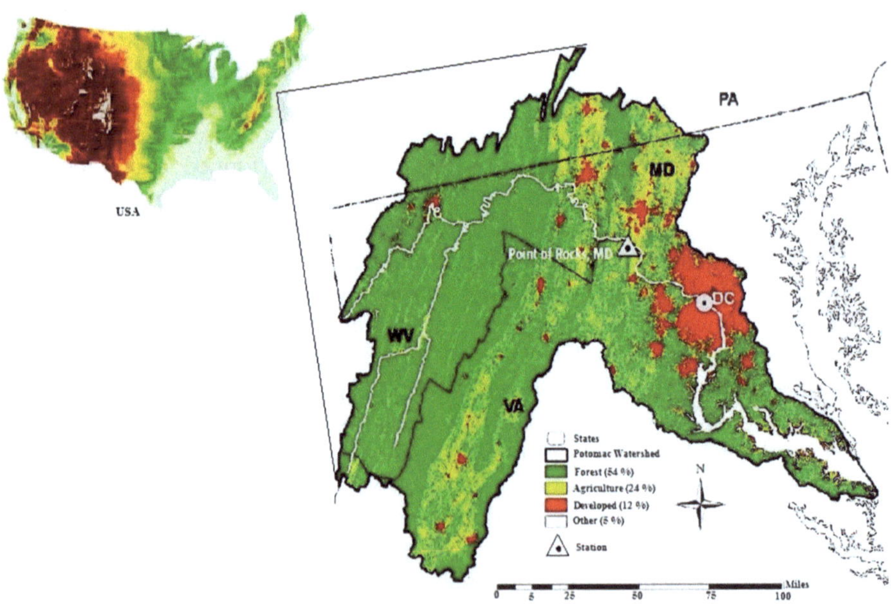

Fig. 12.1 The study area map, Potomac River at Point of Rocks at Frederick County

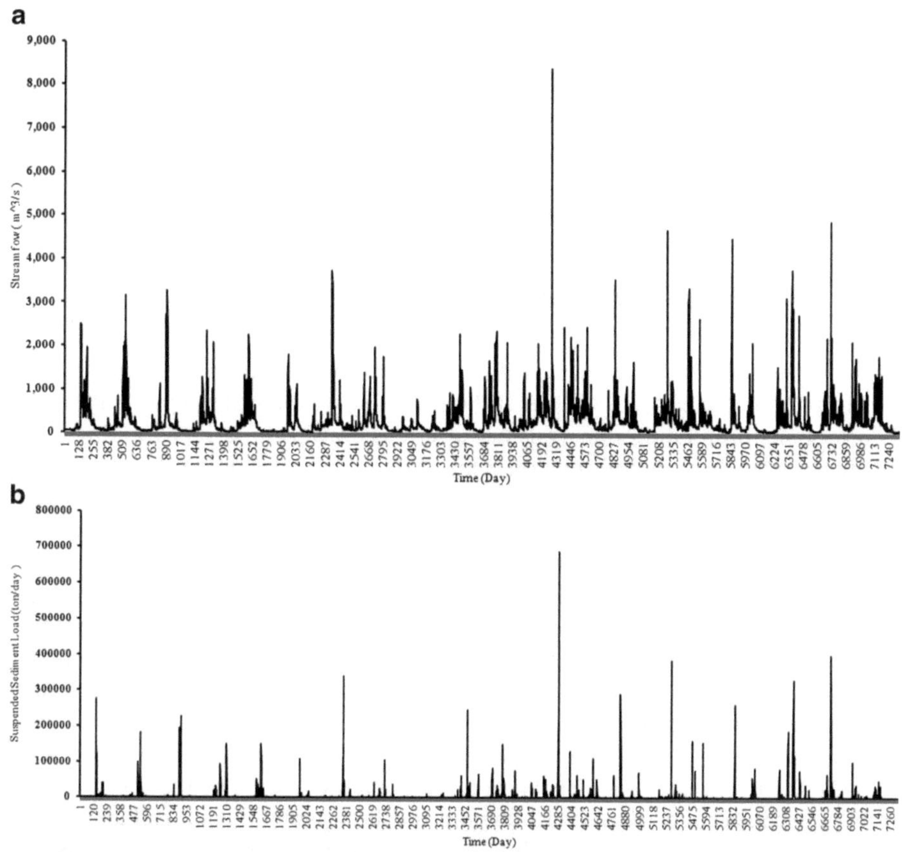

Fig. 12.2 (**a**) Stream-flow time series and (**b**) SSL time series observed at outlet of the Potomac River from 1960 to 1980

Table 12.1 Statistics of stream-flow and SSL time series for whole data as well as calibration and verification data

	Data		Calibration		Verification	
Statistical parameters	Stream-flow (m^3/s)	SSL (ton/day)	Stream-flow (m^3/s)	SSL (ton/day)	Stream-flow (m^3/s)	SSL (ton/day)
X_{mean}	276.42	3,144	266.94	3,182.03	304.84	41,885.46
X_{max}	8,288	689,000	8,288	689,000	4,816	400,000
X_{min}	15.316	2	15.316	2	30.8	3.7
S_d	2,324.3	19,082.3	2,271	18,541	2,469.8	20,622

The time series are presented in Fig. 12.2a, b for stream-flow and SSL, respectively. The statistical parameters of the stream-flow and sediment data such as the mean, standard deviation, and maximum and minimum values (i.e., X_{mean}, S_d, X_{max}, and X_{min}, respectively) are given in Table 12.1. Due to the training and verification goals, data sets were divided into two parts. The first division was 75 % of total data (from 1960 to 1975) which included the training set and the rest 25 % (from 1975 to 1980) was used for verification purpose.

12.3 Materials and Methods

12.3.1 Wavelet De-noising Procedure

The wavelet transform has increased in usage and popularity in recent years since its inception in the early 1980s, yet still does not enjoy the widespread usage of the Fourier transform. Fourier analysis has a serious drawback. In transforming to the frequency domain, time information is lost. When looking at a Fourier transform of signal, it is impossible to tell when a particular event took place but wavelet analysis allows the use of long time intervals where more precise low-frequency information and shorter regions are necessary where high-frequency information is wanted. In the field of earth sciences, Grossmann and Morlet (1984), who worked especially on geophysical seismic signals, introduced the wavelet transform application. A comprehensive literature survey of wavelet in geosciences can be found in Foufoula-Georgiou and Kumar (1995) and the most recent hydrological contributions have been cited by Labat (2005) and Sang (2013). As there are many good books and articles introducing the wavelet transform, this chapter will not delve into the theory behind wavelets and only the main concepts of the transform are briefly presented; recommended literature for the wavelet novice includes Mallat (1998) or Labat et al. (2000). The time-scale wavelet transform of a continuous time signal, $x(t)$, is defined as (Mallat 1998):

$$T(a,b) = \frac{1}{\sqrt{a}} \int_{-\infty}^{+\infty} g^{*}\left(\frac{t-b}{a}\right) x(t) \cdot dt \tag{12.1}$$

where * corresponds to the complex conjugate and $g(t)$ is called wavelet function or mother wavelet. The parameter a acts as a dilation factor, while b corresponds to a temporal translation of the function $g(t)$, which allows the study of the signal around b. The main property of wavelet transform is to provide a time-scale localization of process, which derives from the compact support of its basic function. This is opposed to the classical trigonometric function of Fourier analysis. The wavelet transform searches for correlations between the signal and wavelet function. This calculation is done at different scales of a and locally around the time of b. The result is a wavelet coefficient ($T(a, b)$) contour map known as a scalogram. In order to be classified as a wavelet, a function must have finite energy, and it must satisfy the following "admissibility conditions" (Mallat 1998):

$$g_{m,n}(t) = \frac{1}{\sqrt{a_0^m}} g^* \left(\frac{t - nb_0 a_0^m}{a_0^m} \right) \tag{12.2}$$

where m and n are integers that control the wavelet dilation and translation, respectively; a_0 is a specified fined dilation step greater than 1; and b_0 is the location parameter and must be greater than 0. The most common and simplest choice for parameters is $a_0 = 2$ and $b_0 = 1$. This power-of-two logarithmic scaling of the dilation and translation is known as the dyadic grid arrangement. The dyadic wavelet can be written in more compact notation as (Mallat 1998):

$$g_{m,n}(t) = 2^{-m/2} g(2^{-m}t - n) \tag{12.3}$$

Discrete dyadic wavelets of this form are commonly chosen to be orthonormal, i.e. (Mallat 1998):

$$\int_{-\infty}^{+\infty} g_{m,n}(t) g_{m',n'}(t) dt = \delta_{m,m'} \delta_{n,n'} \tag{12.4}$$

which δ is Kronecker delta.

This allows for the complete regeneration of the original signal as an expansion of a linear combination of translates and dilates orthonormal wavelets.

For a discrete time series, x_i, the dyadic wavelet transform becomes (Mallat 1998):

$$T_{m,n} = 2^{-m/2} \sum_{i=0}^{N-1} g(2^{-m}i - n) x_i \tag{12.5}$$

where $T_{m,n}$ is wavelet coefficient for the discrete wavelet of scale $a = 2^m$ and location $b = 2^m n$. Equation (12.5) considers a finite time series, x_i, $i = 0, 1, 2, \ldots,$ $N - 1$; and N is an integer power of 2: $N = 2^M$. This gives the ranges of m and n as, respectively, $0 < n < 2^M - m - 1$ and $1 < m < M$. At the largest wavelet scale (i.e., 2^m where $m = M$) only one wavelet is required to cover the time interval, and only one coefficient is produced. At the next scale (2^{m-1}), two wavelets cover the time interval; hence, two coefficients are produced, and so on down to $m = 1$. At $m = 1$, the a scale is 2^1; that is, 2^{M-1} or $N/2$ coefficients are required to describe the signal at this scale. The total number of wavelet coefficients for a discrete time series of length $N = 2^M$ is then $1 + 2 + 4 + 8 + \cdots + 2^{M-1} = N - 1$.

In addition to this, a signal-smoothed component, $\overline{T}$, is left, which is the signal mean. Thus, a time series of length N is broken into N components, i.e., with 0 redundancy. The inverse discrete transform is given by (Mallat 1998):

$$x_i = \overline{T} + \sum_{m=1}^{M} \sum_{n=0}^{2^{M-m-1}} T_{m,n} 2^{-m/2} g(2^{-m} i - n) \tag{12.6}$$

or in a simple format as (Mallat 1998):

$$x_i = \overline{T}(t) + \sum_{m=1}^{M} W_m(t) \tag{12.7}$$

where $\overline{T}(t)$ is called approximation sub-signal at level M and $W_m(t)$ are detail sub-signals at levels $m = 1, 2, \ldots, M$.

The wavelet coefficients, $W_m(t)$ ($m = 1, 2, \ldots, M$), provide the detail signals, which can capture small features of interpretational value in the data; the residual term, $\overline{T}(t)$, represents the background information of data. Because of simplicity of $W_1(t), W_2(t), \ldots, W_M(t), \overline{T}(t)$, some interesting characteristics, such as period, hidden period, dependence, and jump, can be diagnosed easily through wavelet components.

Wavelet de-noising method based on threshold application was proposed for acquiring correct de-noised results by Donoho (1995). This method, which is now the most common method of wavelet de-noising, is performed as follows:

1. An appropriate mother wavelet and number of resolution level are chosen. The original one-dimensional time series x_i is decomposed into an approximation at resolution level M and detailed signals at various resolution levels up to level M.
2. The absolute values of detailed signals that exceed certain threshold are treated as the difference between the values of detailed sub-signals and threshold by Eq. (12.8), otherwise, are set to 0, which gives the threshold quantifications used to obtain the processed detailed signals at each resolution level during wavelet de-noising. The approximation usually does not perform threshold quantifications.

$$\hat{d}_j(t) = \begin{cases} sgn(d_j(t))\left(|d_j(t)| - T'\right) & |d_j(t)| > T' \\ 0 & |d_j(t)| \le T' \end{cases} \tag{12.8}$$

In Eq. (12.8), T' and $dj(t)$ ($j = 1, 2, \ldots, M$) denote certain threshold and the absolute values of detailed signals for the jth resolution level, respectively.

3. Wavelet reconstruction can derive the de-noised data from the approximation at resolution level M and processed detailed signals ($\hat{d}_j(t)$) at all resolution levels.

Donoho (1995) derived a general optimal universal threshold for the white Gaussian noise under a mean square error criterion and its side condition that with high probability, the enhanced signal is at least as smooth as the clean signal. In this method, threshold is selected as (Donoho 1995):

$$T' = \hat{\sigma}\sqrt{2\ln(n)} \tag{12.9}$$

where n is number of samples in the noisy signal and $\hat{\sigma}$ is the standard deviation of noise that is estimated by Donoho (1995):

$$\hat{\sigma} = \left[\frac{median\left(\left|d_j(t)\right|\right)}{0.6745}\right] \tag{12.10}$$

in which $|d_j(t)|$ is the first-level detail coefficient of wavelet transform of the signal.

In the current study, the soft threshold wavelet-based de-noising was done by global method (i.e., all the detailed signals shrunk just with the same threshold value). In comparison to conventional de-noising methods such as moving average, wavelet de-noising technique is a robust technique. It means that in wavelet-based de-noising technique time series is decomposed to several sub-series each in a specified scale (i.e., multi-resolution property); thereafter, the thresholding is applied to each detail to remove the involved noises proportional to the scale of sub-series. The moving average is a smoothing scheme which conducts smoothing by applying uniform weights to the values of time series without any thresholding, provided, in the wavelet transform the relevant weights are applied nonuniformly via mother wavelet. On the other hand well-matching of the de-noising approach with the nature of the phenomena plays a crucial role in de-noising techniques employment, particularly in hydrological time series. The involved noises in time series are necessitated to be eliminated via a method that encompasses the extremely non-stationary and nonlinear characters; such characters are dominant in hydrological time series.

12.3.2 Artificial Neural Network and Efficiency Criteria

ANN is widely applied in hydrology and water resource studies as a forecasting tool. In ANN, FFBP network models are common to engineers. It has been proved that BP network model with three layers is satisfied for forecasting and simulating any engineering problem (Hornik 1988; Nourani et al. 2008). Three-layered feed forward neural networks (FFNNs), which have been usually used for forecasting hydrologic time series, provide a general framework for representing nonlinear functional mapping between a set of input and output variables. Three-layered FFNNs are based on a linear combination of the input variables, which are transformed by a nonlinear activation function as expressed by Eq. (12.11) where

i, j, and k denote input layer, hidden layer, and output layer neurons, respectively, and w is the applied weight by the neuron. The term "feed forward" means that a neuron connection only exists from a neuron in the input layer to other neurons in the hidden layer or from a neuron in the hidden layer to neurons in the output layer and the neurons within a layer are not interconnected to each other. The explicit expression for an output value of a three-layered FFNN is given by Nourani et al. (2008):

$$\hat{y}_k = f_0\left[\sum_{j=1}^{M_N} w_{kj} \cdot f_h\left(\sum_{i=1}^{N_N} w_{ji}x_i + w_{j0}\right) + w_{k0}\right] \tag{12.11}$$

where $w_{i,j}$ is a weight in the hidden layer connecting the ith neuron in the input layer and the jth neuron in the hidden layer, w_{j0} is the bias for the jth hidden neuron, f_h is the activation function of the hidden neuron, w_{kj} is a weight in the output layer connecting the jth neuron in the hidden layer and the kth neuron in the output layer, w_{k0} is the bias for the kth output neuron, f_0 is the activation function for the output neuron, x_i is ith input variable for input layer, and $\hat{y}_k$, y are computed and observed output variables, respectively. N_N and M_N are the number of the neurons in the input and hidden layers, respectively. The weights are different in the hidden and output layers, and their values can be changed during the process of the network training. Since ANN training sequences involve tuning the values of random-selected weights and biases of the network to optimize network performance, it is possible to get different results in each training procedure. However, in order to reduce the impact of the initial weights randomness in the BP algorithm the training procedures were repeated several times for each input with the specific network structure and target data. Since the reinitialized parameters were adopted according to the same variables of network (i.e., input data, network architecture, target data) the random allocation of parameters had no significant effect on output results, in as much as the average result of all runs was approximately equal to each of the trials.

The model that yields the best results in terms of determination coefficient (R^2) as Eq. (12.12) and RMSE as Eq. (12.13) in the training and verifying steps can be determined through trial and error process (Nourani et al. 2009a).

$$R^2 = 1 - \frac{\sum_{i=1}^{N}\left(O_{\text{obs}_i} - O_{\text{com}_i}\right)^2}{\sum_{i=1}^{N}\left(O_{\text{obs}_i} - \overline{O}_{\text{obs}}\right)^2} \tag{12.12}$$

$$\text{RMSE} = \sqrt{\frac{\sum_{i=1}^{N}\left(O_{\text{obs}_i} - O_{\text{com}_i}\right)^2}{N}} \tag{12.13}$$

where R^2, RMSE, N, O_{obs_i}, O_{com_i}, and $\overline{O}_{\text{obs}}$ are determination coefficient, RMSE, number of observations, observed data, computed values, and mean of observed data, respectively.

The RMSE is used to measure forecast accuracy, which produces a positive value by squaring the errors. The RMSE increases from 0 for perfect forecasts through large positive values as the discrepancies between forecasts and observations become increasingly large. Obviously high value for R^2 (up to 1) and small value for RMSE indicate high efficiency of the model.

The utilized data were normalized due to the fact that the network training process could be speeded up by normalizing the input and target data before training (Rogers 1996). In this study, the input and target data were normalized to scale data between 0 and 1 with Eq. (12.14):

$$y_i = \frac{x_i - x_{\min}}{x_{\max} - x_{\min}} \tag{12.14}$$

Regarding Eq. (12.14) x_i is the desired variable value, and $x_{\min}$ and $x_{\max}$ are the minimum and maximum values, respectively. y_i is the normalized variable.

Utilization of normalizing approach and transferring data between [0,1] cause a small change in $y_{\max}$, i.e., the upper bound of the normalized interval, and they impact on a normalized input in the mentioned range; subsequently the normalized input has a greater influence on the output; also normalizing makes the training of the ANN quicker (Nourani and Fard 2012).

12.4 Results and Discussion

The impact of de-noised and noisy data on model forecasting performance was compared with application of the ad hoc ANN and hybrid ANN-wavelet with de-noising technique over stream-flow and SSL forecasting models. Firstly, the single ANN models with raw and noisy data sets were established to forecast stream-flow and SSL amounts one day ahead. Secondly, the input data were de-noised by threshold-based global wavelet de-noising method and reimposed into ANN. The schematic diagram of wavelet de-noising is presented in Fig. 12.3. The architecture of ANN models was arranged according to the antecedents of stream-flow and sediment processes, whereas, stream-flow and sediment time series usually behave as strong Markovian processes, so that the values of parameters in the current time step may be related to the previous time step condition. Therefore, ten combinations of stream-flow and SSL value antecedents were used as inputs to forecast the stream-flow and SSL at the input layer of ANN.

Input combinations for stream-flow forecasting were as:

Comb. (1-1): Q_{t-1}
Comb. (1-2): Q_{t-1}, Q_{t-2}
Comb. (1-3): Q_{t-1}, Q_{t-2}, Q_{t-3}
Comb. (1-4): Q_{t-1}, Q_{t-2}, Q_{t-3}, Q_{t-4}

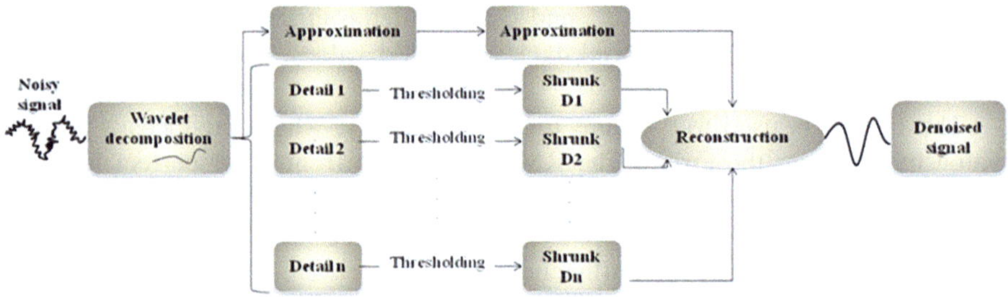

Fig. 12.3 The schematic diagram of thewavelet de-noising

and input combinations for SSL prediction were as:

Comb. (2-1): SSL_{t-1}
Comb. (2-2): SSL_{t-1}, SSL_{t-2}
Comb. (2-3): Q_{t-1}, Q_{t-2}
Comb. (2-4): SSL_{t-1},Q_{t-1}
Comb. (2-5): SSL_{t-1}, SSL_{t-2},Q_{t-1}
Comb. (2-6): SSL_{t-1}, SSL_{t-2},Q_{t-1},Q_{t-2}
Comb. (2-7): SSL_{t-1}, SSL_{t-2}, SSL_{t-3}, Q_{t-1}, Q_{t-2}, Q_{t-3}

In all cases, t represented current time step. The output layer comprised only one neuron, which was stream-flow of current day (Q_t) for the stream-flow forecasting and SSL at current time step (SSL_t) for the SSL forecasting.

12.4.1 Ad Hoc ANN

12.4.1.1 Stream-Flow Forecasting

In order to get appropriate one-day-ahead forecast of stream-flow, the input layer should be arranged in a way that could enjoy all pertinent information on target data. Based on sensitivity analysis, the input layer was optimized with only the most important time memories. In this regard, aforementioned combinations including sliding window on stream-flow time series up to four lag time steps were considered as model inputs. For each input combination, the ANN architecture was a three-layered ANN included of input layer, hidden layer, and finally the output layer. In hidden layer the numbers of neurons were altered up to ten neurons to choose the appropriate value as the hidden layer neuron number. After determination of proper ANN architecture due to the adequate performance criteria, training became over and weights were saved in order to be used in the verification step. Daily stream-flow forecasting results are presented in Table 12.2 for training and validation data sets. According to Table 12.2, the combination with three antecedents performed properly in terms of evaluation criteria; thus, there was no need of more complex ANN models with further time memories of stream-flow.

Table 12.2 Ad hoc ANN results for stream-flow forecasting model with various input combinations

Input combination	Output	Network structure[a]	RMSE (normalized)[b]		R^2	
			Calibration	Verification	Calibration	Verification
Comb. 1-1	Q_t	(1-3-1)	0.021	0.060	0.786	0.762
Comb. 1-2	Q_t	(2-9-1)	0.015	0.020	0.883	0.822
Comb. 1-3	$\mathbf{Q_t}$	**(3-3-1)**	**0.017**	**0.020**	**0.861**	**0.838**
Comb. 1-4	Q_t	(4-6-1)	0.013	0.020	0.911	0.835

[a]The mentioned values on the network structure stand to the number of input neurons, hidden neurons, and output neuron, respectively

[b]The results have been presented for the best structure and the RMSE values in all tables are dimensionless

Table 12.3 Ad hoc ANN results for SSL forecasting model with various input combinations

Input combination	Output	Network structure	RMSE (normalized)		R^2	
			Calibration	Verification	Calibration	Verification
Comb. 2-1	SSL_t	(1-5-1)	0.020	0.022	0.396	0.365
Comb. 2-2	$\mathbf{SSL_t}$	**(2-10-1)**	**0.014**	**0.018**	**0.692**	**0.601**
Comb. 2-3	$\mathbf{SSL_t}$	**(2-10-1)**	**0.016**	**0.020**	**0.635**	**0.543**
Comb. 2-4	SSL_t	(2-10-1)	0.019	0.022	0.449	0.374
Comb. 2-5	SSL_t	(3-7-1)	0.015	0.018	0.685	0.630
Comb. 2-6	$\mathbf{SSL_t}$	**(4-4-1)**	**0.015**	**0.017**	**0.674**	**0.651**
Comb. 2-7	SSL_t	(6-6-1)	0.015	0.018	0.676	0.618

12.4.1.2 SSL Forecasting

For the SSL forecasting the same procedure of ANN modeling was conducted, in which for each combination of input data, the hidden layer neuron values were selected through trial–error procedure and finally the output layer neuron revealed the forecasted SSL value. Based on the R^2 results at verification step, the most appropriate architecture of ANN was selected for SSL prediction. The input layer neurons comprised stated combinations for SSL forecasting, which could be considered in two groups; firstly, the inputs that contained only SSL or stream-flow data (i.e., combinations with one type of variable, stream-flow or SSL) and secondly, combinations included both SSL and stream-flow data in order to forecast SSL amount (i.e., combinations with two types of variables, stream-flow and SSL). Considering the fact that financial and technical problems in the exact measurement of the SSL turn into a barrier and usually after several years measuring would be stopped, the employed available stream-flow data in order to estimate the SSL values could be a reliable choice. According to Table 12.3, poor results in terms of R^2 which were obtained for the Comb. 2-1 and Comb. 2-4 denoted that in order to achieve a good performance of modeling, input data combination with more than one time step lag was required. On the other hand, the increase of lag time in combinations with one type of variables (i.e., Combs. 2-2 and 2-3) led to more reasonable consequences for SSL forecasting. Moreover, while both variables (stream-flow and SSL) were located in the input neurons, the performance of ANN enhanced and Comb. 2-6 revealed the best performance. This might be relevant to the underlying impacts between stream-flow and SSL phenomena. The selected sufficient combinations due to the proper performance were used in choosing parameters for the threshold-based wavelet de-noising approach.

12.4.2 Hybrid ANN-Wavelet

12.4.2.1 Wavelet-Based De-noising Approach

Beyond a shadow of a doubt, efficiency of ANN modeling as a data-driven method depends on the quality and quantity of data. Indeed, most of the hydrological time series like stream-flow and SSL consist of noise. Therefore, the necessity of removing the noise appears in the modeling of hydrological processes. Noise reduction methods are ordinarily chosen on the basis of their robustness, the soft threshold wavelet-based de-noising approach is an applicable and powerful noise reduction method, which can perform more efficiently if reasonable mother wavelet, proper time-scale levels, and eventually appropriate threshold value could be picked out. In order to solve wavelet de-noising concerns of stream-flow and SSL time series in this study, the assessment on precise selection of three so-called parameters was conducted. The brief description of each parameter and the effective factors was presented as well as hybrid models.

The selection of an appropriate mother wavelet is a challenge in wavelet de-noising issue, since the type of utilized mother wavelet can affect the modeling results remarkably (Nejad and Nourani 2012). The essence of wavelet transform is to discover the similarity between the analyzed time series and utilized wavelet prototype. In this way, Daubechies family of wavelets (*Haar*, *db2*, *db3*, *db4*, *db5*) were examined as the mother wavelets in current research (Fig. 12.4). The Daubechies wavelets have associated minimum-phase scaling filters, are both orthogonal and biorthogonal, and do not have an explicit analytic expression except for the *db1* (or *Haar*) form.

In discrete wavelet analysis, Nourani et al. (2009a) and Sang et al. (2009a) suggested some methods to choose adequate decomposition level in hydrological modeling. In this study, the minimum decomposition level for the wavelet-based preprocessing technique was selected due to the historical data length and the importance of seasonalities regarding the process entity (i.e., the periods that stream-flow or SSL happens dominantly). It is obvious that wavelet transform decomposes a signal into approximation and detailed sub-signals, in which the approximation signal that states the general aspect of stream-flow or SSL time series reveals positive integers; otherwise, the negative values in the approximation signal cannot state any physical interpretation for hydrological processes.

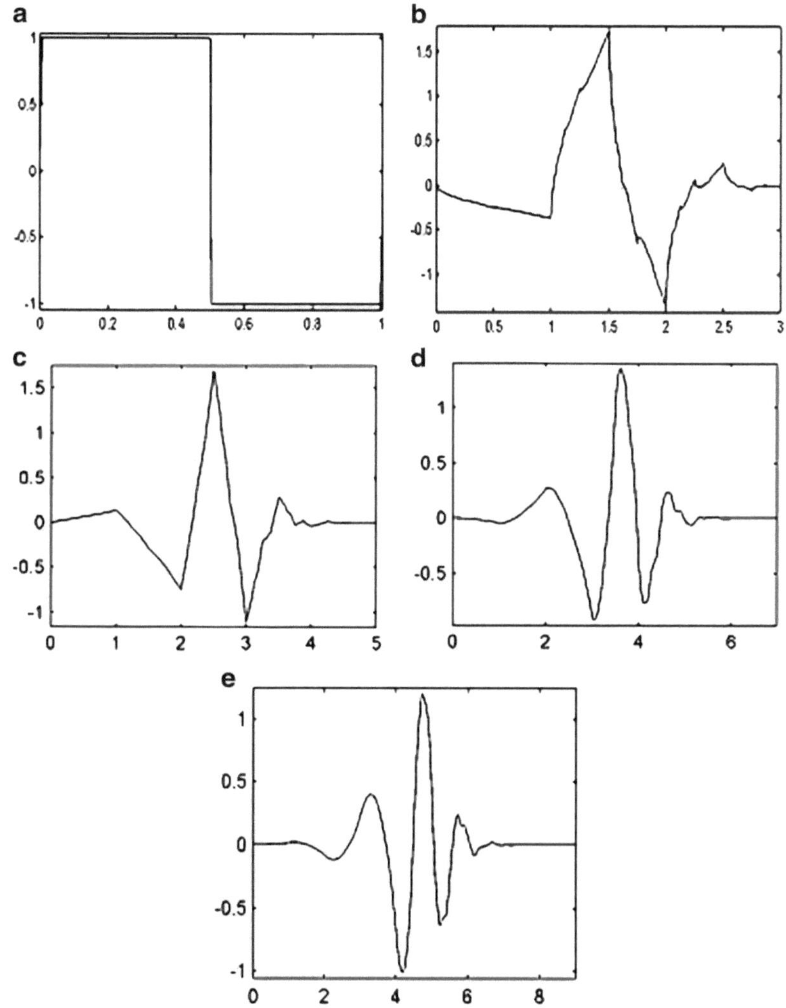

Fig. 12.4 (**a**) Haar, (**b**) *db2*, (**c**) *db3*, (**d**) *db4*, and (**e**) *db5* mother wavelets

Decomposition level *n* for daily data yields *n* detailed sub-series; as an example if *n* is considered 7, decomposition contains details as 2^1-day mode, 2^2-day mode, 2^3-day mode (i.e., nearly weekly mode), 2^4-day mode and 2^5-day mode (i.e., almost monthly mode), and 2^6-day mode and 2^7-day mode (i.e., period of occurrence is nearly per 4 months). In the application of decomposition levels less than 8 only a few seasonalities of main time series might be taken into account, but according to the hydrological base of the processes, there might be other dominant seasonalities with longer periods. Therefore, decomposition levels 8, 9, and 10 were also examined to obtain optimum decomposition level. The decomposition level 8 contains one more detail in comparison to decomposition level 7, which is about yearly

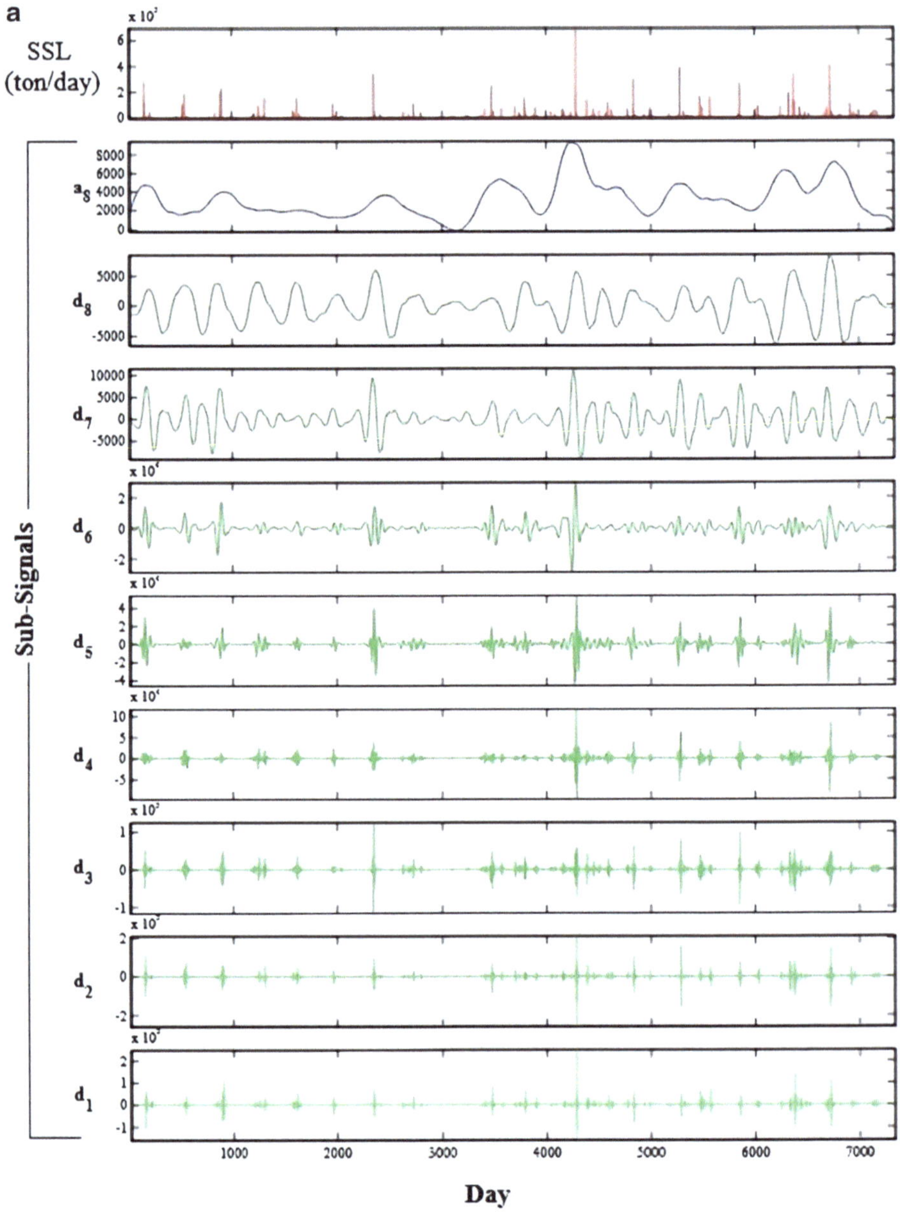

Fig. 12.5 (**a**) Decomposed SSL time series at level 8 via *db5* and (**b**) decomposed stream-flow time series at level 8 via *db3* (a_8: approximation signal; d_1, d_2, ..., d_8: detailed signals)

mode and decomposition levels 9 and 10 yield one and two more details, respectively (i.e., 2^9-day mode and 2^{10}-day mode). Figure 12.5a, b shows approximation and detailed sub-series of SSL and stream-flow time series decomposed by *db5* and *db3* mother wavelets at level 8, respectively. Eventually, based on stream-flow and

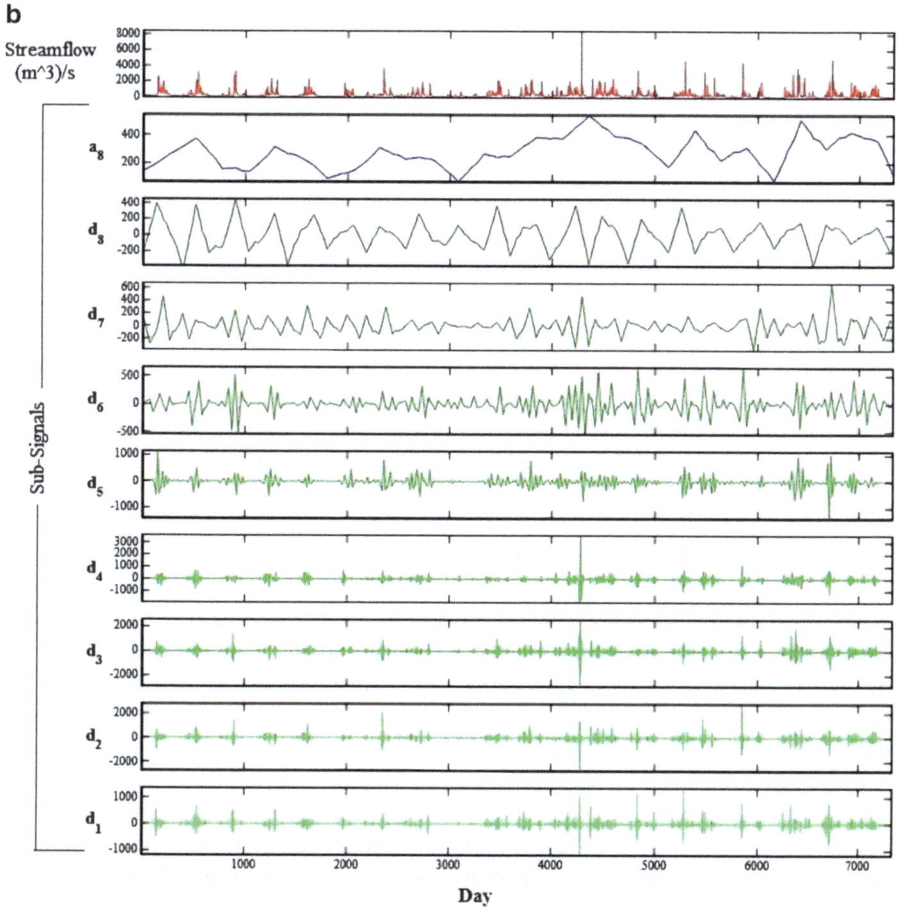

Fig. 12.5 (continued)

SSL nature, decomposition levels 8, 9, and 10, which might get to rational outcomes on de-noising approach, were compared in order to survey the effect of decomposition level on the model performance. Thresholding is the third and the most important step in wavelet de-noising procedure, which de-noises detailed signals by shrinking them as Eq. (12.8). In the current study, the soft threshold wavelet-based de-noising was performed by global method (i.e., all the detailed signals shrank just with the same threshold value). The validity of threshold determination in the hydrological time series was a great challenge and many conventional methods such as Fixed Threshold, SURE, and MINIMAX for determining the thresholds of wavelet detailed signals encountered the disadvantages. The first disadvantage was that the conventional threshold determination methods in hydrological time series were generally suitable for analysis of normal distribution of noises; the second problematic condition was that the noises separated from original hydrologic series were usually auto-correlated well; and finally, for different methods, the threshold values of wavelet detailed signals were assorted (Sang et al. 2009b). But the proposed method in current research not only covered all the noise distributions by threshold variation but also made the detailed signal thresholds equal with global method. It should be mentioned that Eq. (12.9) has been developed on the basis of the Gaussian noise involved in the signal. Moreover, the noise included in a complex hydrological process might not obey Gaussian noise (Sang et al. 2009b). Consequently, more trial–error procedure was needed to determine the appropriate threshold value without considering how the noise distribution was. The threshold value influenced de-noising procedure of hydrological time series in two different aspects. First, if threshold value was fine, noise reduction would be skimpy; inversely if threshold value was great, as a result, noise would be removed, but main information of the time series would be also removed. Therefore, the appropriate threshold not only removed the noise but also kept the main part of the time series. Thus, determination of the appropriate threshold value led to achieve higher performance in the forecasting phase.

12.4.2.2 Stream-Flow Forecasting

To develop the hybrid ANN-wavelet model, the ANN model was fed by de-noised sub-series in order to forecast stream-flow values one day ahead for daily data. In this way, firstly the stream-flow time series decomposed to various levels by different mother wavelets; subsequently the threshold value was found through universal threshold method (Eq. 12.9), and time series were de-noised via global wavelet-based de-noising; eventually the de-noised data were imposed to ANN as inputs. After determination of proper mother wavelet and decomposition level for the time series, the appropriate threshold value was determined via trial–error procedure on threshold values. In order to employ the mentioned methodology, the Comb. 1-3 of stream-flow data, which was selected in the ad hoc ANN modeling section, was utilized as the input time series. Through the universal threshold method the threshold value was obtained for stream-flow time series; afterwards, the threshold value was employed in the wavelet-based global de-nosing method for each detailed sub-signal and various mother wavelets to gain the ultimate clear and de-noised stream-flow time series. At the last step the de-noised time series were arranged in the form of Comb. 1-3 to be utilized in training of ANN. Among the several structures of ANN the proper ones were inserted to Table 12.4. According to the evaluation criteria the most appropriate mother wavelet and decomposition level were determined. The consequences revealed that based on R^2 evaluation criterion, mother wavelet *db4* with decomposition level of 8 performed more efficiently than other Daubechies at other decomposition levels. The better performance of the *db4* wavelet might originate from the form of wavelet (Fig. 12.4d), which was in coincidence to the stream-flow signal.

The other purpose of the current chapter was to find the appropriate threshold value to enjoy more de-noised time series. To this end, different threshold values around the neighborhood of the universal threshold computed by Eq. (12.9) were

Table 12.4 Results of hybrid ANN-wavelet model for stream-flow forecasting using Comb. 3-1

Mother wavelet	Decomposition level	Network structure	RMSE (normalized)		R^2	
			Calibration	Verification	Calibration	Verification
Haar	8	(3-9-1)	0.014	0.019	0.897	0.846
	9	(3-2-1)	0.017	0.019	0.862	0.841
	10	(3-7-1)	0.014	0.019	0.897	0.841
db2	8	(3-8-1)	0.014	0.019	0.900	0.844
	9	(3-8-1)	0.013	0.019	0.911	0.849
	10	(3-9-1)	0.014	0.019	0.899	0.848
db3	8	(3-6-1)	0.014	0.019	0.900	0.850
	9	(3-10-1)	0.013	0.019	0.913	0.849
	10	(3-6-1)	0.013	0.018	0.913	0.855
db4	**8**	**(3-10-1)**	**0.013**	**0.018**	**0.913**	**0.867**
	9	(3-6-1)	0.014	0.019	0.902	0.848
	10	(3-7-1)	0.014	0.019	0.897	0.850
db5	8	(3-8-1)	0.014	0.018	0.906	0.857
	9	(3-5-1)	0.013	0.018	0.919	0.860
	10	(3-10-1)	0.013	0.019	0.918	0.852

Table 12.5 Appropriate threshold determination of stream-flow time series in stream-flow forecasting

Threshold value (m^3/s)	Network structure (inputs: Q_{t-1}, Q_{t-2}, Q_{t-3})	RMSE (normalized)		R^2	
		Calibration	Verification	Calibration	Verification
2.832	(3-4-1)	0.014	0.020	0.900	0.831
19.822	(3-8-1)	0.013	0.019	0.916	0.853
28.316	(3-9-1)	0.014	0.019	0.898	0.854
42.475	(3-4-1)	0.015	0.018	0.879	0.864
56.633	(3-8-1)	0.013	0.018	0.910	0.862
84.950	(3-7-1)	0.013	0.017	0.908	0.875
113.267	(3-9-1)	0.015	0.017	0.891	0.883
169.900	**(3-8-1)**	**0.012**	**0.016**	**0.921**	**0.888**
226.534	(3-8-1)	0.013	0.017	0.913	0.882
283.168	(3-6-1)	0.014	0.017	0.894	0.879
339.802	(3-10-1)	0.014	0.018	0.893	0.855
424.752	(3-10-1)	0.016	0.020	0.875	0.825
453.068	(3-8-1)	0.016	0.021	0.865	0.818
458.732	(3-8-1)	0.016	0.021	0.870	0.817
481.385	(3-9-1)	0.017	0.022	0.851	0.800
509.702	(3-9-1)	0.017	0.022	0.855	0.794
566.336	(3-8-1)	0.018	0.023	0.831	0.777
707.92	(3-9-1)	0.020	0.024	0.804	0.762
849.504	(3-9-1)	0.021	0.026	0.775	0.725

examined for discharge time series in stream-flow modeling to find the appropriate threshold value (see Table 12.5). As depicted in Table 12.5, the more increased threshold value of the de-noising, the higher R^2 value was obtained, albeit after a special threshold value, R^2 descended. This special threshold value could be considered as the "appropriate threshold" which in stream-flow forecasting was 169.9 (m^3/s). In Table 12.5, it was seen that R^2 was increased from 0.838 for the noisy time series to 0.888 for the de-noised one, in stream-flow forecasting modeling.

12.4.2.3 SSL Forecasting

Similar to stream-flow forecasting, effective mother wavelet and decomposition level were chosen for each suitable predefined combinations of SSL (Combs. 2-2 and 2-3) through sensitivity analysis on ANN model with de-noised data obtained from global wavelet de-noising procedure using the universal threshold method (Table 12.6). The consequences of Table 12.6 revealed that mother wavelets *db3* and *db5* with decomposition levels $M = 10$ and $M = 8$ for the Combs. 2-3 and 2-2, respectively, showed the efficient performance in terms of R^2 values.

The proposed de-noising scheme was also applied for the Comb. 2-6, which was the effective combination among ad hoc ANN models and comprised both stream-flow and SSL antecedents. Since the combination 2-6 was the cumulative form of Combs. 2-3 and 2-2, and on the other hand the effective mother wavelet and decomposition level values were determined for each of Combs. 2-3 and 2-2, two different ANN-wavelet models were proposed for Comb. 2-6 as: (1) Both de-noised stream-flow and SSL time series were considered as inputs. (2) Only one de-noised time series (stream-flow or SSL) was considered as input. The outcomes in terms of efficiency criterion R^2 proved that the model with de-noised time series on both variables exhibited high performance in comparison to the models with just individual de-noised variable (see Table 12.7). With respect to the obtained results in Table 12.7, it was obvious that de-noised stream-flow time series had fundamental effect on final results compared with de-noised SSL time series.

After choosing appropriate mother wavelets and decomposition level for each combination, appropriate threshold value should be computed. The range of threshold values within local neighborhood of universal threshold value was obtained from Donoho's equation (Eq. 12.9) to determine "appropriate threshold value" for discharge and SSL time series in SSL forecasting. The results revealed that 453.068 (m^3/s) and 9,000 (ton/day) threshold values were appropriate values for stream-flow and SSL time series, respectively, in terms of R^2 as an evaluation criterion (see Tables 12.8 and 12.9).

Table 12.6 Results of hybrid ANN-wavelet model for SSL forecasting using Combs. 2-2 and 2-3

				RMSE (normalized)		R^2	
Mother wavelet	Input comb.	Decomposition level	Network structure	Calibration	Verification	Calibration	Verification
Haar	2-3	8	(2-4-1)	0.016	0.019	0.619	0.590
		9	(2-8-1)	0.016	0.019	0.663	0.593
		10	(2-10-1)	0.015	0.018	0.682	0.607
	2-2	8	(2-5-1)	0.016	0.018	0.639	0.628
		9	(2-7-1)	0.015	0.018	0.684	0.616
		10	(2-9-1)	0.014	0.018	0.709	0.630
db2	2-3	8	(2-4-1)	0.016	0.019	0.621	0.584
		9	(2-4-1)	0.016	0.019	0.621	0.594
		10	(2-4-1)	0.016	0.018	0.635	0.615
	2-2	8	(2-10-1)	0.016	0.021	0.639	0.603
		9	(2-10-1)	0.014	0.018	0.699	0.620
		10	(2-6-1)	0.015	0.018	0.668	0.608
db3	**2-3**	8	(2-7-1)	0.016	0.019	0.641	0.572
		9	(2-10-1)	0.015	0.018	0.662	0.597
		10	**(2-8-1)**	**0.015**	**0.017**	**0.676**	**0.669**
	2-2	8	(2-8-1)	0.016	0.018	0.651	0.607
		9	(2-6-1)	0.016	0.018	0.620	0.601
		10	(2-7-1)	0.016	0.018	0.625	0.614
db4	2-3	8	(2-7-1)	0.016	0.018	0.610	0.600
		9	(2-5-1)	0.015	0.018	0.668	0.636
		10	(2-7-1)	0.016	0.019	0.625	0.591
	2-2	8	(2-6-1)	0.016	0.018	0.608	0.608
		9	(2-7-1)	0.015	0.018	0.684	0.617
		10	(2-7-1)	0.015	0.018	0.648	0.626

(continued)

Table 12.6 (continued)

Mother wavelet	Input comb.	Decomposition level	Network structure	RMSE (normalized)		R^2	
				Calibration	Verification	Calibration	Verification
db5	2-3	8	(2-4-1)	0.016	0.018	0.602	0.599
		9	(2-7-1)	0.015	0.019	0.650	0.583
		10	(2-6-1)	0.016	0.018	0.615	0.601
	2-2	**8**	**(2-4-1)**	**0.014**	**0.018**	**0.651**	**0.632**
		9	(2-7-1)	0.014	0.018	0.698	0.611
		10	(2-6-1)	0.016	0.018	0.608	0.607

Table 12.7 Results of hybrid ANN-wavelet model for SSL forecasting using Comb. 2-6

				RMSE (normalized)		R^2	
Time series type	Mother wavelet	Decomposition level	Network structure	Calibration	Verification	Calibration	Verification
SSL	*db5*	8	(4-4-1)	0.013	0.016	0.756	0.704
Stream-flow	Noisy data						
SSL	Noisy data		(4-9-1)	0.013	0.014	0.764	0.726
Stream-flow	*db3*	10					
SSL	*db5*	8	(4-4-1)	0.012	0.013	0.802	0.802
Stream-flow	*db3*	10					

Table 12.8 Appropriate threshold determination of stream-flow time series in SSL forecasting

Threshold value (m^3/s)	Network structure (inputs = Q_{t-1}, Q_{t-2})	RMSE (normalized)		R^2	
		Calibration	Verification	Calibration	Verification
2.832	(2-9-1)	0.016	0.020	0.635	0.529
19.822	(2-10-1)	0.015	0.017	0.685	0.653
28.316	(2-10-1)	0.015	0.019	0.672	0.581
42.475	(2-5-1)	0.016	0.018	0.633	0.628
56.633	(2-7-1)	0.016	0.018	0.628	0.609
84.950	(2-10-1)	0.015	0.08	0.669	0.614
113.267	(2-8-1)	0.016	0.017	0.663	0.662
169.900	(2-4-1)	0.015	0.016	0.691	0.689
226.534	(2-4-1)	0.015	0.016	0.699	0.698
283.168	(2-10-1)	0.014	0.016	0.712	0.698
339.802	(2-7-1)	0.014	0.016	0.713	0.698
424.752	(2-10-1)	0.015	0.016	0.710	0.700
453.068	**(2-9-1)**	**0.014**	**0.016**	**0.713**	**0.708**
458.732	(2-7-1)	0.014	0.016	0.710	0.702
481.385	(2-8-1)	0.015	0.016	0.704	0.702
509.702	(2-9-1)	0.015	0.016	0.702	0.700
566.336	(2-4-1)	0.015	0.017	0.688	0.683
707.92	(2-7-1)	0.015	0.017	0.665	0.659
849.504	(2-5-1)	0.016	0.018	0.636	0.631

Table 12.9 Appropriate threshold determination of SSL time series in SSL forecasting

Threshold value (ton/day)	Network structure (inputs = S_{t-1}, S_{t-2})	RMSE (normalized)		R^2	
		Calibration	Verification	Calibration	Verification
50	(2-4-1)	0.015	0.018	0.681	0.632
4,000	(2-6-1)	0.015	0.017	0.695	0.679
5,000	(2-8-1)	0.014	0.017	0.708	0.686
5,500	(2-9-1)	0.015	0.017	0.704	0.687
6,000	(2-7-1)	0.014	0.017	0.706	0.678
6,500	(2-8-1)	0.015	0.017	0.704	0.686
7,500	(2-9-1)	0.015	0.017	0.701	0.685
8,500	(2-7-1)	0.015	0.017	0.702	0.689
9,000	**(2-9-1)**	**0.015**	**0.017**	**0.699**	**0.694**
10,500	(2-8-1)	0.015	0.017	0.685	0.671
11,000	(2-6-1)	0.015	0.017	0.697	0.684
12,000	(2-9-1)	0.015	0.017	0.694	0.682
13,000	(2-8-1)	0.015	0.017	0.693	0.693
13,500	(2-6-1)	0.015	0.017	0.699	0.681
14,500	(2-8-1)	0.015	0.018	0.672	0.648
16,000	(2-7-1)	0.014	0.017	0.709	0.679
16,500	(2-7-1)	0.015	0.017	0.696	0.672
17,000	(2-5-1)	0.014	0.017	0.708	0.686

Table 12.10 Comparative results of ad hoc and hybrid ANN models

Input comb.	Noise condition[a]	Output	Network structure	RMSE (normalized) Calibration	RMSE (normalized) Verification	R^2 Calibration	R^2 Verification
1-3	Noisy	Q_t	(3-3-1)	0.017	0.020	0.861	0.838
	De-noised 1		(3-10-1)	0.013	0.018	0.913	0.867
	De-noised 2		(3-8-1)	0.012	0.016	0.921	0.888
2-2	Noisy	SSL_t	(2-10-1)	0.014	0.018	0.692	0.601
	De-noised 1		(2-4-1)	0.014	0.018	0.651	0.632
	De-noised 2		(2-9-1)	0.014	0.016	0.713	0.708
2-3	Noisy	SSL_t	(2-10-1)	0.016	0.020	0.635	0.543
	De-noised 1		(2-8-1)	0.015	0.017	0.676	0.669
	De-noised 2		(2-9-1)	0.015	0.017	0.699	0.694
2-6	Noisy	SSL_t	(4-4-1)	0.015	0.017	0.674	0.651
	De-noised 1		(4-6-1)	0.013	0.016	0.713	0.706
	De-noised 2		(4-4-1)	0.012	0.013	0.802	0.802

[a]The terms de-noised 1 and de-noised 2 stand for the de-noising time series via universal threshold and appropriate threshold, respectively

12.4.3 Comparison of Models

In order to obtain an intuitive understanding of ad hoc ANN modeling with hybrid ANN-wavelet modeling linked to de-noising approach, comprehensive comparison is presented in Table 12.10. Through single ANN approach noisy time series of stream-flow and SSL were imposed to ANN intended for predicting both stream-flow and SSL one day ahead. On the other hand, de-noised data via threshold-based wavelet de-noising method were employed as inputs again for stream-flow and SSL forecasting. The comparison results revealed that noise removal using the wavelet de-noising approach improved the evaluation criteria of stream-flow and SSL forecasting models. Specifically ascending results of Table 12.10 at each combination denoted that the efficiency criteria values improved while the modeling method altered from ad hoc ANN to hybrid ANN-wavelet with universal threshold approach and appropriate threshold.

Figure 12.6a, b presents scatter plots between observed and computed stream-flow values using ad hoc ANN for noisy and de-noised data, respectively, at verification step. Similarly Fig. 12.7a, b presents scatter plots between observed and computed SSL for single ANN with noisy and de-noised data, respectively, at verification step. Comparison of scatter plots demonstrates that the improvement by

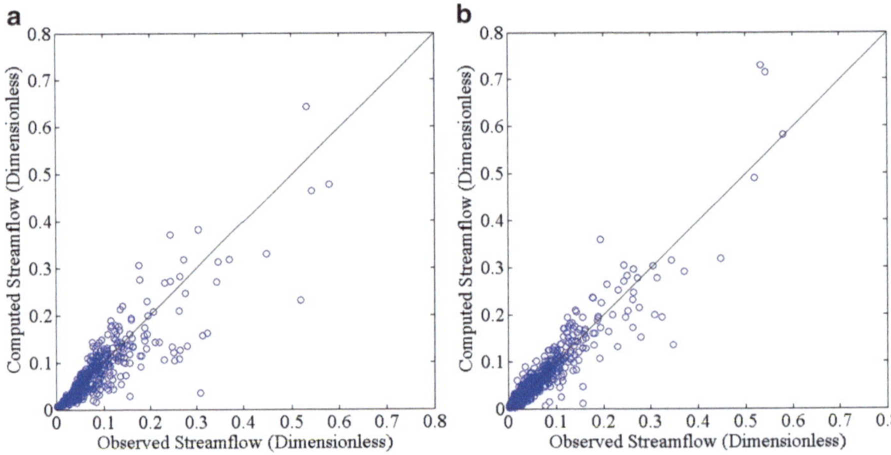

Fig. 12.6 Scatter plot for observed and computed stream-flow using ANN for (**a**) noisy and (**b**) de-noised data at verification step

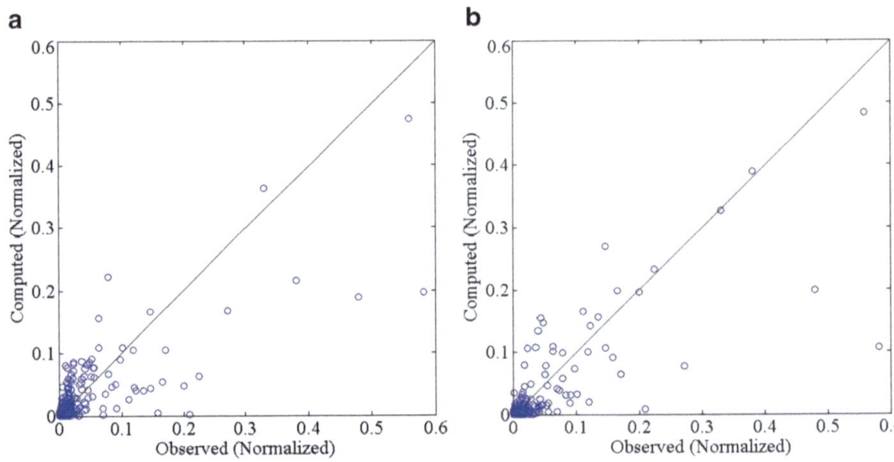

Fig. 12.7 Scatter plot for observed and computed SSL using ANN with (**a**) noisy and (**b**) de-noised data at verification step

using the ANN-wavelet de-noising scheme is not so significant than ad hoc ANN for low stream cases. It is justified due to the entity of noise; noise usually comes out quite strong in high vacillations, since low values of stream-flow include trivial fluctuations in proportion to medium and high stream (Fig. 12.2a, b); therefore, de-noising exposes less significant effect in improvement of ANN for low stream cases. In Fig. 12.8a, b observed noisy data of stream-flow and SSL were compared to de-noised time series, respectively. A visual assessment of the predicted and

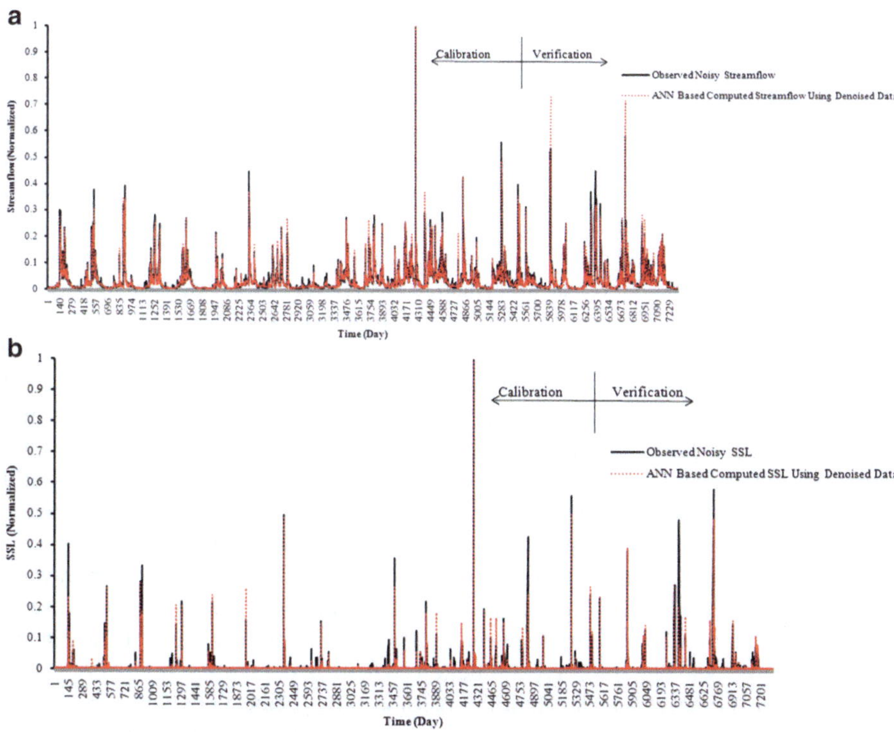

Fig. 12.8 Comparison of observed noisy data of (**a**) stream-flow and (**b**) SSL with de-noised time series of stream-flow and SSL, respectively

observed stream-flow and SSL showed that ANN modeling via de-noised data had the best fit to observed time series.

It was seen that varying the output from Q_t to SSL_t led to change in the chosen mother wavelet for stream-flow de-noising from *db4* to *db3*. Therefore, it was derived that in the threshold-based wavelet de-noising, appropriation of the mother wavelet depended on the type of the information extracted from the time series based on the target and allowed the de-noising procedure to remove unnecessary parts as a noise and kept the main part of the time series.

The appropriate threshold value for discharge time series in stream-flow forecasting was greater than that in SSL prediction and as well in universal threshold method based on Eq. (12.9) and also it was obvious that by increasing the threshold, more information was deleted from the signal as noise. Hence, it was deduced that in hydrological processes modeling, noise had a relative identification and depended on the target, some information in time series could be considered as a noise, and just with target alternation, information value varied from noise to worth information.

12.5 Concluding Remarks

Understanding the stream-flow and SSL processes in order to propose the effective forecasting models would greatly improve the discernment ability in water resources management. Noise is the important factor embedded in hydrological time series, which influences on modeling results. In this study, the wavelet transform was utilized for de-noising purpose and the ANN was applied to develop a hybrid black box model for forecasting and simulating stream-flow and SSL values in the Potomac River at Maryland, USA. It was proceeded to achieve the purpose of study through several steps: (1) determination of proper combinations for both stream-flow and SSL time series as inputs to ANN, (2) selection of appropriate mother wavelet and optimum decomposition level as well as sufficient threshold value for de-noising goal, and (3) imposition of de-noised time series into ANN to fulfill the forecasting purpose. It should be mentioned that second step included the threshold-based wavelet de-noising method. In addition to hybrid ANN-wavelet model, the ad hoc ANN was employed in order to compare efficiency of proposed methodology in prediction of stream-flow and SSL.

The inception of de-noising procedure was the determination of mother wavelet and decomposition level, which were key factors in effectiveness of the wavelet transform. The visual similarity between time series and mother wavelets facilitated to pick proper wavelet prototype function. Then, decomposed time series was de-noised by the universal thresholding method. The greatest challenge in current study was that a high threshold value would remove useful information from signals and a low threshold value could not emit large amount of noise, in order to determine "appropriate threshold value." The obtained results revealed that in context of universal threshold, Donoho's equation was not acceptable preference to determine threshold value. The challenge was obviated through utilizing the universal threshold method in order to estimate the approximate range of threshold value and subsequently, the appropriate threshold value was achieved through trial–error procedure on the assigned threshold range, since the noise contaminating signals were not adhered to Gaussian noise principle. The results of investigation conducted into threshold-based wavelet de-noising procedure could be presented as:

- Appropriate mother wavelet and threshold value changed parallel with target changes.
- Increasing the threshold value did not specifically mean accurate prediction result.
- Stream-flow time series could be successfully employed as the input in order to predict SSL with applicable accuracy.
- Threshold-based wavelet de-noising method as a preprocessed method enhanced the forecasting model performance of stream-flow and SSL time series.

In order to probe the influence of de-noising on hydrological forecasting models, it is suggested to utilize other noise removal methods such as level-dependent soft threshold-based wavelet de-noising method and compare the results with current values. Furthermore, it is suggested to carry out the forecasts via autoregressive integrated moving average and support vector machine models for the monthly scales as well.

References

Abrahart R, Anctil F, Coulibaly P et al (2012) Two decades of anarchy? Emerging themes and outstanding challenges for neural network river forecasting. Prog Phys Geogr 36(4):480–513. doi:10.1177/0309133312444943

Agarwal A, Mishra SK, Ram S et al (2006) Simulation of runoff and sediment yield using artificial neural networks. Biosyst Eng 94(4):597–613. http://dx.doi.org/10.1016/j.biosystemseng.2006.02.014

Alp M, Cigizoglu HK (2007) Suspended sediment load simulation by two artificial neural network methods using hydrometeorological data. Environ Model Softw 22(1):2–13. doi:10.1016/j.envsoft.2005.09.009#doilink. http://dx.doi.org/10.1016/j.envsoft.2005.09.009

Cannas B, Fanni A, See L et al (2006) Data preprocessing for river flow forecasting using neural networks: wavelet transforms and data partitioning. Phys Chem Earth 31(18):1164–1171. doi:10.1016/j.pce.2006.03.020

Dawson C, Wilby R (2001) Hydrological modeling using artificial neural networks. Prog Phys Geogr 25(1):80–108. doi:10.1177/030913330102500104

Dawson CH, See L, Abrahart R et al (2005) A comparative study of artificial neural network techniques for river stage forecasting. Paper presented at the international joint conference on neural networks, Montreal, vol 4, pp 2666–2670

Donoho DH (1995) De-noising by soft-thresholding. IEEE Trans Inf Theory 41(3):613–617. doi:10.1109/18.382009, 10.1109/18.382009#blank

Elshorbagy A, Simonovic SP, Panu US (2002) Noise reduction in chaotic hydrologic time series: facts and doubts. J Hydrol 256(3):147–165. doi:10.1016/S0022-1694%2801%2900534-0#doilink. http://dx.doi.org/10.1016/S0022-1694(01)00534-0

Foufoula-Georgiou E, Kumar P (1995) Wavelets in geophysics. Academic, San Diego

Grossmann A, Morlet J (1984) Decomposition of Hardy functions into square integrable wavelets of constant shape. SIAM J Math Anal 15(4):723–736. doi:10.1137/0515056

Guo J, Zhou J, Qin H et al (2011) Monthly streamflow forecasting based on improved support vector machine model. Expert Syst Appl 38(10):13073–13081. doi:10.1016/j.eswa.2011.04.114#doilink. http://dx.doi.org/10.1016/j.eswa.2011.04.114

Hornik K (1988) Multilayer feed-forward networks are universal approximators. Neural Netw 2 (5):359–366. doi:10.1016/0893-6080(89)90020-8, 10.1016/0893-6080%2889%2990020-8#_self

Hsu KL, Gupta HV, Sorooshian S (1995) Artificial neural network modeling of the rainfall-runoff process. Water Resour Res 31(10):2517–2530. doi:10.1029/95WR01955

Jain SK (2001) Development of integrated sediment rating curves using ANNs. J Hydraul Eng 127 (1):30–37. doi:10.1061/%28ASCE%290733-9429%282001%29127:1%2830%29. http://dx.doi.org/10.1061/(ASCE)0733-9429(2001)127:1(30)

Jain A, Ormsbee LE (2002) Short-term water demand forecast modeling techniques: conventional methods versus AI. J Am Water Works Assoc 94(7):64–72

Jansen M (2006) Minimum risk thresholds for data with heavy noise. IEEE Signal Process Lett 13:296–299. doi:10.1109/LSP.2006.870355, 10.1109/LSP.2006.870355#blank

Kalman RE (1960) A new approach to linear filtering and prediction problems. J Fluids Eng 82 (1):35–45

Kisi O (2009) Daily suspended sediment estimation using neuro-wavelet models. Int J Earth Sci 99:1471–1482. doi:10.1007/s00531-009-0460-2

Labat D (2005) Recent advances in wavelet analyses. Part 1: a review of concepts. J Hydrol 314:275–288. doi:10.1016/j.jhydrol.2005.04.003#doilink. http://dx.doi.org/10.1016/j.jhydrol.2005.04.003

Labat D, Ababou R, Mangin A (2000) Rainfall-runoff relation for Karstic Spring. Part 2: continuous wavelet and discrete orthogonal multi resolution analyses. J Hydrol 238:149–178. doi:10.1016/S0022-1694%2800%2900322-X#doilink. http://dx.doi.org/10.1016/S0022-1694(00)00322-X

Maier HR, Jain A, Dandy GC et al (2010) Methods used for the development of neural networks for the prediction of water resources variables: current status and future directions. Environ Model Softw 25:891–909. doi:10.1016/j.envsoft.2010.02.003#doilink. http://dx.doi.org/10.1016/j.envsoft.2010.02.003

Mallat SG (ed) (1998) A wavelet tour of signal processing. Academic, San Diego

Martyn PC, David ER, Ross AW et al (2008) Hydrological data assimilation with the ensemble Kalman filter: use of streamflow observations to update states in a distributed hydrological model. Adv Water Resour 31:1309–1324. doi:10.1016/j.advwatres.2008.06.005#doilink. http://dx.doi.org/10.1016/j.advwatres.2008.06.005

Melesse AM, Ahmad S, McClain ME et al (2011) Suspended sediment load prediction of river systems: an artificial neural network approach. Agric Water Manag 98:855–866. doi:10.1016/j.agwat.2010.12.012#doilink. http://dx.doi.org/10.1016/j.agwat.2010.12.012

Nejad FH, Nourani V (2012) Elevation of wavelet de-noising performance via an ANN-based streamflow forecasting model. Int J Comput Sci Manag Res 1(4):764–770

Nourani V (2009) Using artificial neural networks (ANNs) for sediment load forecasting of Talkherood River mouth. J Urban Environ Eng 3(1):1–6. doi:10.4090/juee.2009.v3n1.001006

Nourani V, Fard MS (2012) Sensitivity analysis of the artificial neural network outputs in simulation of the evaporation process at different climatologicr. Adv Eng Softw 47(1):127–146. doi:10.1016/j.advengsoft.2011.12.014#doilink. http://dx.doi.org/10.1016/j.advengsoft.2011.12.014

Nourani V, Mogaddam AA, Nadiri AO (2008) An ANN based model for spatiotemporal groundwater level forecasting. Hydrol Process 22:5054–5066. doi:10.1002/hyp.7129

Nourani V, Komasi M, Mano A (2009a) A multivariate ANN-wavelet approach for rainfall–runoff modeling. Water Resour Manag 23:2877–2894. doi:10.1007/s11269-009-9414-5

Nourani V, Alami MT, Aminfar MH (2009b) A combined neural-wavelet model for prediction of Ligvanchai watershed precipitation. Eng Appl Artif Intell 22:466–472. doi:10.1016/j.engappai.2008.09.003#doilink. http://dx.doi.org/10.1016/j.engappai.2008.09.003

Nourani V, Komasi M, Alami MT (2012a) A hybrid wavelet-genetic programming approach to optimize ANN modeling of rainfall-runoff process. J Hydrol Eng 17(6):724–741. doi:10.1061/%28ASCE%29HE.1943-5584.0000506#_blank#Opens new window. http://dx.doi.org/10.1061/(ASCE)HE.1943-5584.0000506

Nourani V, Kalantari O, Hosseini Baghanam A (2012b) Two semi-distributed ANN-based models for estimation of suspended sediment load. J Hydrol Eng 17(12):1368–1380. doi:10.1061/%28ASCE%29HE.1943-5584.0000587#_blank#Opensnew window. http://dx.doi.org/10.1061/(ASCE)HE.1943-5584.0000587

Nourani V, Hosseini Baghanam A, Adamowski JF et al (2013) Using self-organizing maps and wavelet transforms for space–time pre-processing of satellite precipitation and runoff data in neural network based rainfall–runoff modeling. J Hydrol 476:228–243. doi:10.1016/j.jhydrol.2012.10.054#doilink. http://dx.doi.org/10.1016/j.jhydrol.2012.10.054

Porporato A, Ridolfi L (1997) Nonlinear analysis of river flow time sequences. Water Resour Res 33(6):1353–1367. doi:10.1029/96WR03535

Rai RK, Mathur BS (2008) Event-based sediment yield modeling using artificial neural network. Water Resour Manag 22:423–441. doi:10.1007/s11269-007-9170-3; 10.3390/e11041123

Rajaee T, Mirbagheri SA, Nourani V et al (2010) Prediction of daily suspended sediment load using wavelet and neuro-fuzzy combined model. Int J Environ Sci Technol 7(1):93–110

Reichel RH, Walker JP, Koster RD et al (2002) Extended versus ensemble Kalman filtering for land data assimilation. J Hydrometeorol 3:728–740

Rogers R (1996) Neural networks: a systematic introduction. Springer, Berlin

Salas JD, Delleur JW, Yevjevich V et al (eds) (1980) Applied modeling of hydrologic time series. Water Resources Publications, Highlands Ranch, CO

Sang YF (2013) A review on the applications of wavelet transform in hydrology time series analysis. Atmos Res 122:8–15. doi:10.1016/j.atmosres.2012.11.003#doilink. http://dx.doi.org/10.1016/j.atmosres.2012.11.003

Sang YF, Wang D, Wu JC et al (2009a) Entropy-based wavelet de-noising method for time series analysis. Entropy 11(4):1123–1147. doi:10.3390/e11041123

Sang YF, Wang D, Wu JC et al (2009b) The relationship between period's identification and noises in hydrological series data. J Hydrol 368:165–177. doi:10.1016/j.jhydrol.2009.01.042#doilink. http://dx.doi.org/10.1016/j.jhydrol.2009.01.042

Schouten JC, Takens F, Van den Bleek CM (1994) Estimation of the dimension of a noisy attractor. Phys Rev E 50(3):1851–1861. http://link.aps.org/doi/10.1103/PhysRevE.50.1851

Sivakumar B, Phoon KK, Liong SY et al (1999) A systematic approach to noise reduction in chaotic hydrological time series. J Hydrol 219:103–135. doi:10.1016/S0022-1694%2899%2900051-7#doilink. http://dx.doi.org/10.1016/S0022-1694(99)00051-7

Solomatine DP, Price RK (2004) Innovative approaches to flood forecasting using data driven and hybrid modeling. In: Liong S.Y. et al., (ed)Proceedings of the 6th International Conference on Hydroinformatics, Singapore, 2004

Tayfur G (2002) Artificial neural networks for sheet sediment transport. Hydrol Sci J 47(6):879–892

Tokar AS, Johnson PA (1999) Rainfall-runoff modeling using artificial neural network. J Hydrol Eng 4(3):232–239

Wiener N (1964) Extrapolation, interpolation, and smoothing of stationary time series: with engineering applications. Technology Press of the Massachusetts Institute of Technology, Cambridge, MA

Zhang G, Patuwo E, Hu M (1998) Forecasting with artificial neural networks: the state of the art. Int J Forecast 14:35–62. doi:10.1016/S0169-2070%2897%2900044-7#doilink. http://dx.doi.org/10.1016/S0169-2070(97)00044-7

Chapter 13
Evaluation of Mathematical Models with Utility Index: A Case Study from Hydrology

Dr. Renji Remesan and Dr. Dawei Han

Abstract Conventional error-based statistical parameters like the Nash–Sutcliffe efficiency index are popular among hydrologists to check the accuracy of hydrological models and to compare the relative performance of alternative models in a particular modelling scenario. A major drawback of those traditional indices is that they are based on only one modelling attribute, i.e. the modelling error. This study has identified an overall model utility index as an effective error-sensitivity-uncertainty procedure which could serve as a useful quality indicator of data-based modelling. This study has also made an attempt to answer the question—should the increasing complexity of the existing model add any benefit to the model users? The study evaluates the utility of some popular and widely used data-based models in hydrological modelling such as local linear regression, artificial neural networks (ANNs), Adaptive neuro fuzzy inference system (ANFIS) and support vector machines (SVMs) along with relatively complex wavelet hybrid forms of ANN, ANFIS and SVM in the context of daily rainfall–runoff modelling. The study has used traditional error-based statistical indices to confirm capabilities of model utility index values in identifying better model for rainfall–runoff modelling. The implication of this study is that a modeller may use utility values to select the best model instead of using both calibration and validation processes in the case of data scarcity. The study comprehensively analysed the modelling capabilities of SVM and its waveform in the context of rainfall–runoff modelling.

Keywords Support vector machines • Wavelets • Local linear regression • Artificial neural network • Fuzzy inference system • Hydrology

---

R. Remesan (✉)
Cranfield Water Science Institute, Cranfield University, Vincent Building, College Road, Cranfield, Bedfordshire MK43 0AL, UK
e-mail: r.remesan@cranfield.ac.uk

D. Han
Department of Civil Engineering, University of Bristol, Bristol, UK

T. Islam et al. (eds.), *Computational Intelligence Techniques in Earth and Environmental Sciences*, DOI 10.1007/978-94-017-8642-3_13,

13.1 Introduction

Data-based modelling techniques have been popular in the field of hydrology for several decades. Even in the acclaimed success of different data models in hydrology, there are still many questions that need to be answered. The relevant questions in data-based modelling in hydrology are how useful is a model for predicting a particular component within the hydrological cycle? and does a complex model work better than simple ones? Visual judgment and statistical measures are two common approaches employed to establish the integrity of any data-based mathematical models. The usefulness of any model depends ultimately on its contribution to the success of decision making, not on its ability to generate unassailably correct numerical values (Pepelnjak 2009). It is often difficult in hydrology to decide which model should be used for a particular purpose, and the decision is often made on the basis of familiarity rather than the appropriateness and effectiveness of the model. Legates and McCabe (1999) have reviewed many major statistical measures that are used by the hydrologists to validate models, which includes the Nash–Sutcliffe (NS) efficiency index, the root mean square error (RMSE), the coefficient of correlation, the coefficient of determination, the mean absolute error and many more, out of which the NS-efficiency index is one of the most commonly used indicators for model comparison and performance evaluation in hydrology. A study by Jain and Sudheer (2008) has demonstrated the weakness of the NS-efficiency index in model comparison. Comparing different models just in terms of their better accuracy in simulating the numerical values is often misleading as there are many other aspects that need to be accounted for before declaring that one model with entirely different mathematical concepts is better than the other. It is a known fact that the best model is not necessarily the most complex, or the one which overtly reflects the most sophisticated understanding of the system (Barnes 1995). There is a hypothesis that more complex models simulate the processes in a better way but with high variability in sensitivity and relatively less bias (Snowling and Kramer 2001). On the other hand, a study by Oreskes et al. (1994) argues that there is no strong evidence that simple models are more likely to produce more accurate results than complex models. Snowling and Kramer (2001) have connected the usefulness of the model to model's uncertainty which was assessed through different modelling attributes like model error, model sensitivity and model complexity.

In hydrology and water resources research, there are two major bases of uncertainty attitudes; one is based on stochasticity as a necessary factor and the other is based on deterministic nature of the system. The definition of uncertainty is much more uncertain about the modelled numerical values; it relates to much deeper processes and pertains to the governing mechanisms of the model. Distinguishable uncertainties in hydrology are data uncertainties (mainly associated with measurements), sample uncertainties (e.g. number of data for calibration) and model uncertainty (Plate and Duckstein 1987). Klir (1989) made an attempt to consider uncertainty in terms of the complexity of the model. He found both categories have a conflictive nature, i.e. if complexity decreases, the uncertainty grows. In the last 20 years,

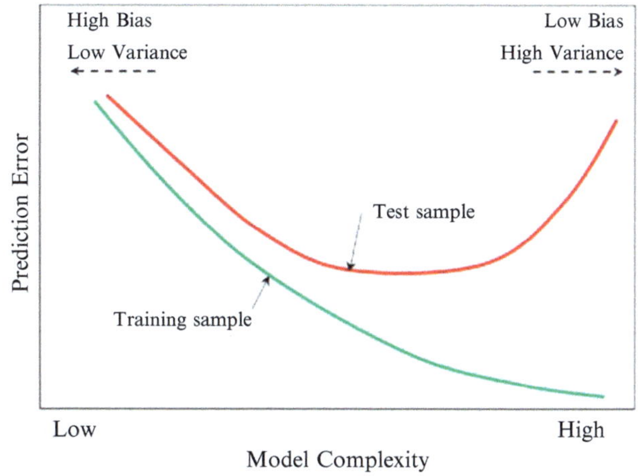

Fig. 13.1 Hypothesis showing effect of complexity during training and testing (Hastie et al. 2001)

the study of complexity in modelling systems has emerged as a recognised field in statistics. Though, the initial attempts to formalise the concept of complexity go back even further to Shannon's inception of information theory (Shannon 1948). The complexity of a model is closely related with the uncertainty of the system which can be defined in terms of model properties like model sensitivity and modelling error. The general hypothesis of model complexity and its influence during training and testing phases is shown in Fig. 13.1. The general hypothesis states that more complex models can simulate reality better than simpler models (i.e. less prediction error), and with a greater variance and low bias during training phase. Less complex models provide a relatively approximate simulation (i.e. more prediction error), but with less variance and high bias. But the case is a bit different in testing phase; highly complex models won't give best test results as the graph is parabolic with a minimum somewhere in the middle.

Figure 13.2 displays the hypothesis which shows the variation of different model parameters particularly with bias–variance interaction during the test phase.

Models of different complexity may show different modelling properties like sensitivity, flexibility, error and data requirements based upon their structure. Figure 13.3 illustrates the hypothetical relationship between model sensitivity, modelling error, model flexibility, training data requirement and model complexity.

The aim of this chapter is to highlight the need to have a statistical comparative index in data-based modelling which considers modelling attributes like model error, model complexity and model sensitivity. The study has made use of modified form of the overall model utility index proposed by Snowling and Kramer (2001) to identify "the best and right" model in data-based hydrological modelling, which was accomplished through a major case study, using the daily information of rainfall and runoff data from the Brue catchment in the United Kingdom. The utility-based results are compared with that of the traditional statistical indices. Another objective of this study is to ascertain if the usefulness of a model changes if one performs the wavelet-based input data splitting.

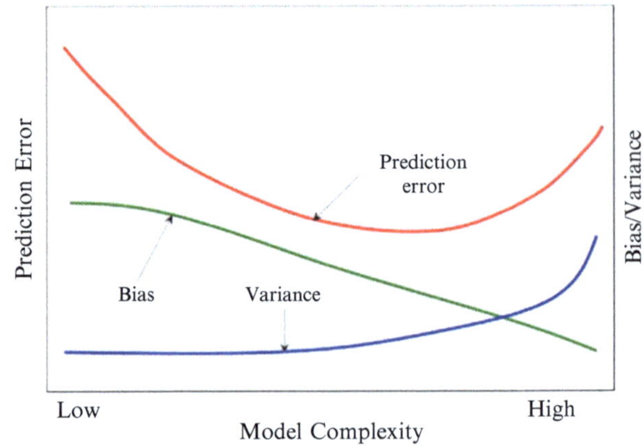

Fig. 13.2 Hypothesis showing effect of complexity on bias–variance interaction (Hastie et al. 2001)

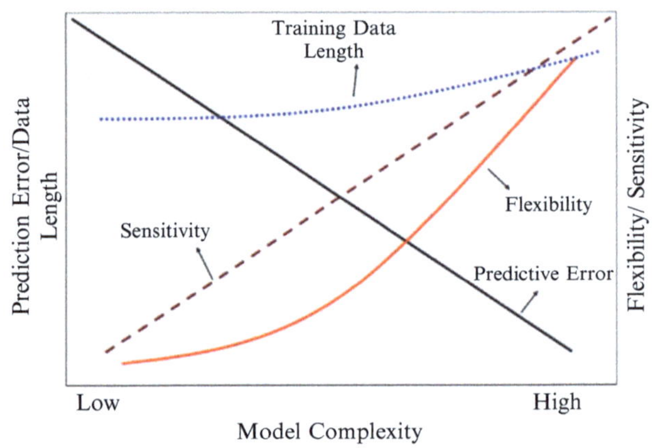

Fig. 13.3 Hypothetical relation of model complexity with sensitivity, flexibility, data requirements and predictive error

13.2 Study Area and Data Used

This study has used daily rainfall and runoff data from the Brue catchment of the United Kingdom. The River Brue catchment is located in Somerset, South West of England. It is considered as one of the best representative catchments to express hydrological responses in England, due to its data quality for a reasonably long time. This catchment has been extensively used in many good quality studies on weather radar, quantitative precipitation and flood forecasting and rainfall–runoff modelling. The location is famous among researchers because of its well-facilitated dense rain gauge network as well as the coverage by three weather radars. The River Brue catchment was the site of the Natural Environment Research Council

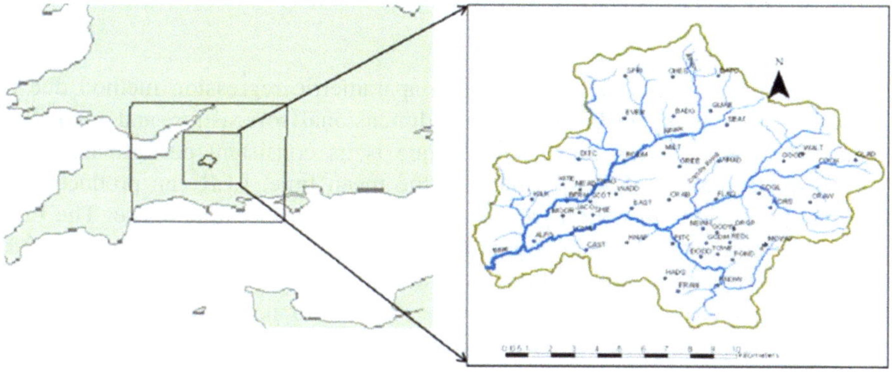

Fig. 13.4 The location map of the study area, the Brue River catchment

(NERC)-funded HYREX project (Hydrological Radar Experiment) from 1993 to 2000. The catchment was chosen for the HYREX project, as its size and relief were seen as representative of many catchments in the United Kingdom to demonstrate the hydrological dynamics and flood forecasting procedures. The catchment has a drainage area of 135 km^2 and an elevation range between 35 and 190 m above sea level. The catchment is located at 51.075°N and 2.58°W (Fig. 13.4). The river gauging point at the catchment is located at Lovington. An automatic weather station (AWS) and an automatic soil water station (ASWS) are located in the catchment which recorded the global solar radiation, net radiation and other weather parameters such as wind speed, wet and dry bulb temperatures and atmospheric pressure in hourly interval. Six years of daily rainfall–runoff data from the Brue catchment, spanning from 1993 to 2000, was used in this study. For the rainfall–runoff modelling, the study has used effective inputs like three-step antecedent runoff values ($Q(t-1)$, $Q(t-2)$, $Q(t-3)$), one-step antecedent rainfall ($P(t-1)$) and current rainfall information ($P(t)$) for hybrid modelling as observed in the previous studies (Remesan et al. 2009). The optimum training data length for this daily rainfall–runoff data set was identified as 1,056 data points (Remesan et al. 2009) which was used as training data set throughout the study and the rest is used for validation.

13.3 Models

The study has used several data-based models such as local linear regression (LLR) model, artificial neural networks (ANNs), Adaptive neuro fuzzy inference system (ANFIS), support vector machines (SVMs) and hybrid wavelet forms of ANN, ANFIS and SVMs in order to cover a wide range of models used in hydrology.

13.3.1 LLR Model

The LLR model is a widely accepted nonparametric regression method due to its better prediction capabilities in low dimensional forecasting and modelling problems. The attraction of LLR technique is its consistent performance even with a small amount of sample data. In the mean time, LLR can produce very accurate predictions in regions of high data density in the input space. The LLR procedure requires only three data points to obtain an initial prediction and then uses all newly updated data as they become available to make further predictions. The only problem with LLR is to decide the size of $p_{\max}$, the number of near neighbours to be included for the local linear modelling.

Given a neighbourhood of $p_{\max}$ points, we must solve a linear matrix equation

$$\mathbf{Xm} = \mathbf{y} \tag{13.1}$$

where $\mathbf{X}$ is a $p_{\max} \times d$ matrix of the $p_{\max}$ input points in d-dimensions, $\mathbf{x}_i (1 \leq i \leq p_{\max})$ are the nearest neighbour points, $\mathbf{y}$ is a column vector of length $p_{\max}$ of the corresponding outputs and $\mathbf{m}$ is a column vector of parameters that must be determined to provide the optimal mapping from $\mathbf{X}$ to $\mathbf{y}$, such that

$$\begin{pmatrix} x_{11} & x_{12} & x_{13} & \cdots & x_{1d} \\ x_{21} & x_{22} & x_{23} & \cdots & x_{2d} \\ \vdots & \vdots & \vdots & \ddots & \vdots \\ x_{x_{p_{\max}}1} & x_{x_{p_{\max}}2} & x_{x_{p_{\max}}3} & \cdots & x_{x_{p_{\max}}d} \end{pmatrix} \begin{pmatrix} m_1 \\ m_2 \\ m_3 \\ \vdots \\ m_d \end{pmatrix} = \begin{pmatrix} y_1 \\ y_1 \\ \vdots \\ y_{p_{\max}} \end{pmatrix} \tag{13.2}$$

The rank $\boldsymbol{r}$ of the matrix x is the number of linearly independent rows, which will affect the existence or uniqueness of the solution for m.

If the matrix $\mathbf{X}$ is square and non-singular then the unique solution to Eq. 13.1 is $\mathbf{m} = \mathbf{X}^{-1}\mathbf{y}$. If $\mathbf{X}$ is not square or singular, we modify Eq. 13.1 and attempt to find a vector $\mathbf{m}$ which minimises

$$|\mathbf{Xm} - \mathbf{y}|^2 \tag{13.3}$$

13.3.2 ANN and ANFIS Models

The theory of ANNs started in the early 1940s when McCulloch and Pitts developed the first computational representation of a neuron (McCulloch and Pitts 1943). The ANNs are nonlinear formations which work based on the function of human neural system. ANNs have become focus of much attention in last few decades in hydrology due to their immense capabilities in the implementation of nonlinear static and dynamic systems. The most commonly used learning algorithm in

ANNs is the back-propagation algorithm. Algorithms like conjugate gradient, quasi-Newton and Levenberg–Marquardt (LM) are considered as some of the faster algorithms, all of which make use of standard numerical optimisation techniques. The Levenberg–Marquardt (LM) learning algorithm was used in this study. There are several types of ANNs like multilayer perceptron, radial basis functions and Kohonen networks. ANN structure defines its structure including number of hidden layers, number of hidden nodes, number of input and output nodes and activation function. For hidden layer the sigmoid activation function and for output layer linear activation function were used in this study. Three-layer feed-forward neural network (one input layer, one hidden layer and one output layer) is the most commonly used topology in hydrology. This topology has proved its ability in modelling many real-world functional problems. The selection of hidden neurons is the tricky part in ANN modelling as it relates to the complexity of the system being modelled and there are several ways of doing it, such as the geometric average between input and output vector dimensions (Maren et al. 1990), the same as the number of inputs used for the modelling (Mechaqrane and Zouak 2004), twice the input layer dimension plus one (Hecht-Nielsen 1990), etc. In this study, the Hecht-Nielsen (1990) approach has been adopted according to our past experimental experience with it.

Adaptive neuro-fuzzy inference system (ANFIS) model is a well-known artificial intelligence technique that has been used in modelling hydrological processes. The ability of neural network to learn fuzzy structure from the input–output data sets in an interactive manner has encouraged many researchers to combine the ANN and the fuzzy logic effectively to organise network structure itself and to adapt parameters of a fuzzy system. The ANFIS model used in this study is based on the Sugeno fuzzy model, which is based on a systematic approach to generate fuzzy rules and membership function parameters for fuzzy sets from a given hydrological time series data set (Sugeno and Kang 1988; Jang 1993). The learning algorithm for ANFIS is a hybrid algorithm, which is a combination between the gradient descent method and the least squares method for identifying nonlinear input parameters and the linear output parameters, respectively. The ANFIS modelling was performed using the "subtractive fuzzy clustering" function due to its good performance with a small number of rules.

For a first-order Sugeno fuzzy model, a typical rule set with two fuzzy If/Then rules can be expressed as

$$\text{Rule 1: If } x \text{ is } A_1 \text{ and } y \text{ is } B_1 \text{ Then } f_1 = p_1x + q_1y + r_1 \tag{13.4}$$

$$\text{Rule 2: If } x \text{ is } A_2 \text{ and } y \text{ is } B_2 \text{ Then } f_2 = p_2x + q_2y + r_2 \tag{13.5}$$

where x and y are the crisp inputs to the node i, A_i and B_i are the linguistic labels (low, medium, high, etc.) characterised by convenient membership functions and p_i, q_i and r_i are the consequence parameters ($i = 1$ or 2). In the ANFIS, nodes in the same layer have similar functions as described below.

(a) *Layer 1* (*input nodes*): Nodes of this layer generate membership grades of the crisp inputs which belong to each of the convenient fuzzy sets using the membership functions. The generated bell-shaped membership function given below was used:

$$\mu_{A_i}(x) = \frac{1}{1 + ((x - c_i)/a_i)^{2b_i i}} \tag{13.6}$$

where μ_{A_i} is the appropriate membership function for A_i fuzzy set, and $\{a_i, b_i, c_i\}$ is the membership function's parameter set (premise parameters) that changes the shape of membership function from 1 to 0.

(b) *Layer 2* (*rule nodes*): In this layer, the rule operator (AND/OR) is applied to get one output that represents the results of the antecedent for a fuzzy rule. The outputs of the second layer, called as firing strengths O_i^2, are the products of the incoming signals obtained from the layer 1, named as w below:

(c) *Layer 3* (*average nodes*): In this layer, the nodes calculate the ratio of the ith rule's firing strength to the sum of all rules' firing strengths

$$O_i^3 = \overline{w}_i = \frac{w_i}{\sum_i w_i}, \quad i = 1, 2 \tag{13.8}$$

$$O_i^2 = w_i = \mu_{A_i}(x)\mu_{B_i}(y), \quad i = 1, 2 \tag{13.7}$$

(d) *Layer 4* (*consequent nodes*): In this layer, the contribution of ith rule towards the total output or the model output and/or the function is calculated as follows:

$$O_i^4 = \overline{w}_i f_i = \overline{w}_i(p_i x + q_i y + r_i), \quad i = 1, 2 \tag{13.9}$$

where $\overline{w}_i$ is the output of Layer 3 and $\{p_i, q_i, r_i\}$ are the coefficients of a linear combination in Sugeno inference system. These parameters of this layer are referred to as consequent parameters.

(e) *Layer 5* (*output nodes*): This layer is called the output nodes. This layer's single fixed node computes the final output as the summation of all incoming signals.

$$O_i^5 = f(x, y) = \frac{\sum_i w_i f_i}{\sum_i w_i} \tag{13.10}$$

13.3.3 Support Vector Machines

Just like ANNs, SVMs can be represented as two-layer networks (where the weights are nonlinear in the first layer and linear in the second layer).

Mathematically, a basic function for the statistical learning process is:

$$y = f(x) = \sum_{i=1}^{M} \alpha_i \varphi_i(x) = w\varphi(x) \tag{13.11}$$

where the output is a linearly weighted sum of M. The nonlinear transformation is carried out by $\varphi(x)$.

The decision function of SVM is represented as:

$$y = f(x) = \left\{ \sum_{i=1}^{N} \alpha_i K(x_i, x) \right\} - b \tag{13.12}$$

where K is the kernel function, α_i and b are parameters, N is the number of training data, x_i are the vectors used in training process and x is the independent vector. The parameters α_i and b are derived by maximising their objective function.

The role of the kernel function simplifies the learning process by changing the representation of the data in the input space to a linear representation in a higher dimensional space called a feature space. A suitable choice of kernels allows the data to become separable in the feature space despite being non-separable in the original input space. Four standard kernels are usually used in classification problems and also used in regression cases: linear, polynomial, radial basis and sigmoid:

Linear: $u' \times v$
Polynomial: $(\gamma \times u'v + \text{coef})^{\text{degree}}$
Radial basis: $e^{-\gamma|u-v|^2}$
Sigmoid: $\tanh(\gamma \times u'v + \text{coef})$

The SVM software used in this study was LIBSVM developed by Chih-Chung Chang and Chih-Jen, and supported by the National Science Council of Taiwan. The basic algorithm is a simplification of both SMO by Platt and SVMLight by Joachims. The source code is written in C++. The choice of this software was made on its ease of use and dependability. It has been tried and tested in several research institutions worldwide including the Computer and Information Sciences Department, University of Florida, USA, and the Institute for Computer Science, University of Freiburg, Germany. The LIBSVM is capable of C-SVM classification, one-class classification, ν-SV classification, ν-SV regression and ε-SV regression. The model first trains the SVM with a list of input vectors describing the training data. It outputs a model file, which contains a list of support vectors and hence the description of the hypothesis for the particular regression problem (Bray and Han 2004).

13.3.4 Wavelet Hybrid Models

This study has used three different types of wavelet hybrid models namely neuro-wavelet (NW) models, wavelet-adaptive-network-based fuzzy inference system

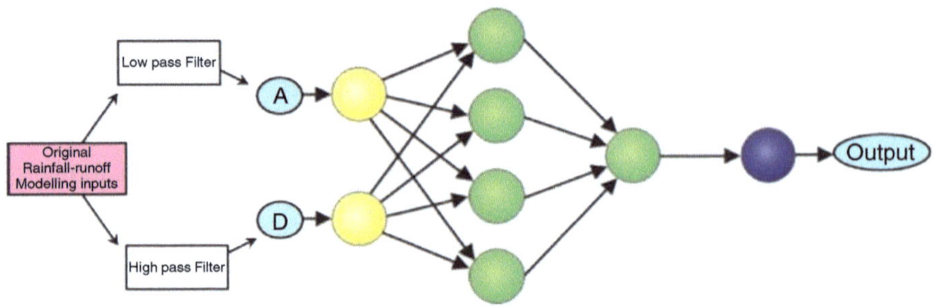

Fig. 13.5 The proposed hybrid scheme for NW model

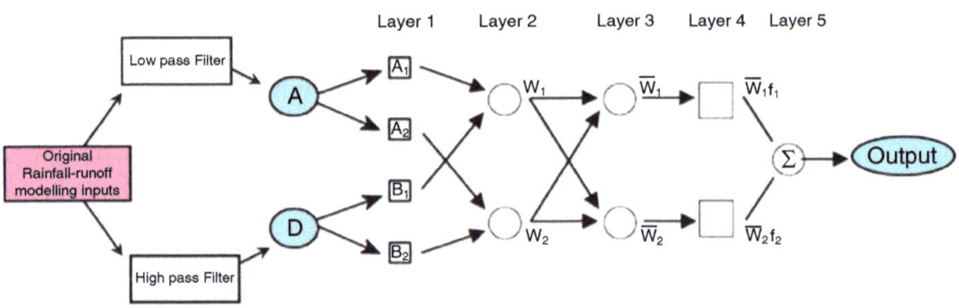

Fig. 13.6 The proposed hybrid scheme for W-ANFIS model

(W-ANFIS) model and wavelet-support vector machine (W-SVM) model in conjunction with ANNs, ANFIS and SVMs, respectively. A multilayer feed-forward network type of ANN and discrete wavelet transfer (DWT) model were combined together to obtain a neuro-wavelet (NW) model. The DWT model is functioned through two set of filters, viz. high-pass and low-pass filters, which decompose the signal into two set of series namely detailed coefficients (*D*) and approximation (*A*) sub-time series, respectively. Please refer to Remesan et al. (2009) for further details of wavelet model construction. In the proposed NW model, these decomposed sub-series obtained from DWT on the original data directly are used as inputs to the ANN model. This study has used another conjunction model: wavelet-neuro-fuzzy is applied in subsequent sections for daily rainfall–runoff modelling. The W-ANFIS model utilises the time–frequency representation ability of DWT to display the data in the time domain in conjunction with a conventional ANFIS model. The input antecedent information data considered are decomposed into wavelet sub-series by discrete wavelet transform and then the neuro-fuzzy model is constructed with appropriate wavelet sub-series as input, and desired time step of the target time series as output. In the case of W-SVM, the DWT model is combined with the SVM model with the best kernel function and best regressor and several trial and error evaluations. The detailed model structure and proposed specifications are given in Figs. 13.5, 13.6 and 13.7 which correspond to NW, W-ANFIS and W-SVM models.

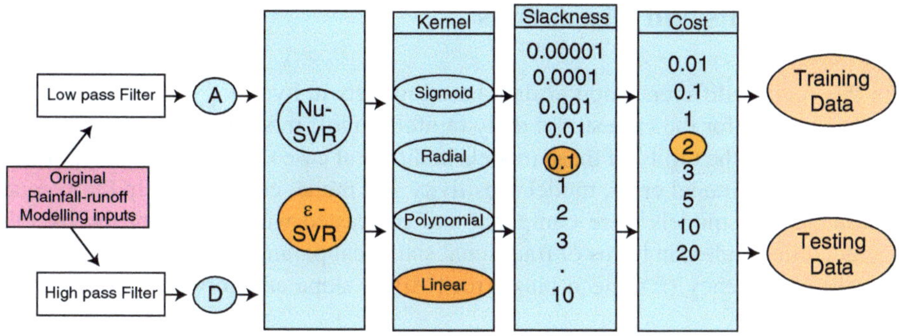

Fig. 13.7 The proposed hybrid scheme for W-SVM model

13.3.5 Index of Model Utility (U)

This study has adopted an index of model utility to make a decision about which is the "best and right" model for a hydrological modelling exercise. The adopted approach is a modified version of Snowling and Kramer (2001) for the suitability in data-based modelling. Statistically the proposed "index of model utility" of a model can be defined as scaled distance from origin on a graph of sensitivity vs. modelling error of different models to the point corresponding to that model in the graph. Mathematically it can be written as

$$U_i = 1 - \sqrt{\frac{K_s S_i^2 + K_e E_i^2}{(K_s + K_e)}} \tag{13.13}$$

where

U_i is the utility index for model i

S_i, sensitivity value for model i (relative to the maximum sensitivity). In this study the value is obtained from the mean value of slope of all sensitivity curves obtained from all inputs

E_i, error value for model i (relative to the maximum error; this study has adopted RMSE as the indicator of model error)

K_s and K_e are weighting constants for sensitivity and error, respectively

U value varies between 0 and 1 and if the value of U is larger the model has higher utility. The values of S and E for each model should be normalised to satisfy the equation; that's the reason for dividing all values by the maximum sensitivity and error value. The values of K_s and K_e depend on how the model values error and sensitivity. If error and sensitivity are valued equally, then K_s and K_e should both be set to 1. In this study both values were set to 1, and the model utility indexes (U) were calculated for this case study using different models detailed above.

13.4 Results and Discussions

In this section different popular data-based models in hydrology were established and compared for the case study: daily rainfall–runoff modelling. The first subsection evaluates the utility of these models in different case studies, in terms of model attributes like model error, model sensitivity and model complexity. In the second subsection the models were compared and contrasted with finding of the overall model utility index, in terms of traditional statistical parameters like RMSE, Nash–Sutcliffe efficiency (*E*), mean bias error (MBE), slope and correlation coefficient (CORR).

In this case study, several models were constructed and tested for predicting daily runoff values (using models ranging from relatively simple LLR model to relatively complex W-SVM). The nonparametric procedure based on LLR models does not require training in the same way as that of neural network models. The optimal number of nearest neighbours for LLR (principally dependent on the noise level) was determined by trial and error method and 16 nearest neighbours were implemented.

The adaptive Adaptive neuro fuzzy inference system (ANFIS) model was used for daily rainfall–runoff modelling, in which a set of parameters in ANFIS were identified through a hybrid learning rule combining the back-propagation gradient descent and a least squares method. The ANFIS model in this modelling used a set of fuzzy IF–THEN rules with appropriate membership functions. The subtractive fuzzy clustering was used to establish the rule-based relationship between input data series and output data variable. The subtractive clustering was used to automatically identify the natural clusters in the input–output data pool. In this ANFIS model, there were 32 parameters to determine in the layer 2 because of five input variables. The three rules generated 3^6 nodes in the subsequent layer. The study set the number of membership functions for each input of ANFIS as three with Gaussian (or bell-shaped) and linear membership functions at the inputs and outputs, respectively.

For SVM modelling, C++-based LIBSVM with ν-SV and ε-SV regressions was used as explained in methodology section. Normalisation of input vectors and proper identification of different parameters are very important in SVM modelling. Initial analysis results in rainfall–runoff modelling at the Brue catchment on both ν-SVM and ε-SVM using different kernel functions are shown in Fig. 13.8. The SVM analysis on the Brue catchment daily data has used different kernel functions like linear, polynomial, radial and sigmoid and compared in terms of mean-squared errors. Out of these eight analysis results, the results from two SVM stood out quite considerably from the remaining six. These were ε-SVM and ν-SVM with linear kernel function (Fig. 13.8). The performance of E-SVM with linear kernel was better than that of ν-SVM with linear kernel. However, it was unclear whether this performance was due to the regression algorithm implemented or whether optimising the parameters within each algorithm would enhance the performance of one SVM over the other.

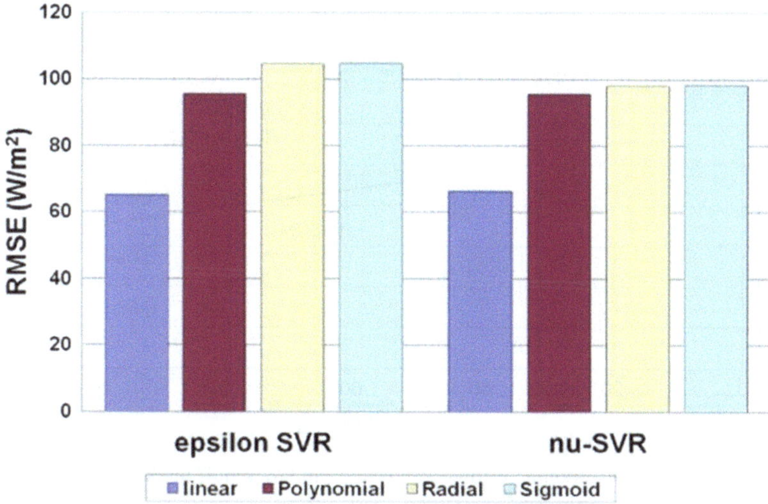

Fig. 13.8 Variation of performance in daily rainfall–runoff modelling at the Brue catchment when applying different support vector machines on different kernel functions

This analysis on daily data from the Brue catchment was performed after fixing the parameters to the default values (degree in kernel function is set as 3, coef0 in kernel function is set as zero, cache memory size is set as 40 Mb, tolerance of termination criterion is set as a default value of 0.001). The SVM hypothesis suggested that the performance of SVM depended on the slack parameter (ε) and the cost factor (C). The study has performed the modelling analysis varying the E values between $\varepsilon = 1$ to $\varepsilon = 0.00001$ and the cost parameters $C = 0.1$–$1,000$. The analysis results have shown that the least error increases rapidly for E greater than 0.1. So the study set the value of E to 0.1 for reliable results and less computation time.

The cost factor of error (C) assigns a penalty for the number of vectors falling between the two hyperplanes in the hypothesis. It suggests if the data is of good quality the distance between the two hyperplanes is narrowed down. If the data is noisy it is preferable to have a smaller value of C which will not penalise the vectors (Bray and Han 2004). So it was important to find the optimum cost value for SVM modelling. The cost value was chosen to be 2 because it produced the least error at that value, with the minimum running time, which was identified after several trial and error analyses.

The study has used a neuro-wavelet (NW) model for modelling; for this purpose, a multilayer feed-forward network type of ANN and DWT model were combined together to obtain an NW model. The DWT model is functioned through two set of filters, viz. high-pass and low-pass filters, which decompose the signal into two set of series namely detailed coefficients (D) and approximation (A) sub-time series, respectively. The present value of runoff has been estimated using the three

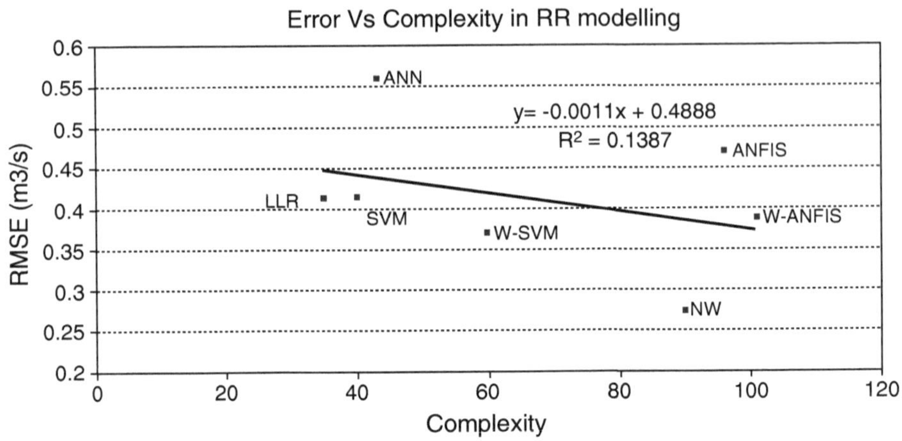

Fig. 13.9 Complexity vs. training error—case study: rainfall–runoff modelling

resolution levels of antecedent runoff and rainfall information (2-day mode (D^j_q1, D^i_p1), 4-day mode (D^j_q2, D^i_p2), 8-day mode (D^j_q3, D^i_p3) and approximate mode A^j_q3, A^i_p3 where q denotes runoff, p denotes rainfall and i and j denote number of antecedent data sets of rainfall and runoff, respectively). The above-mentioned decomposed sub-series were used as inputs in the case of wavelet hybrid forms of ANFIS and SVM (viz. W-ANFIS and W-SVM).

13.4.1 Comparison of Data Models Using Utility Index

The study has used the overall model utility index to evaluate and compare different data-based models from relatively simple LLR model to the hybrid W-SVM model. This index gives a measure of the "utility" of the model in a particular modelling scenario, with respect to modelling uncertainty (assuming that model uncertainty connects to its sensitivity, error and complexity). Thus we have assessed model attributes like model complexity (the study has used the training time as the indicator of complexity), model sensitivity (output response to changes in training input) and model error (closeness of simulation to measurement) of all seven data-based models used for daily rainfall–runoff modelling.

Figure 13.9 shows the variation of error (RMSE) with the model complexity for this case study i.e. daily rainfall–runoff modelling. The RMSE decreases with increasing complexity as this study has hypothesised. However, relatively complex ANN and ANFIS models have shown more erroneous prediction than the relatively simpler LLR model. The better prediction in terms of error was exhibited by the NW model, followed by W-SVM, W-ANFIS and SVM models. Even though, the hypothetical relation is a straight line, we observed a decreasing linear relation with R^2 value of 0.138.

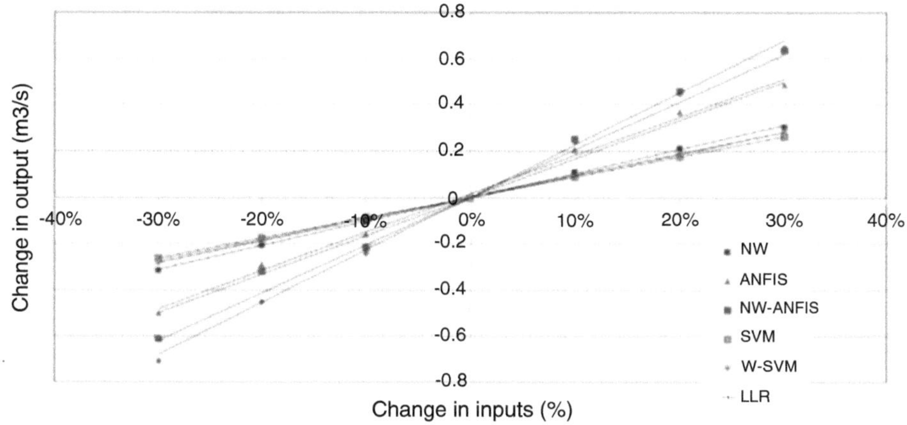

Fig. 13.10 Sensitivity curves for different data-based models

The sensitivity of the model with change in inputs used for training was assessed varying the inputs in the range of certain percentages. To find the training sensitivity of the model to the inputs, the study has changed all the inputs in the range of −30 to +30 % and checked the change in outputs produced in each scenario. These values were averaged to plot sensitivity diagram of each model as shown in Fig. 13.10. The slopes of these sensitivity diagrams were estimated and these slope values were considered as the measure of sensitivity.

Figure 13.11 shows the results of the variation of the sensitivity of different models with the corresponding complexity values. The sensitivity showed an increasing linear trend with increasing complexity with an R^2 value of 0.58. Even though the SVM model was a bit complex in structure the sensitivity value was observed close to that of LLR model. The highest value of sensitivity was observed with NW model, followed by W-ANFIS and ANFIS. The complex W-SVM showed relatively low sensitivity compared to other hybrid wavelet models like NW and W-ANFIS.

Now a modeller can make a decision in terms of uncertainty (expressed in terms of error and sensitivity) and complexity (expressed in terms of modelling time). The overall model utility statistic requires the error and the sensitivity to express in relative ratio to the maximum value. Table 13.1 shows different data models and the corresponding overall utility indices.

The value of the overall model utility index (U) varies between 0 and 1, where the larger the value of U, the greater the model utility considering aspects like uncertainty and complexity. The resultant figure shows the utility values corresponding to different models for the case study: rainfall–runoff modelling is shown in Fig. 13.12. Due to the relatively high sensitivity of NW/W-ANFIS models and relatively high error of ANN models, these three models stand out as the three lowest utility models in the rainfall–runoff modelling study. It means that even

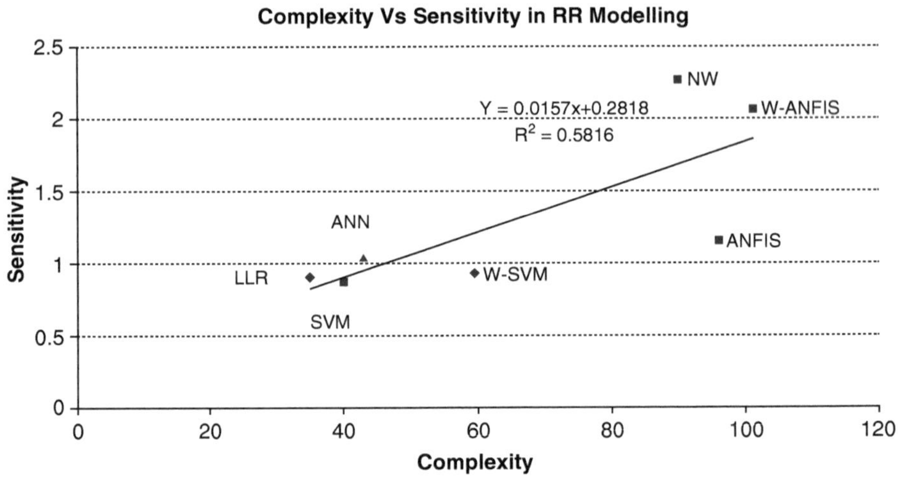

Fig. 13.11 Sensitivity vs. complexity—case study: rainfall–runoff modelling

Table 13.1 Different models and their attributes which decide overall model utility in rainfall–runoff modelling

Model	Complexity (function of modelling time)	Sensitivity (function of slope of sensitivity curves)	RMSE (m^3/s)	RMSE (%)	Sensitivity (%)	Utility (U)
ANN	43	1.04	0.558	1	0.458	0.222
NW	90	2.27	0.274	0.491	1	0.212
ANFIS	96	1.15	0.47	0.842	0.506	0.304
W-ANFIS	101	2.06	0.39	0.698	0.907	0.190
SVM	40	0.88	0.415	0.743	0.387	0.406
W-SVM	60	0.95	0.37	0.663	0.418	0.445
LLR	35	0.9	0.414	0.741	0.396	0.405

though the models are better at predicting numerical values, when considering other attributes, which decide consistency in modelling, complex models may stand out of "better and useful" model groups. Among all seven models, the W-SVM model has the best model utility followed by the models like SVM and LLR. It was interesting to note that the complex model SVM and relatively simple model LLR have very close utility values.

13.4.2 Comparison of Data Models Using Statistical Indices

Though the study has presented the utility evaluation as an alternative method for model comparison, it is essential to have a look into how the traditional statistical

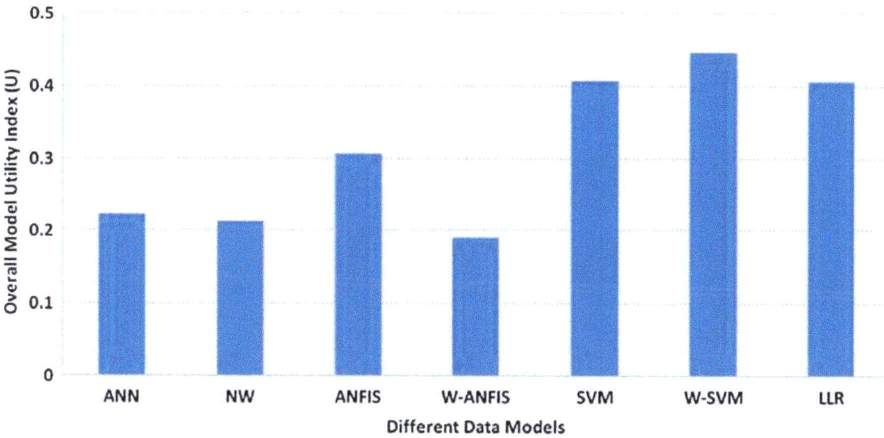

Fig. 13.12 Overall model utility—case study: rainfall–runoff modelling

indices behave in this modelling case study. The performance of the developed LLR technique was compared with other models using major global statistics as shown in Table 13.2. The estimated daily runoff values using the LLR model for 1,056 data points resulted in the overall RMSE value of 0.414 m^3/s which is 20.7 %; compared to observed daily runoff values and the MBE observed as -0.028 m^3/s. The Levenberg–Marquardt algorithm-based ANN underperformed to that of LLR in both the training and validation phases (RMSE value of 0.558 m^3/s (27.9 % of mean observed) and MBE value of -0.085 m^3/s during validation phase). It was seen that the LLR model's performance had a superior efficiency and performance compared with Levenberg–Marquardt ANN model with lower RMSE and higher CORR, for the training period and validation periods.

Table 13.2 implies that the performance of ANFIS model is remarkably better than that of LM-based ANN model in both validation and training data. The ANFIS model showed an efficiency of 88.9 % (increase of 5.45 % from ANN model) for the training data, and a validation efficiency of 77.3 % (an increase of 7.81 % compared to ANN). The correlation coefficients between the computed and observed are found to be 0.88 during training and 0.75 during validation. In order to assess the robustness of the models developed, evaluation criterion like MBE was used. From the MBE value one can deduce that both ANN and ANFIS showed underestimation for the training data and validation data. The underestimation is less for ANFIS during validation phase compared with ANN as a low value of MBE was observed. However, the performance of LLR was observed better than that of ANFIS model during training phase while ANFIS model outperformed LLR in the validation phase.

The analysis results in Table 13.2 have shown that the NW model is superior in predicting runoff values in comparison to all other models. The performance efficiency of the NW model is 4.13 % higher than that of the ANFIS model for validation and the corresponding value for the training data is 8.21 % higher.

Table 13.2 Comparison of some basic performance indices of different models employed in the study for daily rainfall–runoff modelling at the Brue catchment

	Training data (1,056 data points)					Validation data				
Models used	RMSE[a] (m^3/s and %)	CORR	Slope	MBE (m^3/s)	*E*	RMSE[a] (m^3/s and %)	CORR	Slope	MBE (m^3/s)	*E*
LLR	0.414 (20.7)	0.92	0.93	−0.028	0.923	0.922 (37.7)	0.70	0.80	−0.171	0.72
ANN-LM	0.558 (27.9)	0.83	0.90	−0.007	0.843	0.877 (36.4)	0.68	0.80	−0.085	0.717
ANFIS	0.470 (23.4)	0.88	0.93	−0.038	0.889	0.796 (33.2)	0.75	0.85	−0.039	0.773
SVM	0.415 (20.75)	0.89	0.91	−0.062	0.91	0.692 (28.3)	0.79	0.82	−0.012	0.831
W-SVM	0.370 (18.8)	0.90	0.90	−0.051	0.90	0.670 (27.2)	0.75	0.83	−0.112	0.770
W-ANFIS	0.39 (19.5)	0.90	0.89	−0.059	0.905	0.702 (28.6)	0.80	0.89	−0.103	0.802
NW	0.274 (13.6)	0.96	0.97	−0.00025	0.962	0.699 (28.6)	0.81	0.92	−0.0068	0.805

[a]Root mean square error is also shown in percentage of the mean value of observed runoff

Compared with the ANN model, the efficiency values of the NW model are 14.1 % and 12.27 % higher for the training and validation data, respectively. In terms of MBE, the performance of the NW model outperformed all other tested models in both the training and validation phases.

As shown in the above section, the study chose the ε-SVM with linear kernel for modelling applying the value of ε to 0.1 and values of C to 2; the modelling results are shown in Table 13.2. The SVM model made a better modelling with RMSE value of 0.415 m^3/s (20.75 %) and CORR of 0.89 during the training phase. The corresponding values during the validation phase were 0.692 m^3/s (28.3 %) and 0.79, respectively. The SVM model has shown better statistical performance compared to ANN, and ANFIS with an efficiency of 0.91 during training. The ε-SVM is applied with DWT to form a W-SVM model. Likewise, the ANFIS model was combined with DWT to form a hybrid W-ANFIS model. In the case of W-ANFIS, the DWT decomposed the input data sets into three wavelet decomposition levels (2–4–8) as mentioned in the previous sections and are used for rainfall–runoff modelling. The performance analysis of wavelet-based ε-SVM (W-SVM) is shown in Table 13.2 along with its comparison with the W-ANFIS model. The table implies that the performance analysis of both the W-ANFIS model and the W-SVM models was remarkably well in both validation and training data. The W-SVM model showed an efficiency of 90.0 % (increase of 6.76 % from ANN model) for the training data, and a validation efficiency of 77.0 % (an increase of 7.39 % compared to ANN). The correlation coefficient between the computed and observed are found to be 0.90 during training and 0.75 during validation. The RMSE for the LM-based ANN model is lower (0.558 m^3/s (27.9 %)) compared with the W-SVM model (0.37 m^3/s (18.8 %)) during training. From MBE value one can see the significant improvements while using hybrid wavelet forms of SVM models. The performance of W-ANFIS model in predicting runoff values is observed superior to the conventional LM-based ANN model and inferior to hybrid wavelet-based SVM model. The runoff prediction was underestimated by all models including W-ANFIS model for both the training and validation phases as indicated by the MBE values in Table 13.2. The RMSE value in the validation phase obtained by W-ANFIS model was 0.702 m^3/s (28.6 %), which was higher than that of the NW model and W-SVM model. The performance efficiency of W-ANFIS model in the rainfall–runoff modelling was 3.75 % lower than that of W-SVM model for validation and corresponding value for training data was 1.12 % lower. Compared with the ANN model, the efficiency values of the W-ANFIS model are 5.57 % and 11.85 % higher for training and validation data, respectively. However, in terms of MBE values, the performance of NW model outperformed all other tested models including W-ANFIS and W-SVM in both training and validation phases. Though many statistical parameters are used for evaluation of robustness of the model, the major index used for comparison of model performance is the Nash–Sutcliffe efficiency. Figure 13.13 shows comparison of overall model utility index and the Nash–Sutcliffe efficiency (average of training and validation) so in general the comparison using general error-based statistical indices and the N-S efficiency has shown that the rainfall–runoff modelling capabilities of data-based models

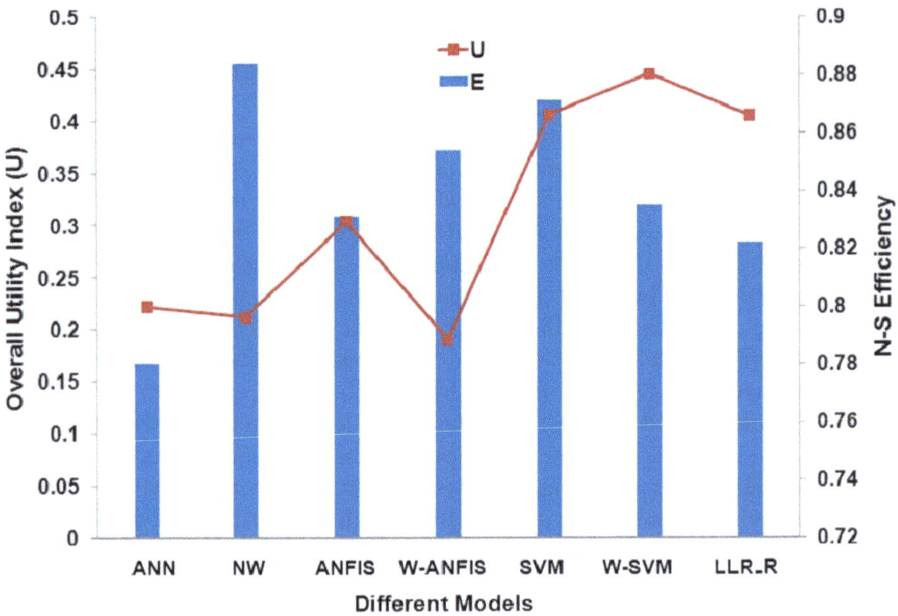

Fig. 13.13 Comparison of *S–E* efficiency and overall model utility index

are in the form of NW > SVM > W-ANFIS > W-SVM > ANFIS > LLR > ANN. However, the results are different when we consider other modelling attributes like model sensitivity and model complexity along with modelling errors. The utility index identified that the usefulness of models in this case study are in the form of W-SVM > SVM > LLR > ANFIS > ANN > NW > W-ANFIS. Both the approaches have acknowledged the better performance of SVM, giving second position in terms of efficiency and utility values. The N-S approach gave high ranking for both W-ANFIS and NW model as the approach couldn't account for the influence of higher sensitivity. The higher utility value of the wavelet-based SVM has shown the capabilities of SVM framework to handle large input space without any difficulty of sensitivity.

13.5 Conclusions

The study adopted a utility index to critically evaluate the acceptance of a model in terms of different modelling properties and contrasted the results with that of traditional statistical indices (particularly the N-S efficiency). This study has demonstrated that such an error-sensitivity-uncertainty procedure could help modellers make effective comparison of different data-based models and it can give an answer on how much the model benefits by increased complexity on data-based models. The study extensively analysed the capabilities of SVM in the context of rainfall–runoff modelling and demonstrated its ability to perform better even in larger wavelet-decomposed input space. The study has demonstrated the weakness of NW and W-ANFIS models. Even though these models had better numerical prediction results, the utility evaluation has shown their limitation in making a useful model for rainfall–runoff modelling due to their inclination towards sensitivity. The overall utility analysis based on the utility index has identified W-SVM as the best model, followed by SVM and LLR models.

The approach would be very useful in the data scarce situation where there is insufficient data for validation. The modeller could use this method for selection of the best possible model for the available data without diverting valuable data away from calibration of model. Though the study presented with useful information, there is room for improvement regarding the sensitivity assessment. The study has assessed local sensitivity of the model with respect to the variation of inputs. The term sensitivity is rather complicated; the local sensitivity slopes of these nonlinear models vary depending on the range of inputs. However the choice of the sensitivity and complexity values of these nonlinear models requires further research to develop the presented utility assessment to a robust method in data-based model comparisons.

References

Barnes CJ (1995) Commentary: the art of catchment modelling: what is a good model? Environ Int 21(5):747–751

Bray M, Han D (2004) Identification of support vector machines for runoff modelling. J Hydroinf 6(4):265–280

Hastie T, Tibshirani R, Friedman J (2001) The elements of statistical learning: data mining, inference, and prediction. Springer, New York

Hecht-Nielsen R (1990) Neurocomputing. Addison-Wesley, Reading, MA

Jain SK, Sudheer KP (2008) Fitting of hydrologic models: a close look at the Nash–Sutcliffe Index. J Hydrol Engrg 13(10):981–986. doi:10.1061/(ASCE)1084-0699(2008)13:10(981)

Jang JSR (1993) ANFIS: adaptive-network-based fuzzy inference systems. IEEE Trans Syst Man Cybern 23(3):665–685

Klir GJ (1989) Methodological principles of uncertainty in inductive modelling: a new perspective. In: Erickson GJ, Smith CR (eds) Maximum-entropy and Bayesian methods in science and engineering, vol 1. Kluwer, Dordrecht

Legates DR, McCabe GJ (1999) Evaluating the use of goodness of fit measures in hydrologic and hydroclimatic model validation. Water Resour Res 35:233–241

Maren AJ, Harston CT, Pap RM (1990) Handbook of neural computing applications. Academic, San Diego, CA

McCulloch WS, Pitts W (1943) A logical calculus of the ideas imminent in nervous activity. Bull Math Biophys 5:115–133

Mechaqrane A, Zouak M (2004) A comparison of linear and neural network ARX models applied to a prediction of the indoor temperature of a building. Neural Comput Appl 13:32–37

Oreskes N, Shrader-Frechette K, Belitz K (1994) Verification, validation, and confirmation of numerical models in the Earth sciences. Science 263:641–646

Pepelnjak JC (2009) Validating the combined approach to online reach prediction. Atlas® Institute Digital Marketing Insights. http://www.atlassolutions.com /uploadedFiles/Atlas/Atlas_Institute/Published_Content/CombinedApproach.pdf
Plate EJ, Duckstein L (1987) Reliability in hydraulic design. In: Duckstein L, Plate EJ (eds) Engineering reliability and risk in water resources. NATO ASI Series, Series E: Applied Sciences, No. 124. Nijhoff, Dordrecht
Remesan R, Shamim MA, Han D, Mathew J (2009) Runoff prediction using an integrated hybrid modelling scheme. J Hydrol 372(1–4):48–60. doi:10.1016/j.jhydrol.2009.03.034
Shannon CE (1948) A mathematical theory of communication. Bell Syst Tech J 27:379–423
Snowling SD, Kramer JR (2001) Evaluating modelling uncertainty for model selection. Ecol Model 138:17–30
Sugeno M, Kang GT (1988) Structure identification of fuzzy model. Fuzzy Set Syst 28:15–33

Index

T. Islam et al. (eds.), *Computational Intelligence Techniques in Earth and Environmental Sciences*, DOI 10.1007/978-94-017-8642-3,

MIX
Papier aus verantwortungsvollen Quellen
Paper from responsible sources
FSC® C105338

If you have any concerns about our products,
you can contact us on
ProductSafety@springernature.com

In case Publisher is established outside the EU,
the EU authorized representative is:
Springer Nature Customer Service Center GmbH
Europaplatz 3, 69115 Heidelberg, Germany

Printed by Libri Plureos GmbH
in Hamburg, Germany